1988

Eukaryote cell recognition

Eukaryote cell recognition: concepts and model systems

edited by

G. P. Chapman
Senior Lecturer in Plant Sciences,
Wye College,
University of London

C. C. Ainsworth
Lecturer in Plant Molecular Genetics,
Wye College,
University of London

C. J. Chatham
Lecturer in Crop Production,
Wye College,
University of London

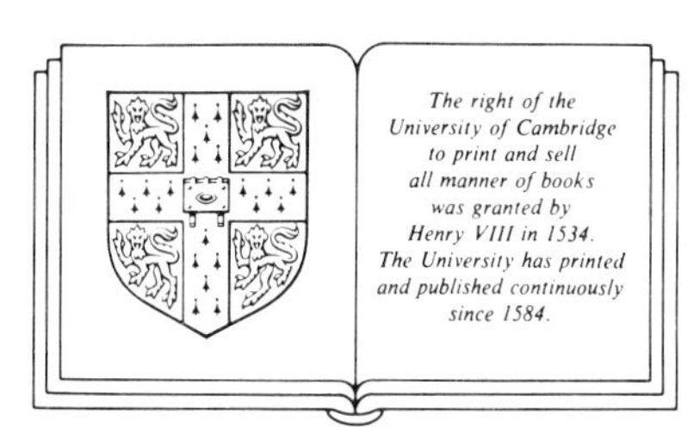

CAMBRIDGE UNIVERSITY PRESS

Cambridge

New York New Rochelle

Melbourne Sydney

Published by the Press Syndicate of the University of Cambridge
The Pitt Building, Trumpington Street, Cambridge CB2 1RP
32 East 57th Street, New York, NY 10022, USA
10 Stamford Road, Oakleigh, Melbourne 3166, Australia

First published 1988

Printed in Great Britain at The Bath Press, Avon

British Library Cataloguing in publication data

Eukaryote cell recognition: concepts and
model systems.
1. Cell interaction
I. Chapman, G. P.   II. Ainsworth, C. C.
Chatham, C. J.
574.87′6   QH604.2

Library of Congress cataloguing in publication data

Eukaryote cell recognition: concepts and model systems/edited by
G. P. Chapman, C. C. Ainsworth, C. J. Chatham.
p.   cm.
Includes index.
ISBN 0 521 34413 1
1. Cellular recognition. 2. Eukaryotic cells. I. Chapman, G. P.
II. Ainsworth, C. C. III. Chatham, C. A.
[DNLM: 1. Cell Interaction. 2. Cells—physiology. QH 604.2 E87]
QH604.2.E94 1988
574.87′6—dc 19
DNLM/DLC
for Library of Congress   87-33850

ISBN 0 521 34413 1

BS

CONTENTS

CONTRIBUTORS

Burger, M. M.
Marine Biological Laboratory, Woods Hole, Massachusetts 02543, USA

Carraway, C. A. C.
Department of Biochemistry, Anatomy & Cell Biology, University of Miami, School of Medicine, Miami, Florida 33101, USA

Carraway, K. L.
Department of Biochemistry, Anatomy & Cell Biology, University of Miami, School of Medicine, Miami, Florida 33101, USA

Cho, A.
Department of Medical Research, Charles H. Best Institute, University of Toronto, Toronto, Ontario, Canada M5G 1L6

Choi, A.
Department of Medical Research, Charles H. Best Institute, University of Toronto, Toronto, Ontario, Canada M5G 1L6

Curtis, A. S. G.
Department of Cell Biology, University of Glasgow, Glasgow G12 8QQ, UK

Dickinson, H. G.
Department of Botany, Plant Science Laboratories, University of Reading, Whiteknights, Reading RG6 2AS, UK

Ende, H. van den
Department of Plant Physiology, University of Amsterdam, Amsterdam, The Netherlands

Hartman, T. D.
Gamete Biology Unit, MRC/AFRC Comparative Physiology Group, Institute of Zoology, Regent's Park, London NW1 4RY, UK

Hiwatashi, K.
Biological Institute, Tohoku University, Sendai, Japan

Hodgkin, T.
Scottish Crop Research Institute, Invergowrie, Dundee DD2 5DA, UK

Holmes, P. V.
Department of Obstetrics & Gynecology, Sahlgrenska Sjukhuset, S-413 45 Gothenburg, Sweden

Kimber, S. J.
Experimental Embryology & Teratology Unit, MRC Laboratories, Woodmansterne Road, Carshalton, Surrey SM5 4EF, UK

Klis, F. M.
Department of Plant Physiology, University of Amsterdam, Amsterdam, The Netherlands

Knox, R. B.
Plant Cell Biology Research Centre, School of Botany, University of Melbourne, Parkville, Victoria 3052, Australia

Kokontis, J. M.
Department of Biochemistry & Molecular Biology, The University of Chicago, Chicago, Illinois, USA

Lindenberg, S.
Department of Obstetrics & Gynecology, Rigshospitalet, Copenhagen, Denmark

Lyon, G. D.
Scottish Crop Research Institute, Invergowrie, Dundee, DD2 5DA, UK

Mansfield, J. M.
Wye College (University of London), Ashford, Kent TN25 5AH, UK

Misevic, G. N.
Biocentre of the University of Basel, Klingelbergstrasse 70, CH-4056, Basel, Switzerland

Moore, H. D. M.
Gamete Biology Unit, MRC/AFRC Comparative Physiology Group, Institute of Zoology, Regent's Park, London NW1 4RY, UK

Musgrove, A.
Department of Plant Physiology, University of Amsterdam, Amsterdam, The Netherlands

Nasrallah, J. B.
Section of Plant Biology, Cornell University, Ithaca, New York 14853, USA

Nasrallah, M. E.
Section of Plant Biology, Cornell University, Ithaca, New York 14853, USA

Perotti, M-E.
Department of General Physiology & Biochemistry, University of Milano, Milan, Italy

Riva, A.
Department of General Physiology & Biochemistry, University of Milano, Milan, Italy

Rowell, P. M.
Dow Chemical Co., Letcombe Laboratory, Letcombe Regis, Wantage, Oxfordshire, OX12 9JT, UK

Ruddat, M.
Department of Molecular Genetics & Cell Biology, Barnes Laboratory, The University of Chicago, Chicago, Illinois, USA

Singh, M. B.
Plant Cell Biology Research Centre, School of Botany, University of Melbourne, Parkville, Victoria 3052, Australia

Smith, C. A.
Gamete Biology Unit, MRC/AFRC Comparative Physiology Group, Institute of Zoology, Regent's Park, London NW1 4RY, UK

Smith, E.
School of Biological Sciences, Macquarie University, Sydney, NSW 2109, Australia

Southworth, D.
Southern Oregon State College, Ashland, OR 97520, USA *and* Plant Cell Biology Research Centre, University of Melbourne, Parkville, Victoria 3052, Australia

Spielman, J.
Department of Anatomy and Cell Biology, University of Miami School of Medicine, Miami, Florida 33101, USA

Stegwee, D.
Department of Plant Physiology, University of Amsterdam, Amsterdam, The Netherlands

Street, P. F. S.
Dow Chemical Co., Letcombe Laboratory, Letcombe Regis, Wantage, Oxfordshire OX12 9JT, UK

Sui, C. H.
Banting & Best Department of Medical Research, Charles H. Best Institute, University of Toronto, Toronto, Ontario, Canada M5G IL6

Tanner, W.
Institut fur Botanik der Universitat Regensburg, Universitatstrasse 31, 8400 Regensburg, FRG

Watzele, M.
Institut fur Botanik der Universitat Regensburg, Universitatstrasse 31, 8400 Regensburg, FRG

Williams, K. L.
School of Biological Sciences, Macquarie University, Sydney, NSW 2109, Australia

Wong, L. M.
Banting & Best Department of Medical Research, Charles H. Best Institute, University of Toronto, Toronto, Ontario, Canada, M5G IL6

Woods, A. M.
Wye College (University of London), Ashford, Kent TN25 5AH, UK

Yanagishima, N.
Biological Institute, Faculty of Science, Nagoya University, Chikusa-Ku, Nagoya 464, Japan

PREFACE

The processes of growth and development throughout the life of all organisms are tightly controlled at the genetic level. Development progresses in an orderly manner in all tissues such that the resultant organism resembles and functions like its parents.

We take it for granted that only gametes of opposite sex can fuse, even in what we call 'isogamy'. This implies one fundamental level of recognition. That gametes are of opposite sex offers no guarantee that they *will* fuse. If their genetic relationship is too distant, or even too close, zygote formation fails. If it succeeds subsequently, development may falter thereafter. Superimposed, therefore, on the initial 'like' versus 'unlike' aspect of gamete recognition are others reflecting distance or closeness of genetic relationship.

Given that zygote formation is accomplished, the intricate processes of differentiation begin. In multicellular organisms cells divide, tissues form and a myriad of recognition events takes place that are integral to normal development. However, development may not be normal. Either some kind of misrecognition event occurs autonomously whereby the organism misdirects its own development or there is intruded from outside some other kind of recognition event. Some pathogen, for example, recognises an acceptable host and the resources of the latter are requisitioned to the needs of the former.

This, then, is the arena of events that confronts the biologist whether his interests are animal or plant orientated and whether his viewpoint is genetic, anatomical or physiological.

For many years an underlying assumption has been that recognition events are mediated at membrane surfaces. Recent advances in molecular biology and the production of monoclonal antibodies have added to the refinement with which this assumption can be explored. Accompanying this of course are the now routine but still improving, techniques of electron microscopy. The combination of these powerful techniques should allow us to define the structure and function of the receptor molecules themselves. The receptors will provide the key to the signalling process in that it is these which receive the external signal and transmit the message to the cell itself.

While recognition remains a central preoccupation, one's perspective can differ. To the sponge biologist, for example, working with an organism that can be readily dispersed and reassembled depending on the composition of the surrounding environment, interacting cell membranes are readily accessible. Conversely, a plant geneticist exploring the curious but normally automatic process of double fertilisation finds inaccessibility of the interacting membranes a major concern. Indeed, convincing isolation of flowering plant 'gametoplasts' has only been accomplished within the last three years. Compare this situation, therefore, with one in animal reproductive biology where animal sperm surfaces have been explored in great detail both structurally and physiologically.

A further dimension is added to the study of recognition phenomena by the life history of slime moulds which have both a social and a non-social mode. Finally, there is the problem of 'anti-recognition' when a host system is over ridden by an invasive one.

An understanding of cell signalling and recognition processes in these different organisms and at various developmental levels will have profound implications in many areas of biology and may ultimately enable these processes to be controlled advantageously.

The Third Wye International Symposium brought together contributors working on a wide range of organisms. Each contributed something to our central concern, the problem of 'recognition' or how cells signal to each other. The papers are arranged here as follows. Under 'concepts' are three papers dealing respectively with 'recognition', 'anti-recognition' as when a recognition system is overridden, and lastly with 'self-recognition'. Single and multicelled organisms are then treated in turn and lastly inter-organism recognition is examined.

It is our hope that the reader will find stimulating this collection of model systems and the ideas they embody whatever his concern with recognition phenomena.

G. P. Chapman
C. C. Ainsworth
C. J. Chatham

Wye College
Ashford
Kent

July, 1987

ACKNOWLEDGEMENTS

Our thanks are due to Professors J. G. Duckett, J. Howard, K. Roberts and W. W. Schwabe and to Dr D. Ingram for their help and advice in the preparation of this volume.

We are also grateful to all the contributors, not least for the readiness with which they met the various editorial requests.

We should also like to thank Mrs Sue Briant and Mrs Margaret Critchley for their excellent secretarial help and the staff of Cambridge University Press who saw this volume through the press.

G. P. C.
C. C. A.
C. J. C.

While this book was in press, the death occurred of one of the contributors, Professor Yanagashima. He was latterly Professor in the Department of Biology at Nagoya University and Director of the Laboratory of Microbiology, and one-time Professor at Osaka City University. His academic activities included membership of the Board of the Botanical Society of Japan, the Japan Society of Plant Physiology and the Japanese Society of Developmental Biologists. Professor Yanagashima died on March 28th, 1987, aged 62.

We extend our condolences to his family and friends.

PART I

Concepts

I Cell–cell interactions: activation or specific adhesion

A. S. G. Curtis

Department of Cell Biology,
University of Glasgow,
Glasgow G12 8QQ, UK

Abstract

After a short review of a classification of recognition systems, consideration is given to the following problems, particularly in relation to studies of cell adhesion.

1. Do activation systems, acting indirectly on cell adhesion, have a major role in adhesion?
2. Do lectins and nectins act as activating systems rather than as direct binding systems?
3. What types of experiment will reveal the existence of direct bonding molecules?

Consideration is also given to CAMs (Cell Adhesion Molecules).

General features of interaction systems

Cell–cell interactions are but a sub-set of the larger field of cell interactions, and thus it is natural to expect that some of the general rules and conditions that apply to that larger field will be reflected in the smaller field. Cell interactions range from the well-understood field of the reactions of hormones with cells (Cuatrecasas & Hollenberg, 1976) to those which still remain relatively obscure, e.g. pattern formation by cell movement in embryos (Garrod & Nicol, 1983). Table 1.1 provides a shortlisting of some of the phenomena.

Those hormones which react with the cell surface are received by receptors with reasonably strong affinity constants ($K_a > 10^5$) and then, usually, the reception of the signal leads to an action as a result. It is perhaps worthwhile considering for a moment the possibility that signals have to be continuously

Table 1.1. *Examples of cell recognition phenomena*

Signal	Receptor	Phenomenon
Soluble		
(f-met-leu-phe)	Leukocyte	Activation of adhesion and chemotaxis
Catecholamines	Many cell types	c-AMP stimulus
'Max factor'	On mouse trigeminal ganglion	Nerve cell attraction (Lumsden & Davies, 1986)
Cell spreading factor*	Fibroblasts	Spreading of cells (Barnes & Silnutzer, 1983)
Insoluble		
Mating type determinants	Yeasts	Mating reaction
NCAM*	Many cell types	Adhesion
?	Sponge cells	Incompatibility reactions (Curtis, Kerr & Knowlton, 1982)
Contact activation of platelets	Blood platelets	Adhesion
?	Many cell types	Sorting out phenomena in cell mixtures

*It is not clear whether these factors must be presented to the cell in an insoluble or a soluble form or in either manner.

received in order to maintain a status quo. A human analogy for this situation might be the reaction of a sentry who queries each approaching person for the password, or who even stands still and merely expects the password to be given. Failure to give the password, or the uttering of the wrong word results in death by rifle fire. We do not know for certain that such systems exist in cell biology, but the continued requirement of various growth hormones for successful cell culture suggests such a possibility.

The conceptual model which derives from hormone action studies is of considerable importance. Nevertheless it should be borne in mind that other types of interaction may be used in cell–cell interactions. For instance:

(i) the sequence in time of events in a cell population may be such that only certain cells are carrying out like events at the same time. Curtis (1961) suggested that this might account for the sorting out of cells in aggregates.

(ii) topographical cues may affect cell interactions (Curtis & Varde, 1964; Dunn, 1982; Clark *et al.*, 1987). For instance, cell movement may be stopped, polarised or even oriented by suitable small-scale features of the environment. Since care was taken by these workers to ensure that there were no alterations in chemical cues in the environment, it seems likely that the cell is reacting by some mechanism other than direct reception of a chemical signal.

(iii) simple non-specific cues such as the quantitative value of adhesiveness of the substrate (Steinberg, 1978). Steinberg has suggested that the sorting out behaviour of embryonic cells may be explained by differences in adhesion driving cells, by simple rules of interfacial tensions, from thermodynamically unfavourable i.e. mixed, into thermodynamically favourable, e.g. sorted out configurations.

It should always be borne in mind that alternative explanations, such as those advanced above should be reviewed in the general light of features of receptor recognition systems. These are simple, namely that:

(i) if continued reaction of the cell is to take place, new unoccupied receptor sites must become available, either by dissociation of the former signal–receptor complexes or by recycling of the occupied receptor sites (Hopkins, Miller & Beardmore, 1985).

(ii) the signal must be present at a reasonable concentration in relation to the number of receptor sites. Ideally the signal will be most effective when half the site is occupied at one time (Cuatrecasas & Hollenberg, 1976; Zigmond, 1982). In turn this carries implication about the affinity constant for binding (or its reciprocal the dissociation constant).

These two points suggest that timing events may be no more than changes in the rate at which receptors are recycled (Smith & Hollers, 1980; Zigmond, 1982) and topographical events nothing more than restrictions on membrane recycling imposed by physical constraint of the cell. Nevertheless such points have not been tested. It should be noted that topographical control of cell movement and behaviour may arise from neighbouring cells. Features of the extracellular matrix in a tissue will offer topographical cues to a given cell; we should not ignore the importance of this type of cue.

The signals on reception are transduced into the cell's interior by a variety of systems which include the cAMP based systems (Lefkowitz, Stradel & Caron, 1983), the diacylglycerol–inositol triphosphate system (Berridge & Irvine, 1984) which probably involves changes in intracellular calcium, and in addition the probable existence of less well-appreciated systems. It is possible, at least in theory, to imagine that some signals are received which do not need to have any effect on the internal economy of the cell. For instance, a molecule, see Fig. 1.1, bridging from one cell to another or from a substrate to a cell might simply be required to interact without internal changes taking place in the cell. Whether such simple situations actually occur is more debatable. In practice such reactions involve intracellular events including cytoskeletal changes, see, for example, Kruskal, Shak & Maxfield (1986).

Adhesion: general considerations

I intend to devote special attention to processes of cell adhesion since this is a major topic of the meeting. Other papers, e.g. those on cell adhesion in embryos by Kimber and by Holmes are obviously directly relevant. Nasrallah and also Knox describe pollen–stigma interactions in self-incompatibility reactions in higher plants which is a process in which adhesion may be of importance. Sperm–egg interactions are discussed by Moore and by Perotti

Fig. 1.1. Possible molecular mechanisms of cell interaction. (a) Homophilic binding (two versions) (b) Heterophilic binding (c) Steric destabilisation (intertangling) (d) DLVO, with a balance between electrodynamic forces of attraction and electrostatic forces of repulsion.

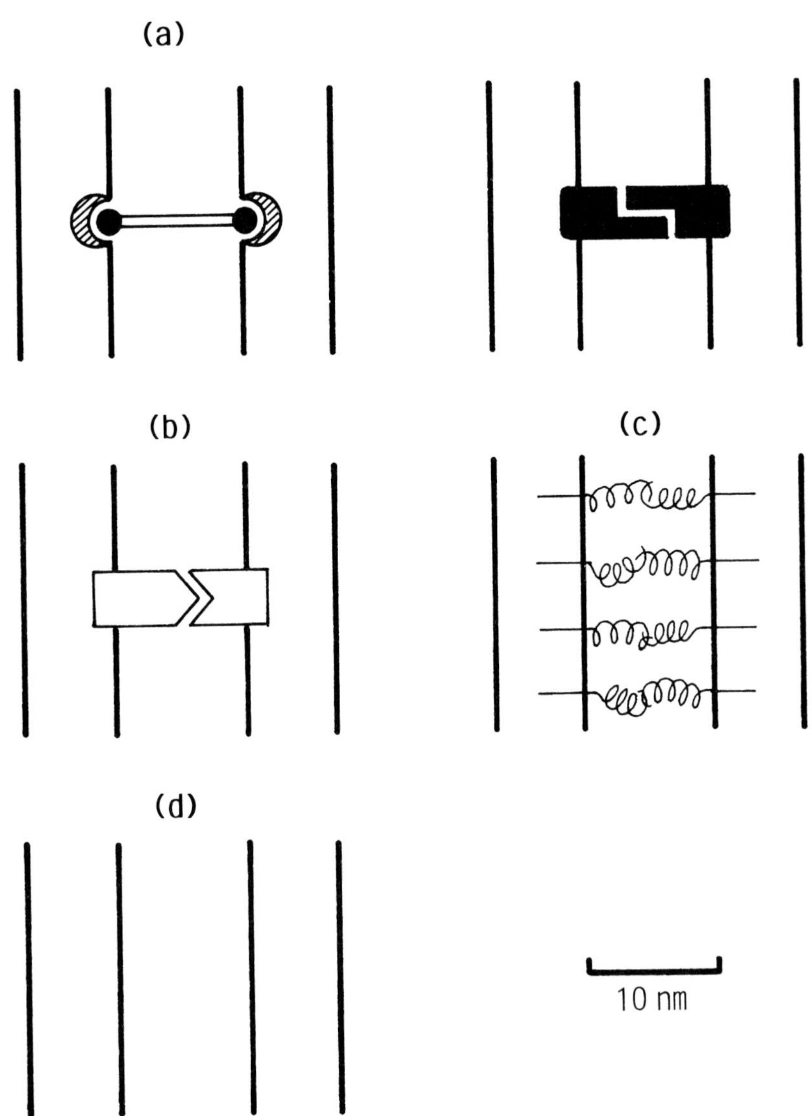

in which adhesion is clearly the major process while Siu describes features of slime-mould adhesion and Misevic aspects of sponge cell adhesion.

A considerable number of reports have appeared describing cell adhesion molecules (CAMs) see Edelman & Thiery, 1986; Curtis, 1987. In general etymological terms the acronym (and its precursor) should perhaps include all molecules directly implicated in effecting adhesion as well as those less directly concerned. These molecules are usually conceived of as being direct molecular bonding agents, perhaps acting as homophilic bonds (see Fig. 1.1) between cells, though other reports on other systems (e.g. Burger, Burkart, Weinbaum & Jumblatt, 1978) have suggested that more heterophile type systems can act. In practice, at present, such concepts are poorly established. In a few cases the particular CAM has been isolated and attached to liposomes (Hoffman & Edelman, 1983) or to derivatised beads (Siu, this symposium), but though such beads show rather increased adhesion compared with controls we do not know whether the molecules concerned are acting directly or in some abnormal manner. Nir, Bentz, Wilschutz & Duzgunes (1983) suggested that liposomes adhere normally by a DVLO (see Rutter, 1980) type system, which, if true, suggests that N-CAM laden liposomes are adhering by a system unlike that suggested for cells. Bell & Torney (1985) have analysed that kinetics of N-CAM liposome adhesion experiments with results which do not support a straightforward interpretation. Furthermore (see below), there may be ambiguities in the interpretation of the Fab experiments by which these molecules were detected and isolated. Clearly there is a need for direct methods which visualise the closeness of approach of molecules on one cell surface to those on another surface, cell or otherwise, to which the first cell is adhering: possibilities in this area are discussed at the end of this paper.

It should be appreciated that there are three main methods for studying cell adhesion:

(i) measurement of the rate of formation of cell adhesions when a cell suspension adheres by collisions induced by shear flow-induced collisions alone, Brownian motion-induced encounters being of little importance for relatively large eukaryote cells. Prokaryote cells will adhere by both processes. If the shear flow is known, viscosity differences due to the incorporation of large soluble macromolecules into the medium may be ignored. Differences in cell volume or population density can only be sensibly compared when quantitative methods which account for these parameters (Curtis & Hocking, 1970; Bongrand, Capo & Depieds, 1982; Duszyk, Kawalec & Doroszweski, 1986) are used. Such methods yield values of the collision efficiency, namely the probability that a collision will result in an adhesion.

(ii) measurement of the extent and rate of cell attachment to a planar surface, e.g. a petri dish. This is a difficult situation for analysis. The approach to the surface under gravity is slow and the cells may well start to flatten before or as they begin to make contact. To some extent such methods may measure the rate at which the cell can establish a large enough contact to resist detachment by subsequent shear, rather than anything else.

(iii) measurement of the force required to detach a cell. This requires very accurate control of the contact area as the force is measured. At present this is probably the least accurate method.

It is important to appreciate that measurement in suspension implies interaction times of 0.01 to 0.1 second between the cells whereas the second method allows interaction times of at least 100 seconds. Thus for cells in suspension in cases in which the collision efficiency is greater than 50%, most cells must form adhesions without any immediate previous effect of cell interaction upon them, though of course soluble signals may have passed between the cells. In situations in which adhesions form more slowly there is considerable opportunity for such events as exchange of small molecules, enzymic effects, etc to modify cell interaction. Active cell movement as in the flagellate and ciliated organisms may enhance the rate at which the cells approach one and another and the energy with which they do so, as well as stirring the medium and destroying gradients of chemicals in the environment.

Molecular interactions between one surface and another will, unless the molecules protrude out far from the cell surface and are present at fairly high number density, be slow because the molecules will have to be brought into alignment and the intervening fluid will have to be drained out of the gap between the cells. However, rather rapid formation of adhesions will tend to imply that rather general large-scale features of the cell surface such as DLVO interactions, hydrogen bonding or molecular entanglement (steric destabilisation, see Rutter, 1980) act. Small freely diffusible signals, on the other hand, may modify adhesion rapidly if they act on generalised processes rather than on specific molecular interactions.

Thus the question should be asked as to whether any agent that affects cell adhesion rapidly is acting as a small soluble molecule that activates cell adhesion in the same way as a hormone activates the cell. Even if it seems that cell adhesion acts by more direct molecular bindings there is still the possibility that large molecules act primarily as activators of cell adhesion rather than as direct bonding molecules, or that the bondings have no specific receptors but have relatively specific steric destabilisation effects. In any event rapid changes in adhesion whether occurring *in vivo* or in experimental situations are suggestive of activation effects.

Of course, collision in sheared suspension is not a process by which adhesions are formed between most cells in nature. In animals, the cells, other than those in circulation in the blood or other fluids, form new adhesions by crawling movement from one cell to another, which is a relatively slow process. What is the real time scale of the adhesive process in such situations, bearing in mind that adhesions may being turned over as cells crawl or spread?

Activation effects in leukocyte and platelet adhesion

Human polymorphonuclear leukocytes and platelets are capable of showing very rapid changes from a relatively non-adhesive to a very adhesive state, see Lackie & Smith (1980). Rough estimates of the change suggest that the increase in adhesion may be of several thousand fold. Since changes in leukocyte adhesion are accompanied by inflammatory events there has been considerable pharmacological interest in the system. Because of the extremely fast reaction of the system there is uncertainty as to what route is followed in the activation. The following main features have emerged (see also Fig. 1.2):

(i) the chemotactic peptide f-met-phe and some of its homologues stimulate adhesion (Smith & Hollers, 1980).
(ii) derivatives of arachidonic acid, in particular leukotriene B4 and thromboxane A2 at levels below 1 nanomole per million cells can stimulate, or are associated by their production with, increased adhesion (Buchanan, Vasquez & Gimbrone, 1983).
(iii) generation of oxygen radicals by the cells may be associated with these changes in adhesion, the 'respiratory burst', Cohen, Chovaniec, Takahashi & Whitin (1986).
(iv) Complement C3bi may activate the cells by another mechanism which does not involve the respiratory burst (Hed & Stendahl, 1982).
(v) Changes in the cell surface glycoprotein complex II/IIIa are detected on activation. This complex has been identified as a possible CAM (Beller, Springer & Schreiber, 1982).

Fibronectin and cell adhesion: binding molecule or activator?

Fibronectin, a 220 kD glycoprotein found in mammalian and avian tissues (see Hynes, 1985 for a good general molecular description) has been implicated in fibroblast adhesion by a number of experimental observations, see Yamada (1983). The most compelling evidence for it being essential for fibroblast adhesion is the work of Grinnell (Grinnell, 1978; Grinnell & Feld, 1979) who showed that mutants of fibroblasts unable to synthesise fibronectin were unable to adhere to a tissue culture dish substrate in the absence of exogenous fibronectin. The soluble form of fibronectin, plasma fibronectin when added to the system adsorbed on the culture dish surface and then the cells were able to adhere and spread.

Akiyama, Yamada & Yamada (1985) isolated a plasmalemmal receptor for avian fibronectin from chick fibroblasts. Thus it would seem at first sight that all the requirements for the demonstration of a cell adhesion system in which fibronectin is a bonding molecule have been achieved. However, some problems remain. Attempts to demonstrate the presence of fibronectin actually in the adhesion plaques of the cells have been, on the whole unsuccessful (Avnur & Geiger, 1981). Cells such as hepatocytes have been shown to have no absolute requirement for fibronectin (Rubin, Johansson, Hook & Obrink, 1981). Curtis, Forrester, McInnes & Lawrie (1983), showed that fibroblasts, under extreme inhibition of protein synthesis with cycloheximide, would adhere and spread, in a morphologically normal manner, provided that whole serum was omitted from the medium.

Fibroblasts are known to resemble leukocytes in at least four features of their adhesion: (i) albumin inhibits adhesion (Curtis & Forrester, 1984). (ii) prostacyclin inhibits cell spreading (and thus probably adhesion) (Ali & Chambers, 1983) in another tissue cell type, the osteoclast. (iii) fibroblasts can produce reactive oxygen species, albeit at a low level (Curtis, Forrester & Clark, 1986). (iv) there is a calcium-dependent feature of adhesion to

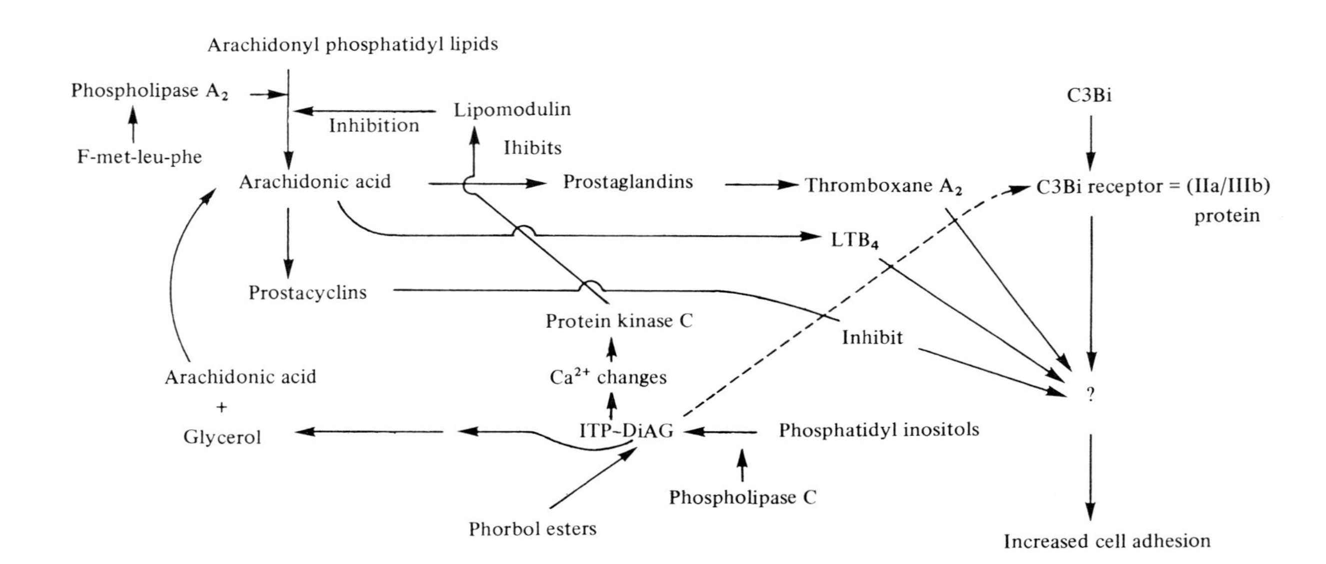

Fig. 1.2. Activation of adhesion in leukocytes. Probable pathways are shown with full lines, possible pathway with broken line. C3bi = activated third component of complement, LTB_4 = Leukotrienes, PG = ITP = Inositol triphosphate, DiAG = Diacylglycerol.

fibronectin operating at about 1×10^{-5} M. This finding is due to work by my colleague Dr J. G. Edwards (Edwards, Robson & Campbell, 1987). I should point out that he inspired much of the work and thought in this section. It is interesting to note that leukocyte attachment and spreading appear to be correlated with a large increase in free-cytoplasmic calcium (Kruskal, Skak & Maxfield, 1986). Thus fibronectin may be activating processes finally resulting in a rise in intracellular calcium, provided that sufficient calcium is available externally for there to be an influx.

Bearing these points in mind, Curtis & McMurray (1986) returned to the question of the adhesion of BHK fibroblasts to polystyrene. The earlier work quoted above had shown that cells would adhere well to hydroxylated polystyrene under conditions in which protein synthesis was inhibited. Since it is possible that the cells were releasing a pool of fibronectin synthesised on an earlier occasion, it was decided to use the tetrapeptide, RGDS, identified by Pierschbacher & Ruoslahti (1984) as the cell binding sequence of fibronectin and shown by them to inhibit cell attachment to fibronectin. If the cells show an absolute requirement for fibronectin, then this tetrapeptide should inhibit attachment under all conditions.

Figure 1.3 shows the results of measurements of cell attachment (BHK cells) to polystyrene bearing either adsorbed fibronectin or bare of exogenous adsorbed fibronectin, and in the presence of cycloheximide. Cell attachment is good (*ca.* 79% of all cells in 30 minutes). The tetrapeptide at 1.1 mM inhibits adhesion to adsorbed fibronectin but has no effect on adhesion in the absence of exogenous fibronectin and when the production of endogenous fibronectin is inhibited with cycloheximide. Cells adhere under these conditions. There is thus no absolute requirement for fibronectin.

This finding supports, though it does not prove, the concept that fibronectin may be acting to activate cell adhesion by some as yet undetected system.

Detection and identification of cell adhesion molecules

Though fibronectin was discovered by a process of testing plasma and cell surface extracts on adhesion to tissue culture polystyrene, it has been possible to demonstrate its probable activity as a molecule with some connection with adhesion by the use of Fab antibody methods. These methods were introduced by Gerisch (Beug, Katz, Stein & Gerisch, 1973; Gerisch, 1980). Though some antibodies against cell surface components inhibit adhesion, the majority, not surprisingly, agglutinate cells. This provides no evidence as to the adhesive role of the epitope with which they react, since a new and non-natural mechanism of adhesion overlies any process which may have been inhibited. Gerisch introduced the idea of using monovalent Fab antibodies against cell surface components. He argued that if the monovalent Fabs isolated from these antibodies inhibited adhesion then they were acting against a cell adhesion molecule. He appreciated two possible difficulties with this interpretation, namely:

(i) controls must be used in which other antigens located on the cell surface are used to raise the antibodies from which the Fabs are derived. These Fabs must not inhibit cell adhesion.

(ii) even if the Fabs inhibited adhesion they might be acting on some component near the adhesion molecule rather than on the adhesion molecule.

Beug, Katz, Stein & Gerisch (1973) were able to obtain a Fab directed against another surface antigen present in large copy number on their slime mould cells which did not inhibit cell adhesion. Unfortunately (Gerische, 1980), found that this antigen, Fabs against which did not lead to adhesion inhibition, lay deeper in the cell surface than the contact site A antigen. Thus it may have been possible that the lack of effect of the Fab was that the antigen lay too deep. In other words would any surface antigen present at sufficient copy number and sufficiently peripheral have bound enough of its appropriate Fab to inhibit cell adhesion by simple steric exclusion? Fab is a relatively large molecule which, once it has used up its single antigen

Fig. 1.3. The lack of inhibition of cell adhesion by the fibronectin tetrapeptide RGDS in the absence of exogenous fibronectin. Histogram of BHK cell attachment (cells cm^{-2}) in 20 minutes at 37 °C to chloric-acid treated polystyrene (Chloric) or to absorbed fibronectin (FN) in the presence or absence of fibronectin tetrapeptide in solution (+ or −) and in the presence of 25 $\mu g\ ml^{-1}$ of cycloheximide (CH) to inhibit protein synthesis. All cell suspensions were prepared with leupeptin to inhibit trypsin action. Error bar: one standard deviation. Reproduced from *Journal of Cell Science* by kind permission of the Company of Biologists.

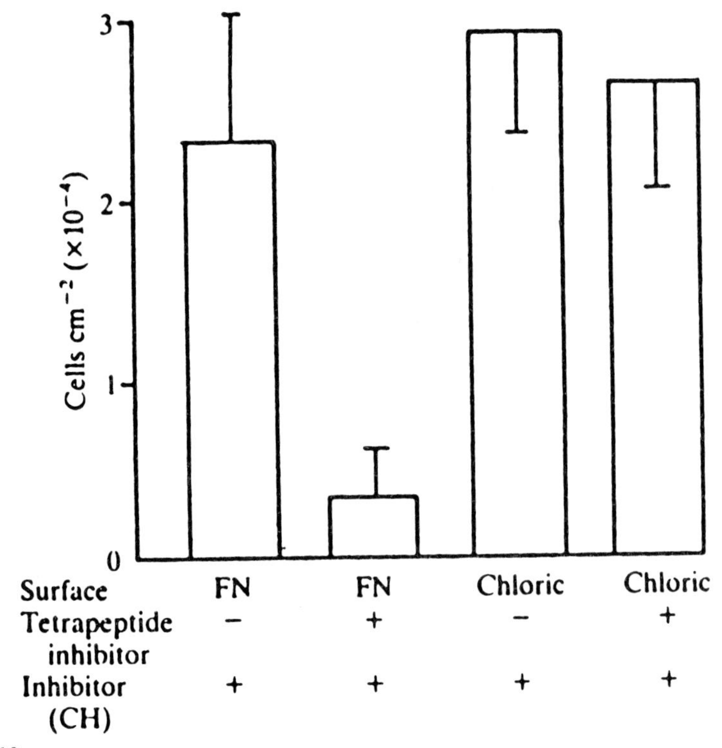

binding site, has little chance of binding by some less specific mechanism to plasmalemmae.

Amongst the controls that should be used in such experiments are the demonstration that several different antigens present in reasonable copy number peripherally on the cell surface elicit Fabs that do not inhibit cell adhesion. Such controls have rarely been carried out to date and the plethora of CAMs being reported (see Edelman & Thiery, 1986; Curtis, 1987) led to the following experimental question.

Would any artificial antigen inserted into the cell surface, which was without effect on adhesion or behaviour, elicit a Fab which would inhibit cell adhesion? If this was so, the possibility of steric exclusion, envisaged by Gerisch, would have been shown to be a real possibility. Two different antigens were used in this work:

(i) 'Monovalent' fluorescein labelled Concanavalin A.
(ii) A 9 Kd dextran stearoylated at a 16 : 1 mole ratio and bearing tetramethyl rhodamine groups at a 10 : 1 mole ratio. This high stearoylation is required to obtain good incorporation in the cell surface.(A 40 kd stearoylated dextran antigen was also used.)

The fluorescent labels are probably haptenic and in any event make it easy to detect cells bearing the antigen. Antibodies were raised in rabbits to these antigens: in the case of the stearolyated TMR dextran the antigen was incorporated in BHK cells which were then used to raise antibodies which were then exhaustively adsorbed out to remove anti-BHK specificities.

Fabs were prepared from these antibodies by standard methods and also from control equine IgG. They were tested on cell-to-cell adhesion and on cell-to-substrate adhesion at various levels, see Figs 1.4 to 1.6. Full details of the experimental conditions are given in the legends to the figures.

Clearly, Fabs raised against an artificial antigen incorporated at fairly high level (*c.* 2 picogram per cell) inhibit cell to cell adhesion and diminish cell to substrate adhesion even though these Fabs have no effect in the absence of the antigen and though the antigen by itself has no effect on cell adhesion. This result should urge caution in the interpretation of experiments using Fab to study cell adhesion. Obviously Fabs will detect cell adhesion molecules, but we must have recourse to other lines of evidence, such as the behaviour of mutants lacking the molecule, or the effects of insertion of the CAM into deficient cells in restoring function, to strengthen the possibility that the effects are not merely due to the type of steric exclusion that presumably operates with the two Fab–antigen systems described above.

The nature of the contact: a need for new methods

The cells adhesions that form with eukaryote cells in the course of an adhesion in suspension experiment or in the course of settling on to substrates appear, from rather limited reports, to be ones in which the plasmalemmae are about 10 nm apart, see discussion by Curtis (1987). Tight junctions and desmosomes form over a period of about 6 hours. Thus the question arises as to whether the molecules involved in adhesion actually traverse this gap

Fig. 1.4. Effects on cell adhesion of Fabs against artificial antigens incorporated into the cell surface. Monovalent FITC Concanavalin A. Attachment of cells to substrate (tissue culture polystyrene) in 20 minutes at 37 °C in upper histogram and aggregation (lower histogram) in presence and absence of monovalent FITC Concanavalin A (Con A) bound to BHK cells (approx 2 pg per cell) and the presence of whole antibody (Ab) 50 $\mu g\ ml^{-1}$ against FITC-succinylated Con A or its Fab (25 $\mu g\ ml^{-1}$. Aggregation (adhesion in suspension) is measured by counting the number of unaggregated cells and expressing adhesion as a percentage of all cells. Dotted lines in upper histograms indicate 1 standard deviation below the means.

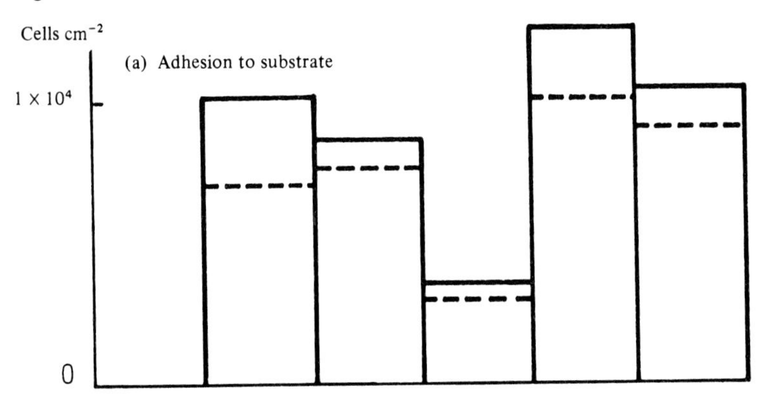

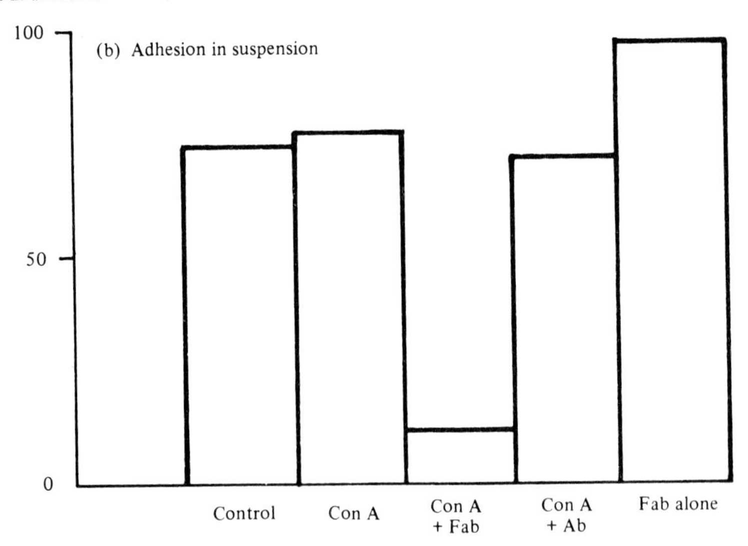

Fig. 1.5. and 1.6. Effects on cell adhesion of Fab against artificial antigens incorporated into the cell surface. The antigen is TRITC labelled stearoylated dextran (40 Kd dextran). Fig. 1.5. Adhesion of BHK cells bearing 2 pg per cell of the antigen (AG on cell) or no antigen to tissue culture polystyrene in the presence of various levels of Fab against the antigen in the medium. Note that considerable inhibition of adhesion can be achieved though this disappears at high levels of Fab. The lack of inhibition at high levels has been found by other workers using polyclonal antibodies. Subsidiary controls showed that whole antibody was without effect on adhesion at these levels. Fig. 1.6. The effect of the same antibody on cell to cell aggregation. Aggregation measured as

$$\ln \frac{\text{cell count at 300 min}}{\text{cell count at 0 min}}.$$

Antigen at the same level as in Fig. 1.5. Error bar: one standard deviation. Note that Fab only inhibits the adhesion of antigen-bearing cells. A non-specific Fab (from whole equine immunoglobulin) was used as a control (Ag + non-sp).

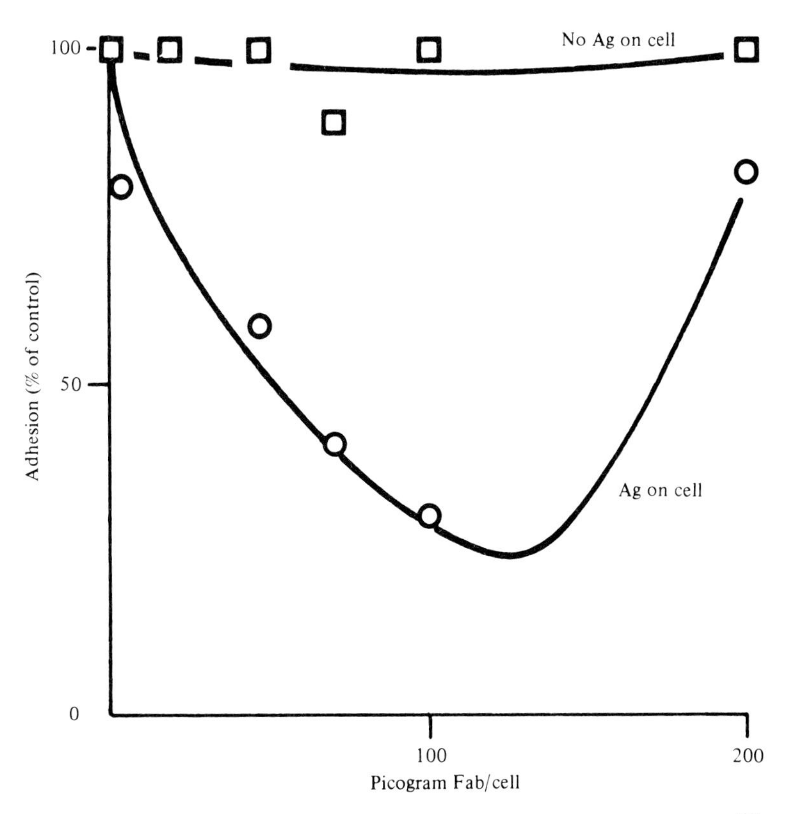

Fig. 1.6.

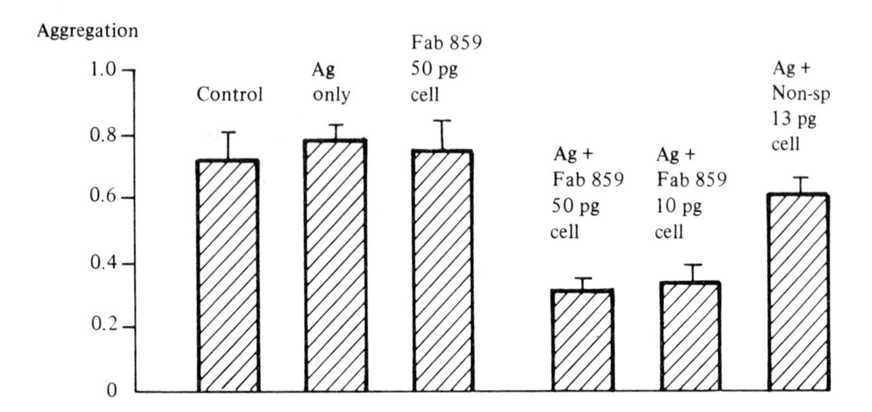

to provide direct molecular contact from one cell surface to another to more or less specific reception sites. Two other possibilities exist: (1) that though there is a degree of molecular entanglement (see Rutter, 1980) in the region between the surfaces there is nothing corresponding to a specific molecular interaction. Nevertheless such sterically destabilised systems can display considerable, if sometimes bizarre specific behaviour; (2) that the surfaces really adhere by long-range force systems, also see Rutter, 1980.

Clearly, techniques are needed to answer this question. Two suggestions about appropriate techniques will be made. The first is Total Internal Reflection Fluorescence Microscopy (TIRFM), (Weis, Balakrishnan, Smith & McConnell, 1982; Axelrod, 1981; Gingell, Todd & Bailey, 1985). In this technique use is made of the fact that at the total internal reflection angle there is a very slight penetration into the second medium of the evanescent wave. Thus, if a cell lies on top of a prism illuminated from below at the internal reflection angle the evanescent wave will penetrate about 100 nm into the cell or into any gap in the contact between the cell and substrate. If a fluorescent tracer molecule can permeate the gap it will then fluoresce. Thus the reality of the gap and its 'exclusion' properties (like a molecular sieve gel) can be investigated.

The second technique is to use Fluorescence Energy Transfer techniques. These have been widely used to investigate intramolecular spacings (Stryer, 1978) but do not seem yet to have been applied to the problem of cell contact structure. Curtis (1985) has proposed that this should be done, because only those molecules located within 5 nm of each other will be detected and quantitative methods should allow closer approaches to be detected. In the method, cells are allowed to attach to a substrate which incorporates a fluorochrome. The cells bear a fluorescent label whose excitation wavelength is the emission wavelength of the first subjacent fluorochrome present beneath the cells. If energy transfer occurs when the first fluorochrome is stimulated, it is detected by appearance of emission from the fluorochrome on the cells.

References

Akiyama, S. K., Yamada, S. S. & Yamada, K. M. (1985). Characterization of a 140 kD avian cell surface antigen as a fibronectin-binding molecule. *J. Cell Biol.*, **102**, 442–56

Ali, N. N. & Chambers, T. J. (1983). The effect of prostaglandin I_2 and 6 alpha-carba-PGI_2 on the motility of isolated osteoclasts. *Prostaglandins*, **25**, 603–14

Avnur, Z. & Geiger, B. (1981). The removal of extracellular fibronectin from areas of cell–substrate contact. *Cell*, **25**, 121–32

Axelrod, D. (1981). Cell–substrate contacts illuminated by total internal reflection fluorescence. *J. Cell Biol.*, **89**, 141–5

Barnes, D. W. & Silnutzer, J. (1983). Isolation of human serum spreading factor. *J. Biol. Chem.*, **258**, 12548–552

Bell, G. I. & Torney, D. C. (1985). On the adhesion of vesicles by cell adhesion molecules. *Biophys. J.*, **48**, 939–47

Beller, D. I., Springer, T. A. & Schreiber, R. D. (1982). Anti-Mac-1 selectively inhibits the mouse and human type three complement receptor. *J. Exp. Med.*, **156**, 1000–9

Berridge, M. J. & Irvine, R. F. (1984). Inositol triphosphate, a novel second messenger in cellular signal transduction. *Nature (Lond.)*, **312**, 315–21

Beug, H., Katz, F. E., Stein, A. & Gerisch, G. (1973) Quantitation of membrane sites in aggregating *Dictyostelium discoideum* by use of tritiated univalent antibody. *Proc. Natl Acad. Sci. USA*, **70**, 3150–4

Bongrand, P., Capo, C. & Depieds, R. (1982). Physics of cell adhesion. *Prog. Surface Sci.*, **12**, 217–86

Buchanan, M. R., Vasquez, M. J. & Gimbrone, M. A. (1983). Arachidonic acid metabolism and the adhesion of human polymorpho-nuclear leukocytes to cultured endothelial cells. *Blood*, **82**, 889–95

Burger, M. M., Burkart, W., Weinbaum, G. & Jumblatt, J. (1987). Cell–cell recognition: molecular aspects. Recognition and its relation to morphogenetic processes in general. In *Cell–cell Recognition* Symp. Soc. Exp. Biol., **32**, 1–23

Clark, P., Connolly, P., Curtis, A. S. G., Dow, J. A. T. & Wilkinson, C. D. W. (1987). Topographical control of cell behaviour. I. Single step cues. *Development*, **99**, 439–48

Cohen, H. J., Chovaniec, M. E., Takahashi, K. & Whitin J. C. (1986). Activation of human granulocytes by arachidonic acid: its use and limitations for investigating granulocyte functions. *Blood*, **67**, 1103–9

Cuatrecasas, P. & Hollenberg, M. D. (1976). Membrane receptors and hormone action. *Adv. Protein Chem.*, **30**, 252–451

Curtis, A. S. G. (1961). Timing mechanisms in the specific adhesion of cells. *Exp. Cell Res. Suppl.*, **8**, 107–22

Curtis, A. S. G. (1985). Cellular and molecular control of cell interactions in embryogenesis. In *Cellular and molecular control of direct cell interactions.* ed. H. J. Marthy. pp. 27–40. NATO Scientific Affairs Division. New York: Plenum Press

Curtis, A. S. G. (1987). The data on cell adhesion (in press)

Curtis, A. S. G. & Forrester, J. V. (1984). The competitive effects of serum proteins on cell adhesion. *J. Cell Sci.*, **71**, 17–35
Curtis, A. S. G., Forrester, J. V. & Clark, P. (1986). Substrate hydroxylation and cell adhesion. *J. Cell Sci.* (in press)
Curtis, A. S. G., Forrester, J. V., McInnes, C. & Lawrie, F. (1983). Adhesion of cells to polystyrene surfaces. *J. Cell Biol.*, **97**, 1500–6
Curtis, A. S. G. & Hocking, L. M. (1970). Collision efficiency of equal spherical particles in a shear flow. The influence of London-van der Waals forces. *Trans. Farad. Soc.*, **66**, 1381–96
Curtis, A. S. G., Kerr, K. & Knowlton, N. (1982). Graft rejection in sponges. *Transplantation*, **33**, 127–33
Curtis, A. S. G. & McMurray, H. (1986). Conditions for fibroblast adhesion without fibronectin. *J. Cell Sci.* (in press)
Curtis, A. S. G. & Verde, M. (1964). Control of cell behaviour – topological factors. *J. Natl Cancer Inst.*, **33**, 15–26
Dunn, G. A. (1982). Contact guidance of cultured tissue cells: a survey of potentially relevant properties of the substratum. In *Cell Behaviour* eds. R. Bellairs, A. S. G. Curtis & G. Dunn. Cambridge University Press: Cambridge, pp. 247–80
Duszyk, M., Kawalec, M. & Doroszewski, J. (1986). Specific cell-to-cell adhesion under flow conditions. *Cell Biophys.*, **8**, 131–9
Edelman, G. M. & Thiery, J-P. (1986). *The cell in contact: adhesions and junctions as morphogenetic determinants.* John Wiley: New York, pp. 507
Edwards, J. G., Robson, R. T. & Campbell, G. (1987). A major difference between serum and fibronectin in the divalent cation requirement for adhesion and spreading of BHK21 cells. *J. Cell. Sci.*, **87**, 657–65
Garrod, D. & Nicol, A. (1983). Cell behaviour and molecular mechanisms of cell–cell adhesion. *Biol. Rev.*, **56**, 199–242
Gerisch, G. (1980). Univalent antibody fragments as tools for the analysis of cell interactions in Dictyostelium. *Curr. Topics Devel. Biol.*, **14**, 243–70
Gingell, D., Todd, I. & Bailey, J. (1985). Topography of cell-glass apposition revealed by total internal reflection fluorescence of volume markers. *J. Cell Biol.*, **100**, 1334–8
Grinnell, F. (1978). Cellular adhesiveness and the extracellular matrix. *Int. Rev. Cytol.*, **53**, 65–144
Grinnell, F. & Feld, M. (1979). Initial adhesion of human fibroblasts in serum-free medium: possible effect of secreted fibronectin. *Cell*, **17**, 117–29
Hed, J. & Stendahl, O. (1982). Differences in the injection mechanism of IgG and C3b particles in phagocytosis by neutrophils. *Immunology*, **45**, 727–36
Hoffman, S. & Edelman, G. M. (1982). Kinetics of homophilic binding by embryonic and adult forms of the neural cell adhesion molecule. *Proc. Natl Acad. Sci. USA*, **80**, 5762–66
Hopkins, C. R., Miller, K. & Beardmore, J. M. (1985). Receptor-mediated endocytosis of transferrin and epidermal growth factor receptors: a comparison of constitutive and ligand-induced uptake. *J. Cell Sci. Suppl.*, **3**, 173–86

Hynes, R. (1985). Molecular biology of fibronectin. *Ann. Rev. Cell Biol.*, **1**, 67–90
Kruskal, B., Skak, S. & Maxfield, F. R. (1986). Spreading of human neutrophils is preceded by a large increase in cytoplasmic free calcium. *Proc. Natl Acad. Sci. USA*, **83**, 2919–23
Lackie, J. M. & Smith, R. P. C. (1980). Interactions of leukocytes and endothelium. In *Cell adhesion and motility* ed. A. S. G. Curtis & J. Pitts. Cambridge University Press: Cambridge, pp. 235–72
Lefkowitz, R. J., Stadel, J. M. & Caron, M. G. (1983). Adenylate cyclase-coupled beta-adrenergic receptors. *Ann. Rev. Biochem.*, **52**, 159–186
Lumsden, A. G. S. & Davies, A. M. (1986). Chemotropic effect of specific target epithelium in the developing mammalian nervous system. *Nature Lond.*, **323**, 538–9
Nir, S., Bentz, J., Wilschutz, J. & Duzgunes, N. (1983). Aggregation and fusion of phospholipid vesicles. *Prog. Surface Sci.*, **13**, 1–124
Pierschbacher, M. D. & Ruoslahti, E. (1984). Cell attachment activity of fibronectin can be duplicated by small fragments of the molecule. *Nature (Lond.)*, **309**, 30–3
Rubin, K., Johansson, S., Hook, M. & Obrink, B. (1981). Substrate adhesion of rat hepatocytes. On the role of fibronectin in cell spreading. *Exp. Cell Res.*, **135**, 127–35
Rutter, P. R. (1980). The physical chemistry of the adhesion of bacteria and other cells. In *Cell adhesion and motility* ed. A. S. G. Curtis & J. Pitts. Cambridge University Press: Cambridge, pp. 103–35
Smith, C. W. & Hollers, J. C. (1980). Motility and adhesiveness in human neutrophils. *J. Clin. Invest.*, **65**, 804–12
Steinberg, M. S. (1978). Cell–cell recognition in multicellular assembly: levels of specificity. In *Cell–cell recognition*. Symposium Soc. Exp. Biol., **32**, 25–49
Stryer, L. (1978). Fluorescence energy transfer as a spectroscopic ruler. *Ann. Rev. Biochem.*, **47**, 819–46
Weis, R. M., Balakrishnan, K., Smith, B. A. & McConnell, H. M. (1982). Stimulation of fluorescence in a small contact region between rat basophil leukemia cells and planar lipid membrane targets by coherent evanescent radiation. *J. Biol. Chem.*, **257**, 6440–5
Yamada, K. M. (1983). Cell surface interactions with extracellular materials. *Ann. Rev. Biochem.*, **52**, 761–99
Zigmond, S. H. (1982). Polymorphonuclear leukocyte response to chemotactic gradients. In *Cell Behaviour* eds R. Bellairs, A. S. G. Curtis & G. A. Dunn. Cambridge University Press: Cambridge, pp. 183–202

2 Molecular mechanisms regulating cell surface organisation and their relationship to tumour survival

Kermit L. Carraway, Julie Spielman and Coralie A. Carothers Carraway

Departments of Anatomy & Cell Biology and Biochemistry,
University of Miami School of Medicine,
Miami, Florida 33101, USA

Abstract

Molecular aspects of the composition and organisation of the cell surfaces of metastatic 13762 mammary ascites adenocarcinoma cells have been investigated to try to understand their ability to escape immune destruction mechanisms. These cells contain a cell surface complex of a large, highly glycosylated sialomucin (ASGP-1) and a second glycoprotein (ASGP-2), representing as much as 1% of the total cell protein. Regulation of the mobility of the ASGP-1/ASGP-2 complex in the plasma membrane has been related to membrane–microfilament interactions through studies of microvilli isolated from the ascites cells. A transmembrane complex containing actin and a third cell surface glycoprotein (CAG) has been isolated from microvilli of cells with mobile cell surface components. In cells with immobile components a third component (58 kDa protein) is isolated with the transmembrane complex. Transmembrane complexes have been shown to be directly associated with microfilaments, thus providing a mechanism for regulating cell surface organisation.

Introduction

The capacity of tumour cells to grow and disseminate depends on their ability to evade the immune recognition and destruction mechanisms of the host. A number of evasion mechanisms have been proposed (Hanfland & Uhlenbruck, 1978), including masking, shedding, and endocytosis of tumour antigens and alterations of antigen mobility in the plasma membrane. Research has focussed on trying to understand the molecular aspects of some of these evasion mechanisms as a means of developing approaches to preventing metastatic spread of tumours. All of these mechanisms depend on the regulation of molecular organisation at the cell surface. Molecular organisation is based on two considerations: composition, which is determined by the synthesis, incorporation and degradation of cell surface molecules, and topography, which is determined by the mobility of molecules in the membrane. Both of these aspects of organisation can be important in evasion mechanisms.

Studies by the authors in this area have focussed on metastatic 13762 rat ascites mammary tumour cells (Carraway *et al.*, 1978). The 13762 tumour was originally obtained as a product of dimethylbenzanthrene carcinogenesis as a solid tumour (Segaloff, 1966), from which ascites forms were derived by the Mason Research Laboratories, Worcester, Massachusetts. Sublines of these ascites cells were obtained and selected for studies on their cell surfaces (Sherblom *et al.*, 1980; Howard *et al.*, 1982). The work described here will concentrate on the MAT-Bl and MAT-Cl sub-lines because they represent the extremes of differences in cell surface properties of interest among the sub-lines that have been examined (Sherblom *et al.*, 1980; Carraway *et al.*, 1979).

Cell surface composition: the dominance of the sialomucin ASGP-1

The cell surface composition of the 13762 ascites cells has a preponderance of the sialomucin ASGP-1 (ascites sialoglycoprotein-1) (Sherblom *et al.*, 1980; Sherblom, Buck & Carraway, 1980), which is the major glycoprotein of these cells and a major cell surface protein. This dominance is best illustrated by metabolic labelling of the cells with glucosamine (Fig. 2.1). Analysis of the labelled glycoproteins by sodium dodecyl sulphate polyacrylamide gel electrophoresis (SDS PAGE) and fluorography shows only one labelled band unless the fluorograms are over-exposed. By slicing and counting the gels, the fraction of the labelled glucosamine in ASGP-1 was determined to be 70–90%, depending on the sub-line analysed (Sherblom, Buck & Carraway, 1980).

ASGP-1 could be readily isolated from the ascites cells by a two-step procedure (Sherblom, Buck & Carraway, 1980). A crude microsome fraction was prepared from the cells by homogenisation and differential centrifugation. These membranes were mixed with 4 M guanidine hydrochloride and centrifuged on a CsCl gradient. ASGP-1 migrated to a density indicative of a high carbohydrate content. By this method ASGP-1 of purity $>95\%$ was obtained for biochemical and biophysical studies. From isolation studies using leucine-labelled cells, it was estimated that ASGP-1 comprised $>0.5\%$ of the *total cell protein* of the ascites cells. From the sialic acid content (Buck, Sherblom & Carraway, 1979; Sherblom, Buck & Carraway, 1980) it was calculated that MAT-Cl cells contain about 6×10^6 molecules of ASGP-1. Table 2.1 shows some of the properties of MAT-Cl ASGP-1. Two features of ASGP-1, the size and the carbohydrate content, are particularly important and are suggestive of a mucin-type glycoprotein. More indicative are the high threonine and serine contents and the high content of galactosamine compared to the other sugars.

From the galactosamine content, carbohydrate content and molecular weight it was estimated that there are about 500 oligosaccharide chains per

ASGP-1 molecule (Carraway & Spielman, 1986). Therefore almost one of every three amino acids in ASGP-1 is glycosylated. This high degree of glycosylation should interfere with folding of the polypeptide and prevent assumption of a more compact conformation. This prediction has been confirmed by electron microscopy of rotary shadowed preparations of isolated ASGP-1, which show an extended rod-like structure terminated on one end by a small, ball-like region (Berman & Hull, unpublished observations). These results were interpreted to indicate that most of the ASGP-1 molecule is in the form of an extended polypeptide chain, presumably carrying the carbohydrate. It is envisioned that the spherical region is the site for association of the ASGP-1 with the plasma membrane.

This type of structure can explain the proposed function of ASGP-1 in masking cell surface antigens. The ASGP-1 glycosylated polypeptide will

Fig. 2.1. Predominance of glucosamine-labelled ASGP-1. SDS PAGE of glucosamine-labelled MAT-B1 and MAT-C1 cells. A, Coomassie blue stain; B, fluorogram. MAT-B1 cells are on left of each pair. Figure modified from Sherblom, *et al.* (1980).

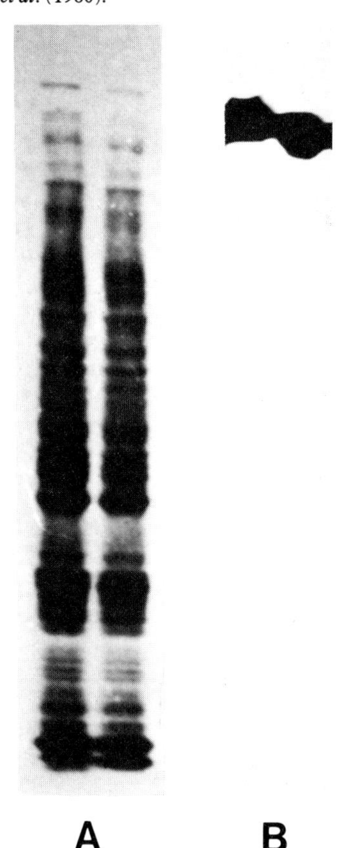

Table 2.1. *Properties of MAT-C1 ASGP-1*

Stokes radius (nm)	13.9
Sedimentation coefficient (S)	8.0
Partial molar volume (cm^3/g)	0.704
Molecular weight	690,000
Polypeptide molecular weight	186,000
Carbohydrate (%)	73
Serine (mole %)	16
Threonine (mole %)	18
oligosaccharides/polypeptide	500
GalNAc : GlcNAc : Gal : NANA	1 : 1.1 : 1.4 : 1.6

extend far beyond the structures of other cell surface molecules (Fig. 2.2). Because of the large number of ASGP-1 molecules on the cell surface, they will cover the cell surface rather efficiently. Electron microscopy of 13762 ascites cells stained with cationised ferritin show some regions of the cell surfaces with hexagonal packing of the ferritin (Carraway & Spielman, 1986). These results clearly show that ASGP-1 will provide an effective screen for the cell surface of the 13762 ascites cells, particularly against the approach of molecules or cells which also carry a negative charge.

Association of ASGP-1 with the cell surface: complex with ASGP-2

This role for ASGP-1 depends on its association with the plasma membrane. When membranes from 13762 ascites cells are extracted with Triton X-100 or deoxycholate and fractionated by gel filtration or velocity sedimentation on sucrose gradients, ASGP-1 is observed as a 1 : 1 complex with a second cell surface glycoprotein, ASGP-2 (Sherblom & Carraway, 1980*a*). ASGP-2 is the major Concanavalin A (Con A)-binding glycoprotein of the ascites cells and has an apparent molecular weight by SDS PAGE of 120,000. Based on its density on CsCl gradients, a carbohydrate content is estimated of about

Fig. 2.2. Model for antigen masking by sialomucins.

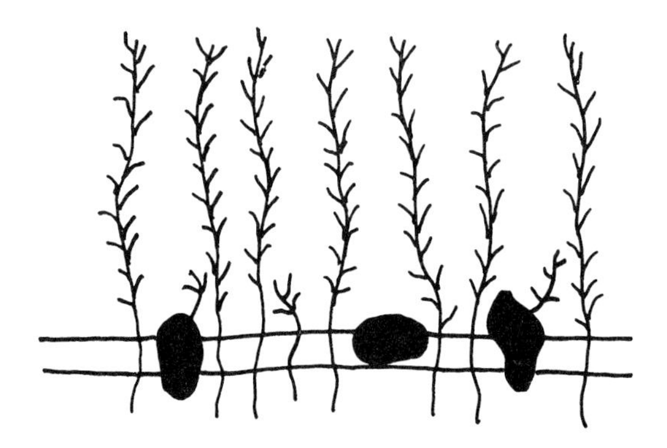

25%. The ASGP-1/ASGP-2 complex from MAT-B1 cells was found to have a sedimentation coefficient of 14.9, density of 1.44 g/cm^3 and a molecular weight of 640,000 (Sherblom & Carraway, 1980*a*).

Further evidence for the presence of the ASGP-1/ASGP-2 complex in the plasma membrane of the ascites cells was obtained by studies of the redistribution of MAT-B1 cell surface glycoproteins in the presence of lectins (Helm & Carraway, 1981). These studies were possible because ASGP-1 is the only detectable MAT-B1 glycoprotein which binds peanut agglutinin (PNA) (Huggins *et al.*, 1980), and ASGP-2 is the major MAT-B1 Con A-binding glycoprotein (Sherblom & Carraway, 1980*a*). When MAT-B1 cells (fixed or unfixed) are incubated with fluorescent PNA, the fluorescence is observed as a 'ring' around the periphery of the cell, indicating that no redistribution of the ASGP-1 has occurred with the PNA treatment (Helm & Carraway, 1981). When MAT-B1 cells are similarly incubated with fluorescent Con A, the fluorescence is observed as a 'cap', indicating that the Con A-binding molecules have been redistributed in the plane of the membrane by the Con A crosslinking. Fixed MAT-B1 cells treated with fluorescent Con A show a 'ring' distribution of fluorescence. When MAT-B1 cells are treated with fluorescent PNA followed by unlabelled Con A or with unlabelled Con A followed by fluorescent PNA, the fluorescence is observed as a 'cap', indicating that ASGP-1 is associated with a Con A-binding molecule in the MAT-B1 cells (Helm & Carraway, 1981). Together with the studies on the isolation of the ASGP-1/ASGP-2 complex, these results clearly demonstrate the association of ASGP-1 and ASGP-2 in the membrane of the MAT-B1 cells.

The mode of association of the complex with the membrane remains uncertain. The fact that ASGP-1 is removed from the membrane by the centrifugation in guanidine hydrochloride and CsCl suggests that it is not a strongly bound integral membrane protein (Sherblom, Buck & Carraway, 1980). Moreover, most of the ASGP-1 can be removed from the cell surface by trypsinisation, indicating that only a small segment is involved in the membrane binding. However, extraction of ascites membranes with 0.1 M sodium carbonate (pH 11) does not remove either ASGP-1 or ASGP-2 (Carraway & Spielman, 1986), suggesting that neither is a peripheral membrane protein. From these results, two possible reasonable explanations for the association of ASGP-1 with the membrane remain, as shown in Fig. 2.3. 1) ASGP-1 is associated strongly with ASGP-2, but is not associated with the bilayer at all. 2) ASGP-1 is associated weakly with the bilayer and strongly with ASGP-2.

ASGP-1 oligosaccharides

Oligosaccharitols can be readily isolated from ASGP-1 after alkaline borohydride elimination to release the carbohydrate chains from the polypeptide (Sherblom, Buck & Carraway, 1980; Hull *et al.*, 1984). All of the oligosaccharides are derivatives of the Gal-GAlNAc disaccharide, which is a common precursor of mucin oligosaccharides and the determinant for PNA binding. The structures of the five major oligosaccharides from MAT-B1 and MAT-C1 cells were found to be the following (Hull *et al.*, 1984):

Structure	
NANA(α2,3)Gal(β1,4)GlcNAc(β1,6)GalNAc \|(β1,3) NANA(α2,3)Gal	I
NANA(α2,3)Gal(β1,4)GlcNAc(β1,6)GalNAc \|(β1,3) Gal	II
Gal(β1,4)GlcNAc(β1,6)GalNAc \|(β1,3) Gal	III
Gal(α1,3)Gal(β1,4)GlcNAc(β1,6)GalNAc \|(β1,3) NANA(α2,3)Gal	IV
Gal(α1,3)Gal(β1,4)GlcNAc(β1,6)GalNAc \|(β1,3) Gal	V

These structures are all based on a common tetrasaccharide core. Surprisingly, in view of the strong binding of PNA by ASGP-1, there is little of the disaccharide Gal-GalNAc, although it is obviously a precursor to all of these oligosaccharides. Moreover, with one exception little or no trisaccharide precursor Gal(GlcNAc)GalNAc is observed. The exception, sulphated oligosaccharides, is an interesting one. About 20% of the MAT-B1 oligosaccharides are sulphated; none of the MAT-C1 oligosaccharides are

Fig. 2.3. Likely models for association of ASGP-1 with plasma membrane.

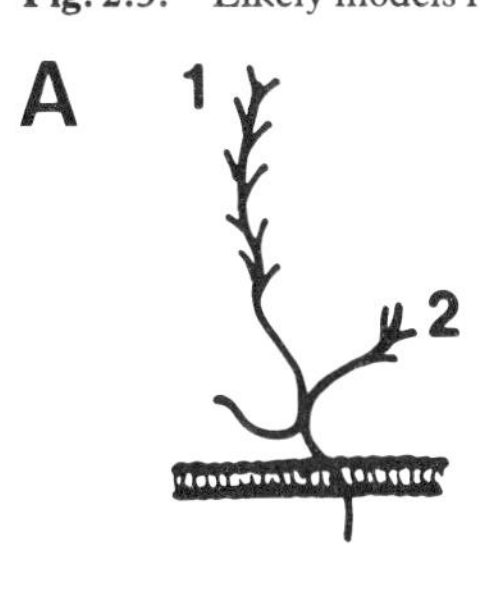

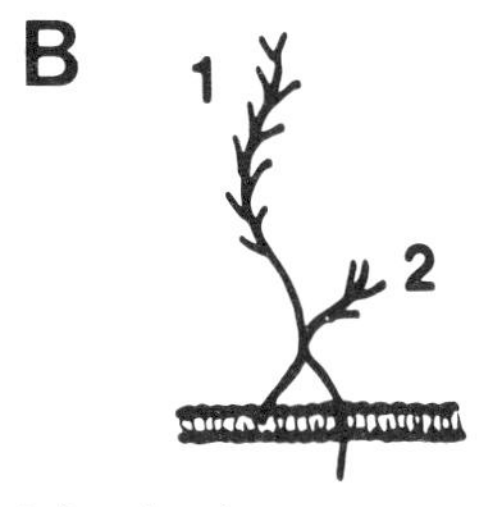

sulphated (Sherblom & Carraway, 1980*b*). Moreover, the major sulphated products are the trisaccaride and its sialated derivative tetrasaccharide. These results suggest that sulphation may act as a chain terminating event and that there is a difference in the regulation of oligosaccharide synthesis between the MAT-B1 and MAT-C1 sub-lines. Fucosylated derivatives of III and V have also been observed, but their exact structures are not yet known.

Expression of ASGP-1: biosynthesis and shedding

The amount of ASGP-1 at the cell surface is regulated by the balance between its biosynthesis and degradation. Expression of ASGP-1 at the cell surface requires four operations: synthesis of the polypeptide, glycosylation, transit through the cell and incorporation into the plasma membrane. To understand the biosynthesis of ASGP-1, the time and location of each of these operations must be determined. The appearance of ASGP-1 at the cell surface can be determined by combined labelling and trypsinisation, since trypsin will cleave >70% of the ASGP-1 at the cell surface (Spielman, Rockley & Carraway, 1987). If cells are labelled with glucosamine or galactose and treated with trypsin at timed intervals after labelling, the time required for label to appear in the trypsin-released fragments will correspond to the minimum time from final glycosylation to appearance at the cell surface. Since the final glycosylation events presumably occur in the *trans* Golgi, this time should also correspond to the minimum transit time from the *trans* Golgi to the plasma membrane. As shown in Fig. 2.4, the time required is about 10 min. By using amino acid label, one can similarly determine the minimum time from polypeptide synthesis to appearance at the cell surface, which should be the minimum transit time from the rough endoplasmic reticulum to the plasma membrane. About 80 min is required for this transit (Fig. 2.4). By analysing the amount of labelled ASGP-1 remaining in the cells after trypsinisation after various times of labelling, the half-time for appearance of ASGP-1 at the cell surface can also be estimated to be >4 h (Spielman *et al.*, 1987).

To determine the minimum time required for glycosylation of ASGP-1, cells were labelled with amino acid and specifically precipitated ASP-1 from cell lysates at timed intervals with soluble PNA (Spielman *et al.*, 1987). If O-glycosylation were initiated in the Golgi apparatus, as proposed by some workers, there should be a lag before appearance of amino acid label in the PNA precipitates. In contrast, if O-glycosylation is co-translational, no lag time should be observed. In fact, no lag was observed. Moreover, the fraction of glycosylated sites in ASGP-1 at any time after polypeptide synthesis could be quantified by using labelled threonine and subjecting the PNA precipitates to alkaline borohydride treatment and amino acid analysis. The alkaline borohydride elimination of the oligosaccharides converts all glycosylated threonine residues to 2-aminobutyric acid. By this method it was shown that 3 and 15% of the ASGP-1 threonines were glycosylated after 5 and 15 min of threonine labelling, respectively. Thus, glycosylation must be co-translational, or the time required to initiate glycosylation after polypeptide synthesis must be very short. These results strongly suggest that initiation of O-glycosylation occurs in the rough endoplasmic reticulum.

To determine whether synthesis of individual oligosaccharides can also be initiated late in the biosynthesis and transit process, the individual amino sugars associated with the ASGP-1 fragments released by trypsin from glucosamine-labelled cells were analysed (Spielman *et al.*, 1987). Since galactosamine is found only attached to the polypeptide, the minimum time for the appearance of labelled galactosamine in the fragments will correspond to the minimum time from the latest initiation event to appearance at the cell surface. Labelled galactosamine was found within 5–10 min, indicating that initiation of oligosaccharide formation can occur late as well as early in the biosynthesis process.

These results support a continuous model of glycosylation, in which oligosaccharide formation can be initiated on the polypeptide at any time from the time of translation to just before appearance at the cell surface. The continuous initiation of oligosaccharides and the long transit time through the cell may be required for the high degree of glycosylation found on ASGP-1.

Turnover of cell surface ASGP-1 is relatively slow, with a half-life of 2–4 days, depending on the subline (Howard *et al.*, 1981). Degradation can occur by two pathways, internalisation or shedding. Since antigen shedding may contribute to tumour survival, the mechanisms by which ASGP-1 is shed were investigated. Two pathways were found. 1) ASGP-1 can be shed in a particulate form, associated with microvilli or vesicles released from the cell surface. 2) ASGP-1 can be shed in soluble form. Shed, soluble ASGP-1 is smaller (by both gel filtration and velocity sedimentation) and

Fig. 2.4. Kinetics of appearance of amino acid- and carbohydrate-labelled ASGP-1 at MAT-B1 cell surface.

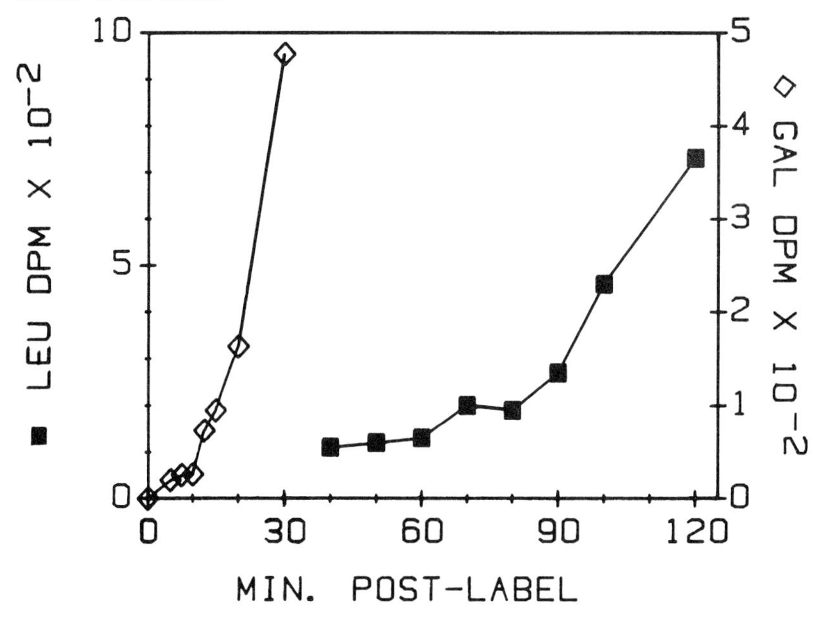

denser (by CsCl centrifugation) than ASGP-1 isolated from membranes, yet has the same relative sialic acid content. These results suggest that the shed ASGP-1 is a fragment of the membrane form, released by proteolysis. The fact that the protease inhibitor aprotinin inhibits ASGP-1 shedding also supports a proteolytic mechanism (Howard *et al.*, 1981).

Expression of ASGP-1 depends on the environment in which the cell is grown. When cells are grown in ascites form, ASGP-1 is abundant. If ascites cells are transferred to culture or grown as solid tumours, the amount of ASGP-1 in the cells decreases (Huggins *et al.*, 1980). When 13762 cultured cells, which have no detectable ASGP-1, are injected intraperitoneally, ascites cells grow which have no ASGP-1. Upon passage of these cells weekly in ascites form, ASGP-1 appears abruptly at about passage 18 (Howard *et al.*, 1980). Results of these studies indicate that ASGP-1 is not required for growth of the tumour cells in ascites form and that the expression of ASGP-1 is regulated in some as yet unknown manner.

Anchorage of the ASGP-1/ASGP-2 complex in the plasma membrane

A major difference between MAT-B1 and MAT-C1 cells is the mobility of the ASGP-1/ASGP-2 complex, Con A-binding proteins or other cell surface antigens (Jung *et al.*, 1984; Carraway *et al.*, 1979). Since receptor mobility may be regulated by attachment of plasma membrane molecules to the cytoskeleton (Geiger, 1983), we examined the anchorage of the ASGP-1/ASGP-2 complex to the cytoskeleton in the presence of Con A. Extraction of either MAT-B1 or MAT-C1 cells with Triton X-100 releases 70–80% of the ASGP-1/ASGP-2 complex from the cells. However, if cells are first treated with Con A, only 25–35% of the complex is released. The remainder is associated with the nuclear/cytoskeleton residue. The same experiment can be performed with microvilli isolated from the two sublines. Microvilli are obtained from the cells by a gentle shearing procedure which does not disrupt the cell bodies. As a result the microvilli have no contaminating nuclei or other cellular organelles (Carraway *et al.*, 1980; C. Carraway *et al.*, 1982). They contain only microfilaments, microfilament-associated proteins and cytoplasmic proteins enclosed in a sealed microvillar membrane. From the microvilli, 75–80% of the complex could be extracted in the absence of Con A, but only 10–20% in its presence. The unextractable complex in the presence of Con A was clearly associated with the microfilament core (Jung *et al.*, 1984).

To determine the mode of association of ASGP-1/ASGP-2 complex with the core, we examined the effects of Con A on the extraction of the complex in microvillar membranes (Jung *et al.*, 1984). These membranes were prepared under conditions which deplete microfilaments (K. Carraway *et al.*, 1982; C. Carraway *et al.*, 1982). MAT-B1 and MAT-C1 microvillar membranes were extracted in the absence and presence of Con A and examined by SDS PAGE. The MAT-B1 membrane residues, obtained in the absence of Con A, contained two major components, actin and a cell surface Con A-binding glycoprotein, designated CAG (cytoskeleton-associated glycoprotein) (Carraway *et al.*, 1983*c*). MAT-B1 membrane residues, obtained in the presence of Con A, also contained ASGP-1, ASGP-2 and Con A. Results with MAT C1 membranes were similar, but both Con A-treated and untreated residues contained an additional 58,000 D polypeptide (58 K) (C. Carraway *et al.*, 1982). The association of these six proteins (ASGP-1, ASGP-2, CAG, actin, 58 K and Con A) was demonstrated directly by their comigration on CsCl gradients in the presence, but not absence, of Con A. The high density of the ASGP-1 sediments the other proteins into the gradient when they are associated (Jung *et al.*, 1984).

Since both ASGP-2 and CAG are Con A-binding glycoproteins, the simplest interpretation of these results is that Con A cross-links ASGP-2 to CAG at the cell surface. If CAG is linked to the microvillar microfilaments, as suggested by the presence of actin, these crosslinks would effectively link the ASGP-1/ASGP-2 complex to the cytoskeleton (Fig. 2.5). These results also show that the difference in ASGP-1/ASGP-2 complex mobility between MAT-B1 and MAT-C1 cells is not due to the anchorage to the cytoskeleton, since complex becomes anchored in both sublines. This difference must relate to events occurring subsequent to anchorage, most likely involving the reorganisation of the microfilaments and/or their association with the membrane.

Isolation and characterisation of the CAG–actin transmembrane complex

The results described above suggest that the ASGP-1/ASGP-2 cell surface complex is linked to the cytoskeleton via a second complex, a transmembrane complex, composed of actin and CAG, with the possible participation of a third protein in the MAT-C1 cells. To investigate the nature of this second putative complex, we fractionated Triton extracts of microfilament-depleted MAT-B1 or MAT-C1 microvillar membranes by gel filtration or velocity sedimentation sucrose density gradient centrifugation. In all cases CAG and actin were found to comigrate, and, in the extracts of MAT-C1 membranes, 58 K also comigrated (Carraway, Jung & Carraway, 1983*a*). By two-dimensional isoelectric focussing-SDS PAGE no other proteins were found present in the complexes. The ratio of the three components was approximately 1 : 1 : 1, as estimated by leucine labelling.

Further evidence for the association of the three components was obtained by Con A treatment of the complex obtained by gel filtration. Con A precipitated the three components in equivalent amounts (C. Carraway, Jung, Metcalf, Andrews & Carraway, submitted for publication). The stability of these associations was illustrated by extraction studies on the microvilli. When microvilli were treated with 0.1 M sodium carbonate (pH 11), most of the microvillar proteins were released, including 90% of the actin. However, ASGP-1, ASGP-2, CAG and 58 K were quantitatively retained with the membranes, indicating that they are strongly bound to the membrane. Moreover, about 10% of the microvillar actin, an amount equivalent to the amount of CAG and 58 K in the microvilli, was also retained. When microvilli were extracted with glycine/EDTA (pH 9.5) in Triton, a residue was obtained which contained primarily CAG, actin and 58 K in equivalent amounts (Carraway, Jung, Metcalf, Andrews, Carraway, submitted). These results indicate that the transmembrane complex is stable to the strong dissociating conditions used for removing peripheral proteins from membranes or for depolymerising microfilaments.

Isolation and characterisation of CAG

The results described above and subsequent studies showed that the CAG-actin complex could only be disrupted by strong dissociating or denaturing conditions. CAG could be separated from the bulk of the actin by incubation of microvilli in 0.1 M sodium carbonate (pH 11) in Triton or in SDS followed by velocity sedimentation sucrose gradient centrifugation. In either case CAG migrated as a peak with a sedimentation coefficient of 20 S (Jung *et al.*, 1985). The size of CAG conflicted with molecular weight values (75,000–

Fig. 2.5. Model for anchorage of ASGP-1/ASGP-2 cell surface complex (CSC) to cytoskeleton by Con A bridge between ASGP-2 and CAG of the microfilament-associated transmembrane complex (TMC).

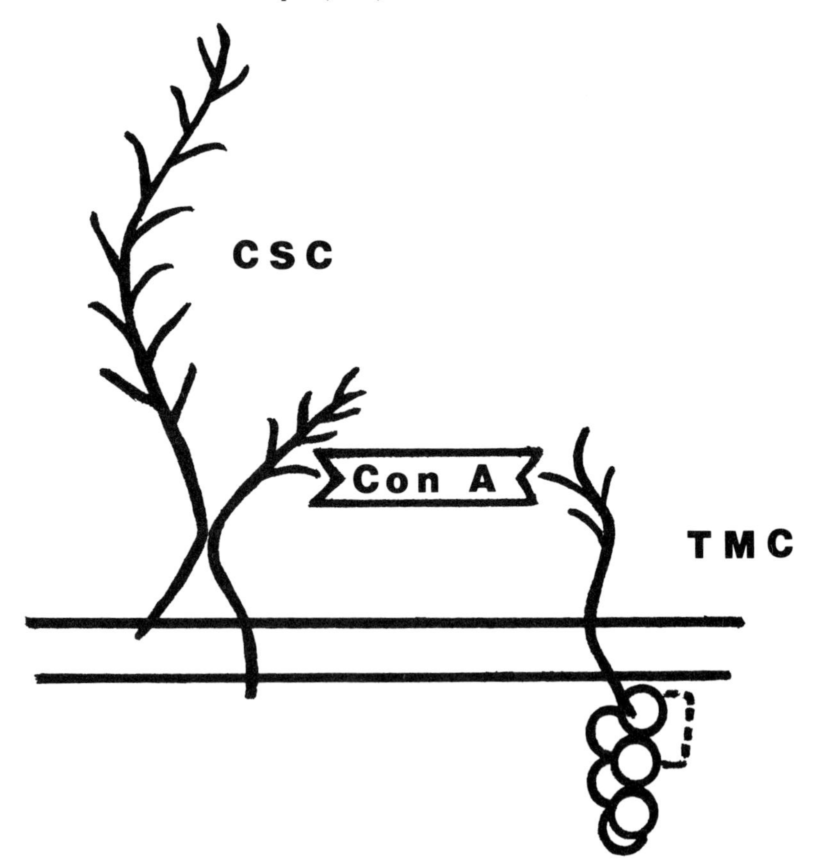

80,000) obtained from SDS PAGE under the usual reducing conditions (Carraway *et al.*, 1983). This discrepancy was resolved when CAG was analysed on velocity sedimentation gradients in the presence of reducing agents. Under reducing conditions CAG sedimented with a sedimentation coefficient of 5 S. These results indicated that CAG exists in the membrane as a large disulphide-linked multimer (Jung *et al.*, 1985).

These findings have permitted conditions to be devised for isolating reduced and unreduced CAG. For the former, microvilli are solubilised under non-reducing conditions in SDS and fractionated on a sucrose gradient. Because of its large size under nonreducing conditions CAG is separated from all of the microvillar proteins except ASGP-1, which overlaps with the CAG peak. The CAG fractions are then run on a second gradient under reducing conditions to separate CAG from ASGP-1 and obtain purified, reduced CAG (Jung *et al.*, 1985). To obtain unreduced CAG, Triton–glycine–EDTA extracts of microvilli are applied to a two-phase discontinuous gradient. The top phase contains Triton/PBS in 4% sucrose, while the bottom phase contains 2% SDS-phosphate. The top phase acts as a 'wash' layer through which microfilament fragments and transmembrane complex are sedimented into the SDS layer. This layer removes ASGP-1 and other membrane and soluble microvillar proteins. The CAG-enriched, ASGP-1-depleted SDS layer is heated briefly to complete dissociation of actin and CAG and applied to a sucrose gradient in SDS. From this second gradient unreduced CAG is obtained in purified form (Carraway, Jung, Metcalf, Andrews & Carraway, submitted).

CAG purified in this manner can be renatured to exhibit its functional capability, i.e. binding of actin. Using an ^{125}I-actin overlay (Snabes, Boyd & Bryan, 1981) on agarose–acrylamide gel, G-actin binding to CAG was demonstrated. The very large size and disulphide-linked structure of CAG led us to compare it to soluble IgM, which is also large and disulphide-linked. By gel filtration and velocity sedimentation analysis CAG was similar to IgM, indicating that it has an unreduced molecular weight of about 10^6. Immunoblot analyses using antibodies directed against rat IgM showed that CAG has at least one epitope in common with the heavy chain of IgM (Metcalf, Andrews, Carraway & Carraway, submitted). However, CAG does not appear to be a heterologous multimer, as is IgM. Results by the authors suggest that it is a multimer of about 12 equivalent, or very similar, subunits. Moreover, the fact that CAG has an external, glycosylated domain, binds tightly to the plasma membrane and has a site for binding actin suggests that it has at least three different domains. Additional studies are underway to define the domain structure of CAG and its relation to function.

Transmembrane complex association with microfilaments

Both CAG and 58 K are associated with microvillar microfilament cores obtained by sedimenting Triton extracts of microvilli at 10,000 g (Carraway *et al.*, 1985). The remainder of the microvillar CAG and 58 K is obtained by sedimenting the 10,000 g supernate at 100,000 g. This fraction containing primarily actin, CAG and 58 K is believed to be transmembrane complex which is unassociated with microfilaments (Carraway, Jung & Carraway,

1983*b*). To show that the transmembrane complex in the 10,000 g pellets is not due to co-pelleting, a procedure for identifying microfilament-associated proteins has been devised (Carraway & Weiss, 1985). When microvilli are extracted with Triton and applied to a velocity sedimentation gradient, the microfilament core migrates as a peak well into the gradient. The sedimentation rate and peak width depend on the stability of the core under the extraction conditions used. If the extraction is performed in the presence of phalloidin, the microfilament-stabilising toxin, the sedimentation rate of the core is increased, resulting in a shift of the microfilaments further into the gradient. Since phalloidin specifically stabilises the microfilaments, any protein which shifts with the microfilaments as a result of phalloidin treatment must be a microfilament-associated protein. The phalloidin shift technique has also been used to demonstrate the association of the cell surface enzyme 5′-nucleotidase with microfilaments in MAT-C1 microvilli (Carraway, Sindler & Weiss, 1986).

Since both CAG and 58 K comigrate with microfilaments in the presence and absence of phalloidin, the transmembrane complex must associate with microfilaments (Carraway, Jung, Metcalf, Andrews & Carraway, submitted). Further evidence for the association of CAG with microfilaments was obtained by polymerising actin in the presence and absence of purified CAG. Low speed sedimentation of the reaction mixtures pelleted actin only in the presence of CAG; CAG was quantitatively pelleted from this mixture. In contrast, actin was not pelleted in the absence of CAG, and CAG was not pelleted under these centrifugation conditions when incubated alone. These results show that CAG is associated with multiple microfilaments, since individual microfilaments will not pellet under these conditions. Moreover, they not only demonstrate directly the association of CAG with microfilaments, but they also suggest that CAG may act as a microfilament-nucleating site at the membrane.

Discussion

Tumour cells escape immune destruction by a variety of mechanisms. Investigations have sought to describe molecular aspects of three of these mechanisms in 13762 mammary ascites cells: masking of tumour antigens, shedding of tumour antigens and alterations of antigen mobility. Because these mechanisms depend on the organisation of particular molecules at the tumour cell surfaces, work has focussed on mechanisms regulating that organisation. These studies emphasise two important general principles of membrane organisation, which have been recognised but not well supported by experimental results until recently. 1) Complexes of membrane polypeptides, either homologous or heterologous, are important features of plasma membrane organisation. 2) Plasma membrane-microfilament interactions, either direct or indirect, are important in regulating the topography of cell surfaces.

ASGP-1/ASGP-2 is an example of a strong heterologous (more than one type of subunit) complex. Examples of such complexes have become increasingly abundant in recent years, particularly from studies of membrane receptors. Included among these are the insulin receptor (Czech & Masságue,

1982), acetylcholine receptor (Conti-Tronconi *et al.*, 1982), immunoglobulin receptors (Hood, Kronenberg & Hunkapiller, 1985) and T-cell antigen receptors (Hood *et al.*, 1985). Other important cell surface functions, such as the histocompatibility antigens (Nathenson *et al.*, 1981), sucrase–isomaltase enzyme complex (Kenny & Maroux, 1982) and sodium ion channel (Costa, Casnellie & Catterall, 1982), also involve heterologous protein complexes. CAG, which is apparently a homologous multimer (having only one type of subunit), is an example of another prevalent cell surface structure, the disulphide-linked complex. Other disulphide-linked plasma membrane components include the transferrin receptor (Geisow, 1986), insulin receptor (Czech & Massague, 1982), histocompatibility antigen (Nathenson *et al.*, 1981), acetylcholine receptor (Karlin *et al.*, 1983), immunoglobulin E receptor (Alcaraz *et al.*, 1984), T-cell antigen receptor (Allison & Lanier, 1985) and T-cell growth factor receptor (Leonard *et al.*, 1985).

The involvement of sialomucins in masking cell surface antigens was first proposed for TA3 mouse mammary adenocarcinoma cells (Sanford, 1967), in which the sialomucin is epiglycanin (Codington & Frim, 1983). The structures of epiglycanin and ASGP-1, which are similar in size and carbohydrate content, appear to be ideal for serving as masking agents (Codington, 1981). The extended rod structure should screen the approach of immune recognition factors from the tumour cell surface antigens. In the case of TA3 cells the presence of epiglycanin appears to prevent antibody binding to the murine histocompatibility antigens (Codington, 1981). Recent studies on 13762 ascites cells suggest that antibody binding to rat class I antigens is not prevented by the presence of ASGP-1 (Spielman, unpublished observations). Thus the involvement of the sialomucin in immune evasion may be different in different tumours. Recent studies by Sherblom & Moody (1986) have suggested an ASGP-1-dependent resistance of 13762 cells to killing by NK cells, suggesting that the sialomucin may be inhibiting cell–cell recognition processes involved in tumour killing.

The efficiency of ASGP-1 as a masking agent must be related to its abundance at the cell surface, which depends on the balance between biosynthesis and turnover. Biosynthesis of ASGP-1 and its transit to the cell surface is a relatively slow process, involving a continuous glycosylation of the ASGP-1 polypeptide throughout transit. Presumably this slow transit and continuous glycosylation are important for the high degree of glycosylation of ASGP-1. Functionally, specificity of the ASGP-1 glycosylation is less important than the density of glycosylation. Therefore, glycosylation is probably a random event dependent only on the presence of the appropriate glycosyltransferases in the organelles through which the ASGP-1 passes on its way to the cell surface. The factors regulating the expression of ASGP-1 are still unknown and will require studies of the ASGP-1 mRNA and gene levels.

The 13762 ascites cells have also provided a system for investigating the regulation of the mobility of cell surface components. The key elements are CAG, which is the site of interaction of multiple microfilaments at the microvillus membrane, and 58 K, which is associated with CAG and actin in the transmembrane complex from MAT-C1 cells. It has been shown that microvilli of MAT-C1 cells are more stable than those of MAT-B1 cells (or most other similar cells). It has been proposed that the presence of 58 K stabilises the association of microfilaments with the microvillar membrane of MAT-C1 cells (Carraway *et al.*, 1983), preventing dissociation of the microfilaments from the membrane and subsequent microfilament depolymerisation and microvillus collapse. It has also been proposed that this stabilisation leads to the formation of the branched microvilli and to restriction of receptor mobility. The former occurs because the MAT-C1 microvilli are not lost as the cell proceeds through the cell cycle, and new microvilli push out from the cell surface behind those already present. Restriction of receptor mobility is envisioned to occur because the cell surface is locked into a relatively static villus form and the dynamics of actin depolymerisation–polymerisation near the plasma membrane is repressed.

Figure 2.6 shows a feasible model for the membrane–microfilament interaction site, and how 58 K might act to stabilise the interaction. It is envisioned that formation of this species occurs by the association of G-actin with CAG to form a transmembrane complex, which can then act as a polymerisation site for microfilament formation and microvillus growth (Carraway & Carraway, 1985). 58 K would then become associated at the transmembrane complex and interact with the nearest actin of the filament as well as the actin directly associated with CAG. By this mechanism 58 K would bridge the near end of the filament to the membrane, locking the filament into its association with the membrane.

The results of these studies and those of others suggest a hierarchy of tumour cell 'escape' or survival mechanisms. The presence of cell surface sialomucin has been related to metastasis in 13762 mammary tumours (Steck & Nicolson, 1983, 1984) and allotransplantability in TA3 mammary tumours (Codington & Frim, 1983). In contrast, sialomucin presence in 13762 cells appears unrelated to xenotransplantability (Sherblom *et al.*, 1980). Studies by the authors suggest xenotransplantability may correlate with the presence of branched microvilli and concomitant reduced mobility of cell surface antigens (Howard *et al.*, 1982). Moreover, the escape mechanism may vary in different tumours. Sialomucin appears to mask Class I histocompatibility antigens in TA3 tumours, but not in similar 13762 tumours. In the latter the effects of the sialomucins may be directed toward cell–cell recognition processes (Sherblom & Moody, 1986). Clearly, very complex cellular phenomena are being investigated. However, recent advances in understanding molecular processes in cellular immunology and structural features of tumour cells offer encouragement that answers to some of the questions concerning tumour survival will begin to be available in the near future.

Acknowledgements

We thank the many colleagues who have contributed to these studies, particularly Drs Anne Sherblom, Goeh Jung, Susan Howard, John Huggins, Rick Helm, Steven Hull and Tom Metcalf. Unpublished work cited was supported by grants from the National Institutes of Health (GM 33975 and CA 31695), National Science Foundation (PCM 8300771) and Papanicolaou Comprehensive Cancer Center of the University of Miami (CA 14395).

Fig. 2.6. Model for association of microfilament core with plasma membrane transmembrane complex.

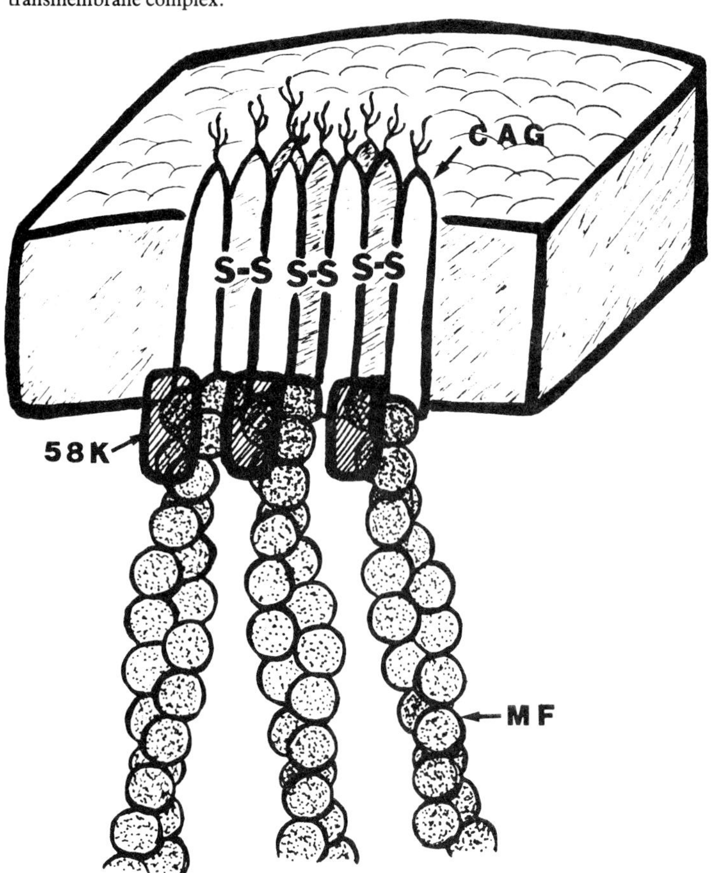

References

Alcaraz, G., Kinet, J.-P., Kumar, N., Wank, S. A. & Metzger, H. (1984). Phase separation of the receptor for immunoglobulin E and its subunits in Triton X-114. *J. Biol. Chem.*, **259**, 14922–7

Allison, J. P. & Lanier, L. L. (1985). Identification of antigen receptor-associated structures on murine T-cells. *Nature (Lond.)*, **314**, 107–9

Buck, R. L., Sherblom, A. P. & Carraway, K. L. (1979). Sialoglycoprotein differences between xenotransplantable and nonxenotransplantable ascites sublines of the 13762 rat mammary adenocarcinoma. *Arch. Biochem. Biophys.*, **198**, 12–21

Carraway, C. A. C., Cerra, R. F., Bell, P. B. & Carraway, K. L. (1982). Identification of a protein associated with both membrane and cytoskeleton fractions from branched but not unbranched microvilli of 13762 rat mammary adenocarcinoma ascites tumor cells. *Biochim. Biophys. Acta*, **719**, 126–39

Carraway, C. A. C., Jung, G. & Carraway, K. L. (1983*a*). Isolation of actin-containing transmembrane complexes from ascites adenocarcinoma sublines having mobile and immobile receptors. *Proc. Natl Acad. Sci. USA*, **80**, 430–4

Carraway, C. A. C., Jung, G. & Carraway, K. L. (1983*b*). Organizational forms of actin in 13762 ascites mammary tumor cell microvilli. *Cell Motil.*, **3**, 491–500

Carraway, C. A. C., Jung, G., Craik, J. R., Rubin, R. W. & Carraway, K. L. (1983*c*). Identification of a cytoskeleton-associated glycoprotein from isolated microvilli of a mammary ascites tumor. *Exp. Cell Res.*, **143**, 303–8

Carraway, C. A. C., Jung, G., Hinkley, R. E. & Carraway, K. L. (1985). Isolation of microvillar microfilaments and associated transmembrane complex from ascites tumor cell microvilli. *Exp. Cell Res.*, **157**, 71–82

Carraway, C. A. C. & Weiss, M. (1985). Phalloidin shift on velocity sedimentation sucrose gradient centrifugation for identification of microfilament-associated proteins. *Exp. Cell Res.*, **161**, 150–60

Carraway, C. A. C., Sindler, C. & Weiss, M. (1986). Demonstration of the association of the cell-surface enzyme, 5′-nucleotidase, with microvillar microfilaments by phalloidin shift on velocity sedimentation gradients. *Biochim. Biophys. Acta*, **885**,68–73

Carraway, K. L. & Carraway, C. A. C. (1985). Plasma membrane-microfilament interaction in animal cells. *Bioessays*, **1**, 55–8

Carraway, K. L., Cerra, R. F., Jung, G. & Carraway, C. A. C. (1982). Membrane-associated actin from the microvillar membranes of ascites tumor cells. *J. Cell Biol.*, **94**, 624–30

Carraway, K. L., Doss, R. C., Huggins, J. W., Chesnut, R. W. & Carraway, C. A. C. (1979). Effects of cytoskeletal perturbant drugs on ecto 5′-nucleotidase, a Concanavalin A receptor. *J. Cell Biol.*, **83**, 529–43

Carraway, K. L., Huggins, J. W., Cerra, R. F., Yeltman, D. R. & Carraway, C. A. C. (1980). α-Actinin-containing branched microvilli from an ascites adenocarcinoma. *Nature (Lond.)*, **285**, 508–10

Carraway, K. L., Huggins, J. W., Sherblom, A. P., Chesnut, R. W., Buck, R. L., Howard, S. P., Ownby, C. L. & Carraway, C. A. C. (1978). Membrane glycoproteins of rat mammary gland and its metastasizing and nonmetastasizing tumors. *Am. Chem. Soc. Symp.*, **80**, 432–45

Carraway, K. L. & Spielman, J. (1986). Structural and functional aspects of tumor cell sialomucins. *Mol. Cell. Biochem.*, **72**, 109–20

Codington, J. F. (1981). The masking of cancer cell surface antigens. *Handb. Cancer Immunol.*, **8**, 171–203

Codington, J. F. & Frim, D. M. (1983). Cell surface macromolecular and morphologic changes related to allotransplantability in the TA3 tumor. *Biomembranes*, **11**, 207–58

Conti-Tronconi, B. M., Hunkapiller, M. W., Lindstrom, J. M. & Raftery, M. A. (1982). Subunit structure of the acetylcholine receptor from *Electrophorus electricus*. *Proc. Natl Acad. Sci. USA*, **79**, 6489–93

Costa, M. R. C., Casnellie, J. E. & Catterall, W. A. (1982). Selective phosphorylation of the a subunit of the sodium channel by cAMP-dependent protein kinase. *J. Biol. Chem.*, **257**, 7918–21

Czech, M. P. & Massague, J. (1982). Subunit structure and dynamics of the insulin receptor. *Fed. Proc.*, **41**, 2719–23

Geiger, B. (1983). Membrane–cytoskeleton interaction. *Biochim. Biophys. Acta*, **737**, 305–41

Geisow, M. J. (1986). Cell-surface receptors: puzzles and paradigms. *Bioessays*, **4**, 149–51

Hanfland, V. P. & Uhlenbruck, G. (1978). Aktuelle immunchemische betrachtungen zum aufbau der tumorzellmembran. *J. Clin. Chem. Clin. Biochem.*, **16**, 85–101

Helm, R. M. & Carraway, K. L. (1981). Evidence for the association of two cell surface glycoproteins of 13762 mammary ascites tumor cells. Concanavalin A-induced redistribution of peanut agglutinin-binding proteins. *Exp. Cell Res.*, **135**, 418–24

Hood, L., Kronenberg, M. & Hunkapiller, T. (1985). T cell antigen receptors and the immunoglobulin supergene family. *Cell*, **40**, 225–9

Howard, S. C., Hull, S. R., Huggins, J. W., Carraway, C. A. C. & Carraway, K. L. (1982). Relationship between xenotransplantability and cell surface properties of ascites sublines of a rat mammary adenocarcinoma. *J. Natl Cancer Inst.*, **69**, 33–40

Howard, S. C., Sherblom, A. P., Chesnut, R. W., Carraway, C. A. C. & Carraway, K. L. (1980). Changes in expression of a major sialoglycoprotein associated with ascites forms of a mammary adenocarcinoma. *Biochim. Biophys. Acta*, **631**, 79–89

Howard, S. C., Sherblom, A. P., Huggins, J. W., Carraway, C. A. C. & Carraway, K. L. (1981). Shedding of the major plasma membrane sialoglycoprotein from the surface of 13762 rat ascites mammary adenocarcinoma cells. *Arch. Biochem. Biophys.*, **207**, 40–50

Huggins, J. W., Trenbeath, T. P., Sherblom, A. P., Howard, S. C. & Carraway, K. L. (1980). Glycoprotein differences in solid and ascites forms of the 13762 rat mammary adenocarcinoma. *Cancer Res.*, **40**, 1873–8

Hull, S. R., Laine, R. A., Kaizu, T., Rodriguez, I. & Carraway, K. L. (1984). Structures of the O-linked oligosaccharides of the major cell surface sialoglycoprotein of MAT-B1 and MAT-C1 ascites sublines of the 13762 rat mammary adenocarcinoma. *J. Biol. Chem.*, **259**, 4866–77

Jung, G., Andrews, D. M., Carraway, K. L. & Carraway, C. A. C. (1985). Actin-associated cell surface glycoprotein from ascites cell microvilli: a disulfide-linked multimer. *J. Cell. Biohem.*, **28**, 243–52

Jung, G., Helm, R. M., Carraway, C. A. C. & Carraway, K. L. (1984). Mechanism of Concanavalin A-induced anchorage of the major cell surface glycoproteins to the submembrane cytoskeleton in 13762 ascites mammary adenocarcinoma cells. *J. Cell Biol.*, **98**, 179–87

Karlin, A., Holtzman, E., Yodh, N., Lobel, P., Wall, J. & Hainfeld, J. (1983). The arrangement of the subunits of the acetylcholine receptor of *Torpedo californica*. *J. Biol. Chem.*, **258**, 6678–81

Kenny, A. J. & Maroux, S. (1982). Topology of microvillar membrane hydrolases of kidney and intestine. *Physiol. Rev.*, **62**, 91–128

Leonard, W. J., Depper, J. M., Kronke, M., Robb, R. J., Waldmann, T. A. & Greene, W. C. (1985). The human receptor for T-cell growth factor. *J. Biol. Chem.*, **260**, 1872–80

Nathenson, S. G., Uehara, H., Ewenstein, B. M., Kindt, T. J. & Coligan, J. E. (1981). Primary structural analysis of the transplantation antigens of the murine H-2 major histocompatibility complex. *Ann. Rev. Biochem.*, **50**, 1025–52

Sanford, B. H. (1967). An alteration in tumour histocompatibility induced by neuraminidase. *Transplantation*, **5**, 1273–9

Segaloff, A. (1966). Hormones in breast cancer. *Recent Prog. Hormone Res.*, **22**, 351–79

Sherblom, A. P., Buck, R. L. & Carraway, K. L. (1980). Purification of the major sialoglycoproteins of 13762 MAT-B1 and MAT-C1 rat ascites mammary adenocarcinoma cells by density gradient centrifugation in cesium chloride and guanidine hydrochloride. *J. Biol. Chem.*, **255**, 783–90

Sherblom, A. P. & Carraway, K. L. (1980*a*). A complex of two cell surface glycoproteins from ascites mammary adenocarcinoma cells. *J. Biol. Chem.*, **255**, 12051–9

Sherblom, A. P. & Carraway, K. L. (1980*b*). Sulphate incorporation into the major sialoglycoprotein of the MAT-B1 subline of the 13762 rat ascites mammary adenocarcinoma. *Biochemistry*, **19**, 1213–19

Sherblom, A. P., Huggins, J. W., Chesnut, R. W., Buck, R. L., Ownby, C. L., Dermer, G. B. & Carraway, K. L. (1980). Cell surface properties of ascites sublines of the 13762 rat mammary adenocarcinoma. Relationship of the major sialoglycoproteins to xenotransplantability. *Exp. Cell Res.*, **126**, 417–26

Sherblom, A. P. & Moody, C. E. (1986). Cell surface sialomucin and resistance to natural cell-mediated cytotoxicity of rat mammary tumor ascites cells. *Cancer Res.*, **46**, 4543–6

Snabes, M. C., Boyd, A. E. III & Bryan, J. (1981). Detection of actin-binding proteins in human platelets by ^{125}I-actin overlay of polyacrylamide gels. *J. Cell Biol.*, **90**, 809–12

Spielman, J., Rockley, N. L. and Carraway, K. L. (1987). Temporal aspects of O-glycosylation and cell surface expression of ASGP–1, the major cell surface sialomucin of 13762 mammary ascites tumor cells. *J. Biol. Chem.*, **262**, 269–75

Steck, P. A. & Nicolson, G. L. (1983). Cell surface glycoproteins of 13762NF mammary adenocarcinoma clones of differing metastatic potential. *Exp. Cell Res.*, **147**, 255–67

Steck, P. A. & Nicolson, G. L. (1984). Cell surface properties of spontaneously metastasizing rat mammary adenocarcinoma clones. *Transplant. Proc.*, **16**, 355–60

3

Cell-specific expression of the S-gene in *Brassica* and *Nicotiana*

Mikhail E. Nasrallah and June B. Nasrallah

Section of Plant Biology,
Cornell University
Ithaca, New York 14853, USA

Abstract

Metabolic labelling of pistil tissue from *Brassica* and from *Nicotiana* with ^{35}S-methionine demonstrates that the self-incompatibility genes are expressed in different cell types of the pistil in the two species. Based on the different pollen inhibition sites, it is further proposed that, in gametophytic systems of the *Nicotiana* type, the S-gene is activated in pollen tubes following pollen germination. The cellular and physiological mechanisms underlying sporophytic and gametophytic incompatibility systems therefore appear to be markedly different, and may reflect a polyphyletic origin of self-incompatibility in dicots. This conclusion is supported by molecular studies of cloned S-sequences in *N. alata* and *B. oleracea*.

Introduction

Self-incompatibility in *Brassica* is genetically controlled by one multiallelic locus designated S. Control of pollen phenotype is sporophytic, with allelic interactions possible in expressing tissues of the anther as well as the stigma. The expression of the S locus can be analysed at various levels. In *Brassica* this expression is temporally and spatially regulated. The developmental regulation of the highly localised inhibition response can be readily and dramatically visualised by microscopic monitoring of self-pollen development on stigmas at different stages of maturity. As shown in Fig. 3.1, pollen tubes are inhibited on stigmas from open flowers (see Fig. 3.1(a)–(c)) and on stigmas from buds at one day prior to anthesis (see Fig. 3.1(d)). On stigmas from younger buds, however, pollen tubes are allowed to develop normally (see Fig. 3.1(e)–(f)). The response of the pollen to the state of the stigma occurs within minutes after the initial contact between the pollen and the papillar cells on the outer surface of the stigma. Pollen grains usually fail to germinate

(Fig. 3.1(a)–(c)), or more rarely germinate to produce pollen tubes that coil at the surface of the papillar cells and are unable to invade the surface layer of stigma cells (Fig. 3.1(d)). The interaction is clearly that between one pollen and one papillar cell, and the transition in the stigma from compatibility to incompatibility is an abrupt all-or-none phenomenon, with all papillar cells becoming incompatible at once.

At the biochemical level, the microscopic observations just described have been correlated with the accumulation, during stigma maturation, of glycoproteins which, on the basis of a number of criteria (Nasrallah *et al.*, 1985), appear to be the products of the S locus, and have therefore been termed S-locus specific glycoproteins (SLSG). Experiments in which stigmas were submerged in ^{35}S-methionine have demonstrated the *de novo* synthesis of SLSG, and correlated changes in the relative incorporation of label into

Fig. 3.1. UV-fluorescence microscopy of pollen behaviour on a sequence of developing stigmas from one inflorescence of *Brassica*: (a), (b), (c) inhibition of self-pollen on stigmas from two-day old flowers, one-day old flowers and at anthesis respectively; (d) inhibition of self-pollen on stigmas from buds at one day prior to anthesis; (e), (f) normal development of self-pollen on stigmas from immature buds at two and three days prior to anthesis respectively.

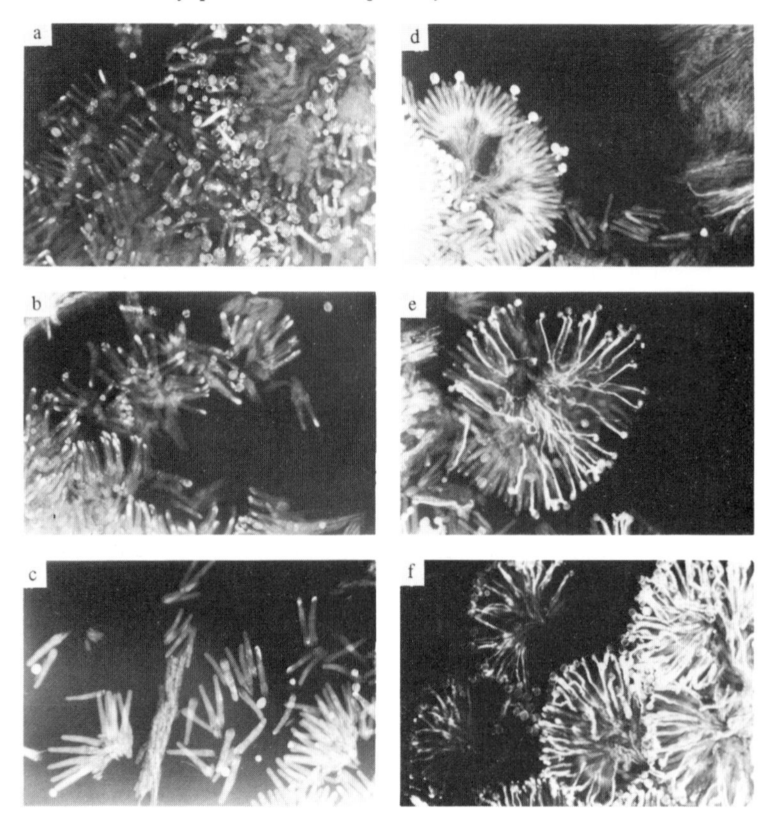

Fig. 3.2. Developmental regulation of S-gene expression during stigma development. *Top:* Relative incorporation of radioactive amino acids into SLSG extracted from stigmas at the −3, −2, −1, 0, +1, +2 developmental stages. Autoradiograms of IEF gels were scanned and percent incorporation calculated (Nasrallah, Doney & Nasrallah, 1985). *Bottom:* level of SLSG mRNA in arbitrary units determined from 'Northern' analysis. Poly A + RNA was obtained from stigmas at the same developmental stages assayed for SLSG (*top*). The Northern blot was probed with cDNA, and the autoradiogram scanned (Nasrallah *et al.*, 1985).

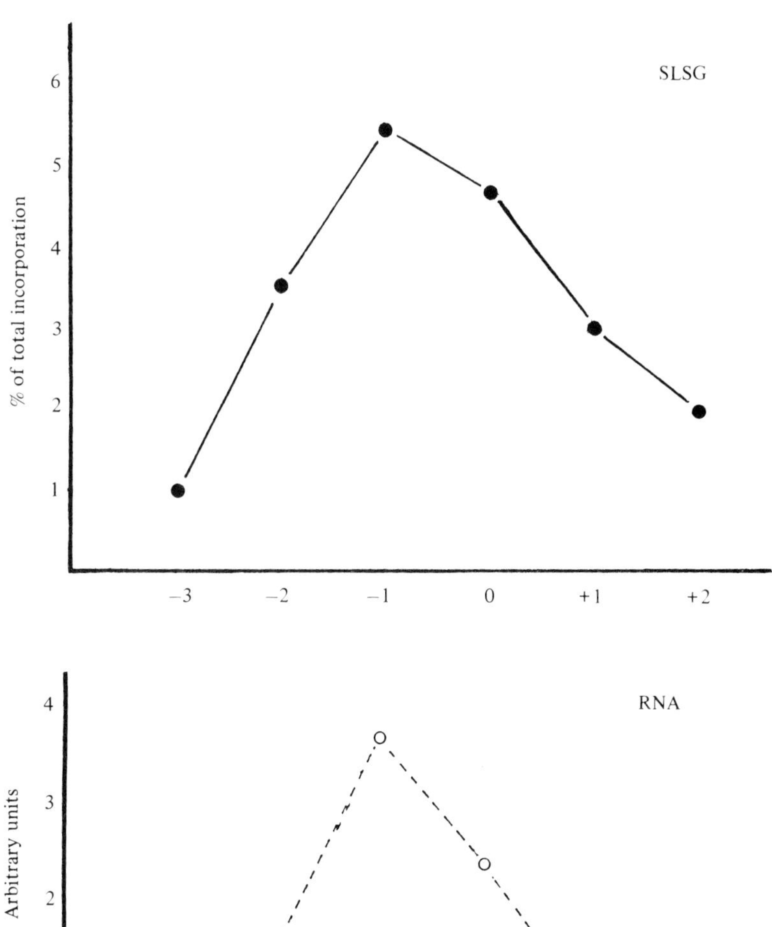

these molecules with the developmental regulation of self-incompatibility (Nasrallah, Doney & Nasrallah, 1985). As shown graphically in Fig. 3.2, relative incorporation into SLSG can be detected early in stigma development and increases in stigmas from buds at one day prior to anthesis coordinately with the onset of self-incompatibility to a maximal level which, in some S genotypes, can account for approximately 6% of total incorporation. Subsequently, and following anthesis, relative incorporation into SLGS decreases. With the isolation of nucleic acid sequences encoding SLSG, these changes in relative synthetic rates were in addition correlated with changes in the steady state levels of SLSG transcripts (see Fig. 3.2 and Nasrallah *et al.*, 1985).

In our attempt to define the spatial regulation of S-gene expression in the stigma, we have carried out experiments to identify the cells in which this gene is turned on. Since it seems likely, in view of the surface inhibition of pollen in *Brassica*, that SLSGs are localised at the stigma surface, the question is whether their synthesis takes place in the papillar cells or, as is also possible, in other cells of the pistil from which they are later translocated to the stigma surface. It is clear that SLSGs are synthesised by the stigma proper, as these molecules cannot be detected in stylar tissue, and since no incorporation into SLSG is evident in stylar extracts even when a large number of styles are examined electrophoretically (see Fig. 3.3). In contrast, labelling of SLSG can be easily demonstrated in extracts from one or a few stigmas (see Fig. 3.3 and Nasrallah, Doney & Nasrallah, 1985). Furthermore, application of ^{35}S-methionine to the surface of the stigma results in efficient and rapid labelling of SLSG, suggesting that superficial stigma cells are responsible for their synthesis (Fig. 3.3). Definitive proof of this conclusion has recently been obtained following microscopic localisation of SLSG transcripts in frozen stigma sections hybridised *in situ* to cloned SLSG-encoding sequences (Nasrallah & Nasrallah, in preparation).

In *Nicotiana*, which exhibits one-locus gametophytic incompatibility, pollen tube inhibition is delayed and occurs during growth in the transmitting tissue of the style. In contrast to the *Brassica* system, early pollen development appears normal, with incompatible pollen germinating and pollen tubes growing into the pistil. The incompatibility response then takes place at different sites within the style, and is characterised by changes in wall morphology, arrest and sometimes bursting of incompatible pollen tubes. S-associated glycoproteins have been reported in *Nicotiana* to be detected along the length of the pistil (Anderson *et al.*, 1986). Metabolic labelling of pistil tissue from *Nicotiana* reveals additional differences between the two species. When submerged in ^{35}S-methionine, segments obtained from the uppermost 3 mm zone of the style show incorporation of label into the S-associated glycoproteins, while dissected stigmas treated in the same way do not (see Fig. 3.4(a)). When applied to the stigma surface, label is translocated down to the style, and substantial incorporation into the S-associated glycoproteins in stylar segments, and not in the stigmas proper, is again demonstrated (see Fig. 3.4(b)). The low levels of these labelled molecules observed in extracts of the stigmas are probably due to contaminating stylar tissue. Furthermore, upon Coomassie Blue staining of the same gels, no S-associated glycoproteins can be detected in the stigma as has been reported, indicating that these molecules are not translocated to the stigma from their site of synthesis in the style. Transcripts encoding S-associated glycoproteins were recently localised to the transmitting tract of the pistil (Anderson *et al.*, 1986). The results of our incorporation experiments predict in addition that such transcripts should be missing from the stigma proper.

Self-incompatibility genes are therefore demonstrably expressed in different cell types of the pistil in *Brassica* and *Nicotiana*. The site of expression of these genes in the male tissue, on the other hand, is only a matter of speculation at this point. It has often been suggested that sporophytic and gametophytic systems may in fact be related, differing mainly in the timing or precise localisation of S-gene expression in the anther. Gametophytic control has been in general attributed to expression in pollen grains following meiosis. Sporophytic control has been ascribed to pre-meiotic S-gene expression in pollen mother cells or, as suggested by Heslop-Harrison *et al.* (1974), to expression in the sporophytic cells of the tapetum followed by transfer

Fig. 3.3. Autoradiogram of an isoelectric focussing separation of extracts from *Brassica* styles (left lane) and stigmas (right lane). Styles from 40 pistils were dissected into 1 mm sections, submerged in ^{35}S-methionine for 12 hours, extracted and applied to one lane. The stigmas, four used in this experiment, were labelled by surface application of ^{35}S-methionine, incubated for 12 hours, extracted and applied to one lane. The arrow points to labelled SLSG band detected only in stigma extracts.

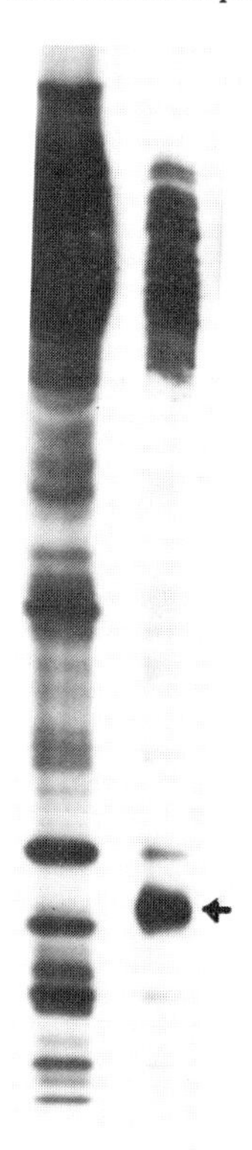

Fig. 3.4. Synthesis of S-associated glycoproteins in *Nicotiana alata*. (a) 15 stigmas and 15 stylar sections (3 mm section proximal to stigma) from two S homozygous genotypes were submerged in ^{35}S-methionine for 24 hours. Isoelectric focussing reveals labelling of S-associated proteins in styles of the two genotypes (arrowheads) but not in the stigmas (middle two lanes). Flower buds were selected at one day prior to anthesis. (b) Labelling by application of ^{35}S-methionine to the surface of stigmas from buds at −1 day prior to anthesis. Label was applied to stigmas which were then incubated at room temperature for 24 hours. Stigmas and style sections were dissected, extracted and subjected to IEF as in (a). The lane on the right consisted of a mixture of stylar extracts from the two homozygous genotypes used in (a), and arrowheads point to the heavily labelled S-associated bands. The left and middle lanes were each loaded with stigma extracts from either one of the two genotypes and show little if any labelling of these bands.

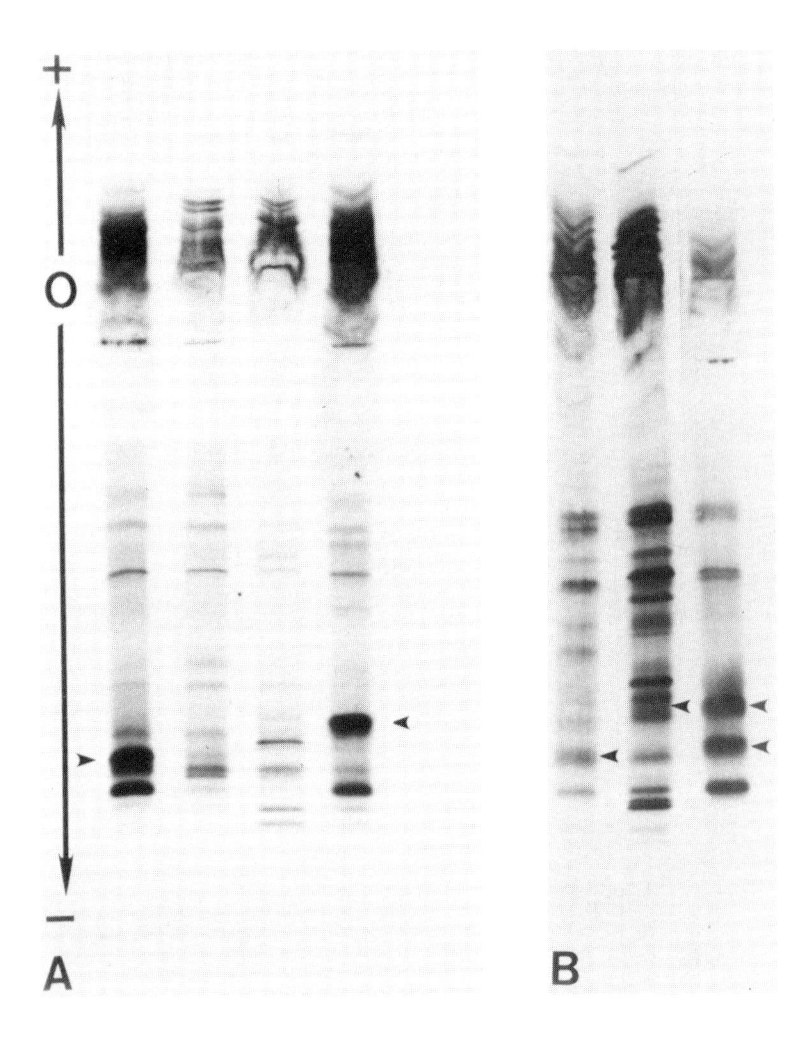

of the S-gene products to the surface of individual pollen grains. While this concept of imprinting of pollen by tapetal cells is quite interesting, the possibility of imprinting from pollen to pollen should also be considered. In the same way, and based on the mode of pollen rejection in gametophytic systems, activation of the S-gene in pollen tubes following germination cannot be dismissed. In any event, now that self-incompatibility sequences have been cloned from both gametophytic and sorophytic systems, the various theories outlined above can be tested, and the S-transcripts localised in time and space by the technique of *in situ* hybridisation to cloned DNA sequences. The cellular and physiological mechanisms underlying sporophytic and gametophytic control of self-incompatibility, although still largely undefined, appear therefore to be markedly different. Together with the lack of homology between the amino acid sequences of the *Brassica* and *Nicotiana* S-proteins as deduced from the sequence of the cloned genes, they point to a polyphyletic origin of incompatibility, and suggest that S loci may have evolved from genes affecting different steps along the pathway of pollen development.

References

Anderson, M. A., Cornish, E. C., Mau., S.-L., Williams, E. G., Hoggart, R., Atkinson, A., Bonig, I., Grego, B., Simpson, R., Roche, P. J., Haley, J. D., Penschow, J. D., Niall, H. D., Tregear, G. W., Coghlan, J. P., Crawford, R. J. & Clarke, A. E. (1986). Cloning of cDNA for a stylar glycoprotein associated with expression of self-incompatibility in *Nicotiana alata*. *Nature (Lond.)*, **321**, 38–44

Heslop-Harrison, J., Knox, R. B. & Heslop-Harrison, Y. (1974). Pollen-wall proteins: exine-held fractions associated with the incompatibility response in *Cruciferae*. *Theor. Appl. Genetics*, **44**, 133–7

Nasrallah, J. B., Doney, R. C. & Nasrallah, M. E. (1985). Biosynthesis of glycoproteins involved in the pollen-stigma interaction of incompatibility in developing flowers of *Brassica oleracea L*. *Planta*, **165**, 100–7

Nasrallah, J. B., Kao, T.-H. Goldberg, M. E. & Nasrallah, M. E. (1985). A cDNA clone encoding an S-locus specific glycoprotein from *Brassica oleracea*. *Nature (Lond.)*, **318**, 263–7

PART II

Recognition in single-celled organisms

4 Sexual interactions in yeast

Naohiko Yanagishima

Biological Institute, Faculty of Science,
Nagoya University, Chikusa-ku,
Nagoya 464, Japan

Abstract

Mating in *Saccharomyces cerevisiae* consists of sexual agglutination, G_1-arrest and sexual cell fusion. Species- and mating-type-specific glycoproteins responsible for sexual agglutination have been isolated from the cell wall and the cytoplasm. These substances are secreted through the yeast secretory pathway and the sex pheromones induce synthesis of their precursors. A hypothesis for the genetic system controlling sexual agglutinability is proposed. In sexual aggregates, gametogenesis leading to zygote formation is induced by sex pheromones. Ascosorogenous yeast mating-types are classified into **a** and α, on the basis of sex pheromones and agglutination substances.

Introduction

The yeast *Saccharomyces cerevisiae* has been successfully used as experimental material for analysis at the molecular level, of mechanisms fundamental to eukaryotes such as cell cycle regulation, sex differentiation and sexual interactions. Genetic control of mating type differentiation and regulation mechanism of mating-type-specific functions in yeast have been analysed in detail (for reviews, see Herskowitz & Oshima, 1981; Sprague, Blair & Thorner, 1983).

The production of sex pheromones, agglutination substances, α-pheromone-inactivating substances and the α pheromone receptor: the mating-type-specific substances responsible for sexual differentiation and interactions, are under the control of the mating type locus (for reviews, see Thorner, 1981; Sprague *et al.*, 1983; Yanagishima, 1986). Hence, sexual differentiation is brought about by the expression of the sex-specific genes whose activity changes through sexual interactions. Thus, the phenomena of sexual interactions and diff-rentiation are interrelated. The mating reaction in the yeast progresses as an integrated sequence of these events eventually leading to the formation of zygotes. This article will consider in detail results of genetic and biochemical studies on the mating-type-specific substances in *S. cerevisiae*. The experimental results are from *S. cerevisiae* unless otherwise specified.

The mating process

In the heterothallic strains of *S. cerevisiae* haploid cells are classified into **a** and α mating types controlled by the mating type alleles, *MAT***a** and *MAT*α respectively located on chromosome III. Fig. 4.1 shows the life cycle of heterothallic *S. cerevisiae*. When **a** and α haploid cells are mixed, sexual agglutination occurs (Fig. 4.2), followed by formation of zygotes in the resultant aggregates (Osumi, Shimoda & Yanagishima, 1974; Shimoda & Yanagishima, 1978; Kawanabe, Yoshida & Yanagishima, 1979) (Fig. 4.3). During the mating process, induction of sexual agglutinability, cell cycle arrest at

Fig. 4.1. Life cycle and mating process of heterothallic *S. cerevisiae*. Mating reaction can occur both in haploid vegetative cells and in spores. In this figure the former case is shown.

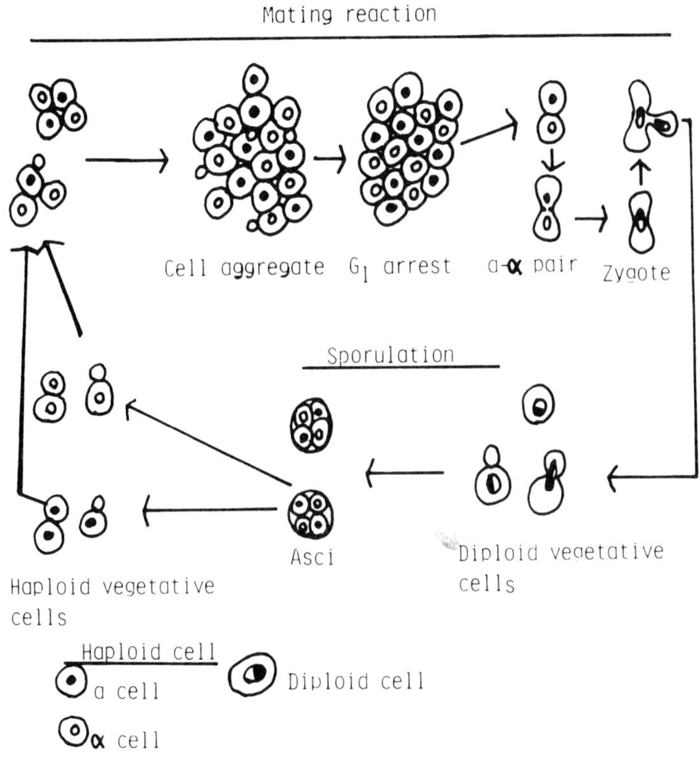

G_1, and formation of large pear-shaped cells occur in order, as shown in Fig. 4.4 (for reviews, see Yanagishima *et al.*, 1977; Yanagishima, 1978; Betz, Manney & Duntze, 1981; Thorner, 1981; Yanagishima & Yoshida, 1981; Yanagishima, 1984; Yanagishima, 1986). Although it is known that each of these events is brought about by a sex pheromone of the opposite mating type, the mechanism determining the sequence is unknown.

Sexual cell–cell recognition

a and α cells recognise each other through mating type-specific substances, resulting in sexual agglutination. Sexual agglutination is of general occurrence in ascosporogenous yeasts (Yoshida & Yanagishima, 1978; Yamaguchi, Yoshida & Yanagishima, 1984*a*). In order to analyse this phenomenon, the degree of the agglutination must first be quantified. An agglutination index (AI) may be calculated, from the decrease in the absorbance at 530 nm (A_{530}) due to the formation of aggregates (Yoshida & Yanagishima, 1978). AI = A_{530} after sonic treatment/A_{530} before sonic treatment. The sonic treatment is too weak to disrupt the cells, but it disperses agglutinating cells completely. The AI is approximately propotional to the mean diameter of the aggregates. The AI, thus defined, is used to quantify agglutination in the experiments described below.

Molecular basis

Sexual agglutination, the first step of the mating reaction, is brought about by complementary binding of the mating-type-specific glycoproteins, agglutination substances on the surface of **a** and α cells (Figs. 4.2 and 4.5) (for

Fig. 4.2. A sexual cell aggregate. α cells were treated with a fluorescent substance before mixing with untreated **a** cells. The aggregate was pressed gently before photographing.

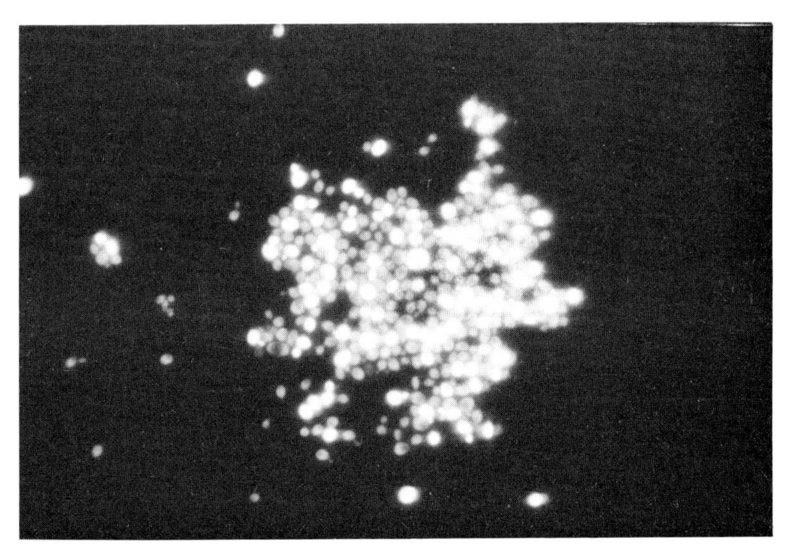

reviews, see Yanagishima, 1978; Yanagishima & Yoshida, 1981; Yanagishima, 1984). The active agglutination substances were solubilised from the cell wall by the two methods, brief autoclaving at 1 kg/cm^2 of **a** and α cells, followed by quick chilling (autoclave method) (Yoshida, Hagiya & Yanagishima, 1976; Hagiya, Yoshida & Yanagishima, 1977) and enzymatic treatment with Glusulase (Endo Lab., Garden City, New York) of cell wall fractions from **a** and α cells (enzyme method) (Shimoda, Kitano & Yanagishima, 1975; Shimoda & Yanagishima, 1976). The agglutination substances solubilised by the autoclave method are released almost exclusively from the cell surface (Hagiya *et al.*, 1977). The active substances released by the autoclave method have molecular weights of about 1×10^4 to 1×10^5 and those by the enzyme method about 1×10^6.

The isolated agglutination substances are glycoproteins with univalent binding activity, that is, the activity to mask sexual agglutinability without causing self-agglutination specifically in the opposite mating type. The **a**

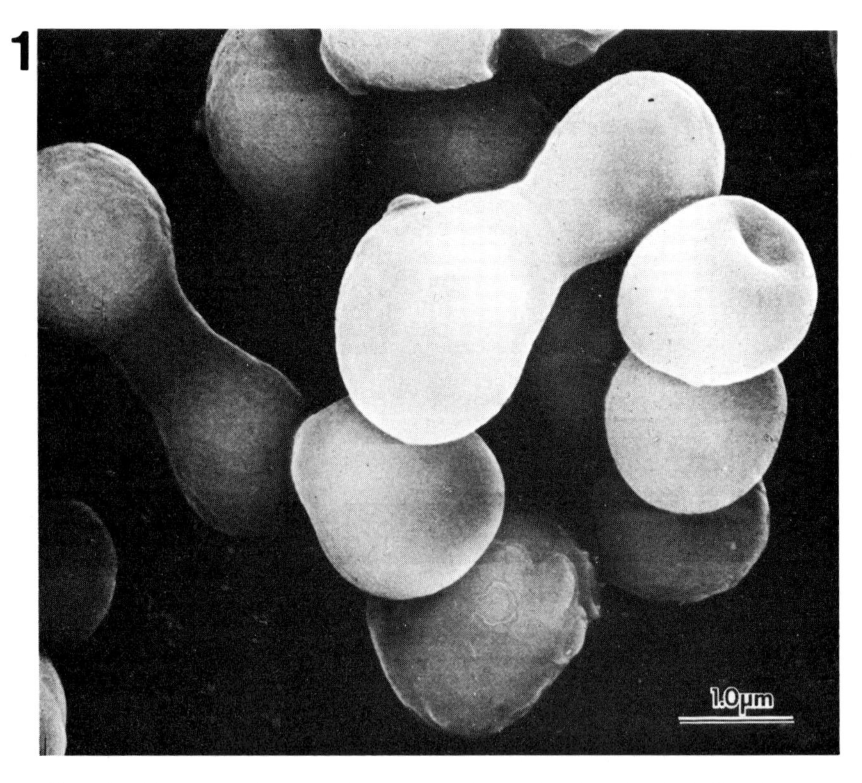

Fig. 4.3. Scanning and transmission electron micrographs of conjugating cells and a zygote. 1, conjugating cells; 2, a conjugating **a** and α cell pair; 3, a zygote. [Reproduced from Yanagishima *et al.*, 1977 (1) and Osumi *et al.*, 1974 (2 and 3)] CW, cell wall; ER, endoplasmic reticulum; M, mitochondrion; NM, nuclear envelope; N, nucleus; V, vacuole.

agglutination substance is more sensitive to 2-mercaptoethanol than the α agglutination substance (Shimoda *et al.*, 1976; Hagiya, 1980; Yamaguchi, Yoshida & Yanagishima, 1982, 1984*a*, *b*). Characteristics of the agglutination substances are shown in Table 4.1.

Both cellular sexual agglutinability and solubilised agglutination substances show the same response to enzymatic and chemical treatments (Shimoda *et al.*, 1975; Shimoda & Yanagishima, 1975; Hagiya, 1980). The amount of **a** agglutination substance extracted by the autoclave method is

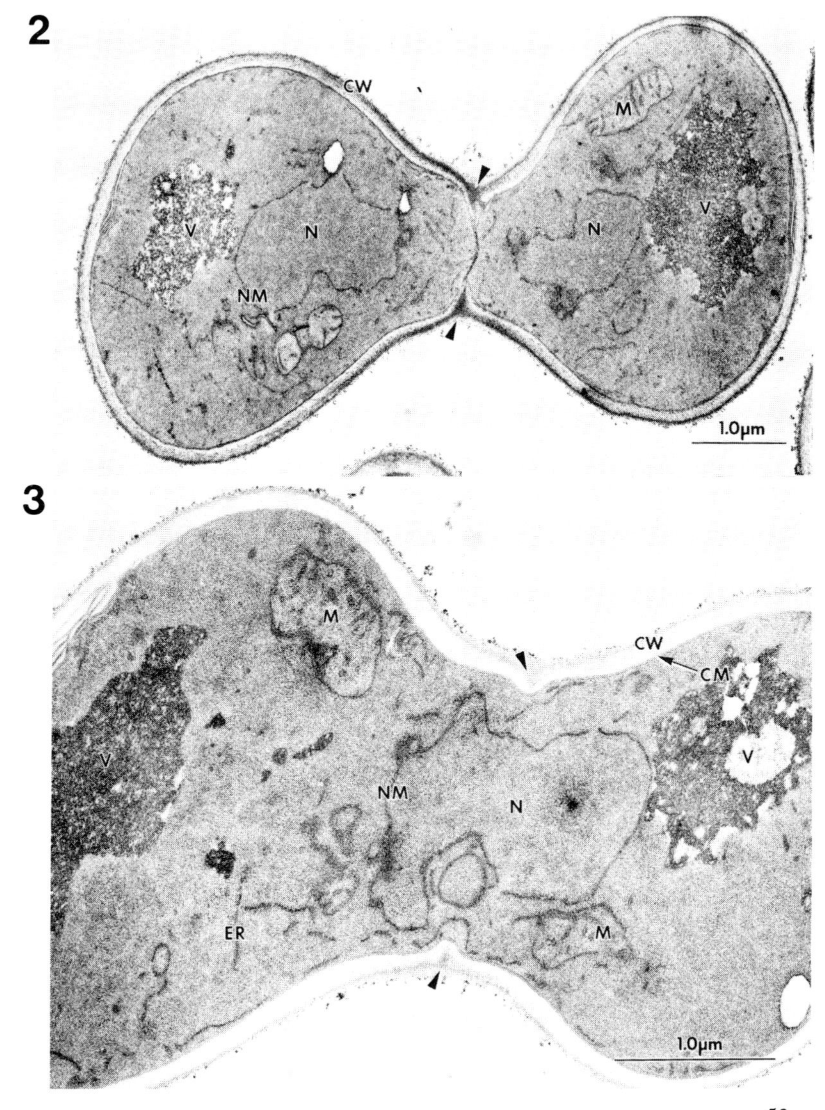

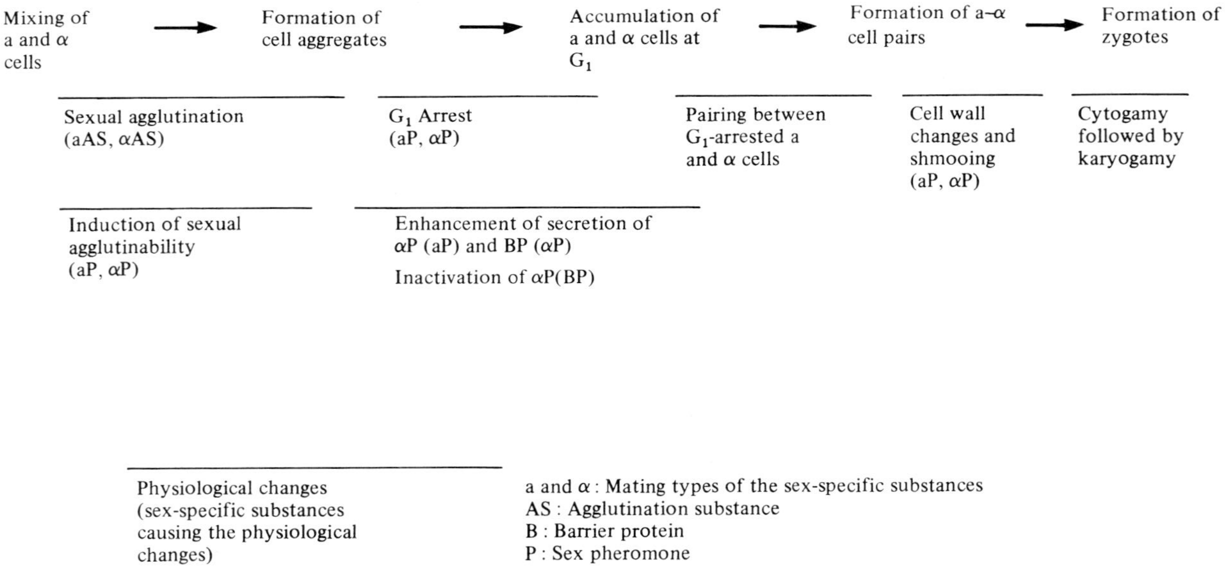

Mixing of a and α cells → Formation of cell aggregates → Accumulation of a and α cells at G_1 → Formation of a–α cell pairs → Formation of zygotes

Sexual agglutination (aAS, αAS)

G_1 Arrest (aP, αP)

Pairing between G_1-arrested a and α cells

Cell wall changes and shmooing (aP, αP)

Cytogamy followed by karyogamy

Induction of sexual agglutinability (aP, αP)

Enhancement of secretion of αP (aP) and BP (αP)

Inactivation of αP(BP)

Physiological changes (sex-specific substances causing the physiological changes)

a and α : Mating types of the sex-specific substances
AS : Agglutination substance
B : Barrier protein
P : Sex pheromone

Fig. 4.4. Physiological changes during the mating reaction in relation to mating-type-specific substances.

approximately proportional to thc sexual agglutinability of the cells (Y. Nakagawa, personal communication). *In vitro* the **a** and *α* agglutination substances, extracted by the autoclave method, form a complementary molecular complex with a concomitant loss of biological activity. The molecular complex dissociates activity recovered when the pH is raised (Yoshida *et al.*, 1976). These results, together with the mating-type-specific masking action, indicate that the agglutination substances extracted by the autoclave method are responsible for sexual cell agglutination.

Active **a** and *α* agglutination substances have also been isolated from the cytoplasm (Yamaguchi *et al.*, 1982, 1984*b*), by disrupting autoclaved cells which had lost the agglutination substance from their cell walls. Autoclaving had little effect on the agglutination substances in the cytoplasm. The same active agglutination substance may be obtained from both autoclaved cells and intact cells even in the case of *α* mating type whose agglutination substance is more sensitive to heat (Yamaguchi *et al.*, 1982). In intact cells, the agglutination substances may be degraded by enzymes, especially proteinases, but this does not occur with autoclaved cells.

The agglutination substance from the cytoplasm of both **a** and *α* cells has a higher molecular weight and carbohydrate content than the substance from the cell wall (Table 4.1). The **a** and *α* agglutination substances from the cytoplasm form complementary molecular complexes with opposite mating type agglutination substances from both the cell wall and the cytoplasm reversibly, depending on the pH values. Sensitivity to enzymatic and chemical treatments is the same between the cell wall and the cytoplasm agglutination substances in the same mating type (Table 4.1). These results suggest that the agglutination substances are processed in the course of export to the cell surface.

Fig. 4.5. Schematic representation of sexual agglutination in relation to agglutination substances. AS, agglutination substance. Isolated agglutination substances mask respective opposite mating type cells and form a complementary complex (molecular complex) with each other. Masked cells and diploid cells have no sexual agglutinability. The complementary complex has no binding activity.

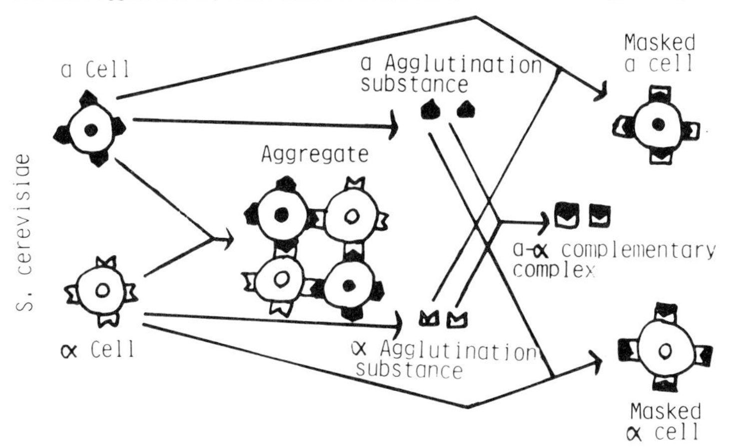

Table 4.1. **a** *and α agglutination substances from the cell wall and the cytoplasm*

	a Mating type		α Mating type	
	Wall	Cytoplasm	Wall	Cytoplasm
Molecular weight	23,000	130,000	130,000	200,000
Characteristic of protein	Glycoprotein	Glycoprotein	Glycoprotein	Glycoprotein
Carbohydrate content	61%	90%	47%	75%
Binding activity	Univalent	Univalent	Univalent	Univalent
pI	4.5	4.5	4.3	4.3
Effects of:				
2-mercaptoethanol	Unstable	Unstable	Stable	Stable
glycosidase	Stable	Stable	Stable	Stable
zymolyase	—	—	Stable	Stable
pronase	Unstable	Unstable	Unstable	Unstable

The cell wall agglutination substances were solubilised by the autoclave method and the cytoplasm agglutination substances were purified by using cytoplasm fractions obtained by disrupting autoclaved cells as starting materials. The agglutination substances released by the autoclave method came almost exclusively from the cell wall and the autoclaved cells contained undetectable amounts of the cell wall agglutination substances (Hagiya *et al.*, 1977).

Pheromonal regulation of sexual agglutinability

The production of the agglutination substances is regulated by various physiological factors such as temperature, carbon source, nitrogen source and sex phermone (Sakai & Yanagishima, 1972; Yanagishima *et al.*, 1976; Doi & Yoshimura, 1977; Tohoyama *et al.*, 1979; Yamaguchi, 1986). Among these factors the sex pheromone is most important as described below.

The α mating type sex pheromone of *S. cerevisiae* was discovered independently through the occurrence of two different phenomena appearing specifically in **a** cells. The first is the arrest of cell division followed by the formation of large pear-shaped cells (shmooed cells) (Levi, 1956; Duntze, McKay & Manney, 1970) (Fig. 4.6) and the second the induction of sexual agglutinability (Sakai & Yanagishima, 1972; Yanagishima *et al.*, 1976) (Figs. 4.7 and 4.8). In 1976, the primary structure of the active substance produced by α-cells and responsible for shmoo induction and that responsible for agglutinability induction were determined independently. It was concluded that the same pheromonal peptide (α factor or α pheromone) has the two modes of action, arresting the cell cycle at G_1 and inducing agglutination (Sakurai *et al.*, 1976, 1977; Stöltzer, Kiltz & Duntze, 1976 (Fig. 4.9)). **a** cells also produce an active substance(s) which arrests cell division at G_1 (Wilkinson & Pringle, 1974) and sexual agglutinability (Yanagishima *et al.*, 1976) specifically in α cells. Determination of the primary structure(s) of the substance(s) showed that the same sex phermone(s) has the two actions (Betz, Manney & Duntze, 1981) (Fig. 4.9).

Fig. 4.6. α-Pheromone-induced large pear-shaped (shmooed) **a** cells. 1, Control (without α pheromone); 2, Shmoo induced by α pheromone.

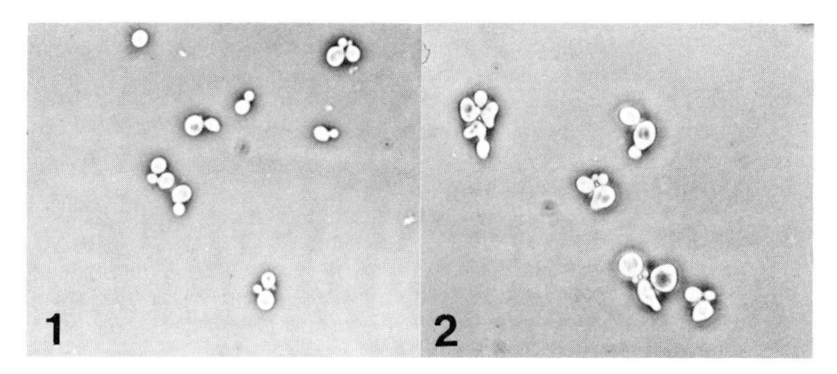

Fig. 4.7. Induction of sexual agglutinability by **a** and α pheromones. Inducible cells having no sexual agglutinability were treated with sex pheromones of the respective opposite mating types and tested for sexual agglutinability by mixing with highly agglutinable cells of the opposite mating types. For genetic and physiological control of inducible cells, see text. Inducible **a** cells (**a**^{i} cells) treated with the same mating type **a** pheromone (1) and the opposite mating type α pheromone (2). Inducible α cells (α^{i} cells) were treated with the same mating type α pheromone (3) and the opposite mating type **a** pheromone (4). Inducible cells can be produced by both physiologically and genetically as described in text. In this experiment genetically inducible strains were used. (Reproduced from Yanagishima & Nakagawa, 1980.)

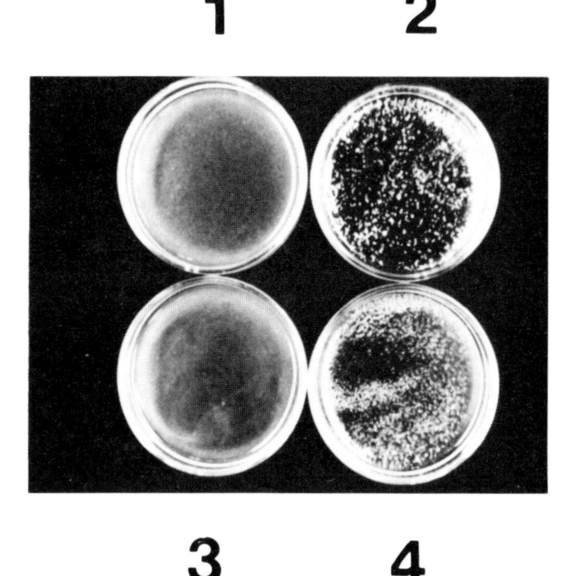

a and α cells incubated at 36 °C or on acetate or glycerol as carbon source produce no detectable amounts of the agglutination substances (Yanagishima *et al.*, 1976; Doi & Yoshimura, 1977; Tohoyama *et al.*, 1979). When these cells are treated with the sex pheromone of the opposite mating type, they begin to produce the agglutination substance of the respective mating type, resulting in expression of sexual agglutinability (Fig. 4.7 and 4.8) (Yanagishima *et al.*, 1976; Tohoyama *et al.*, 1979). This induction of agglutinability is inhibited by inhibitors of protein and RNA synthesis but not by an inhibitor of DNA synthesis (Tohoyama *et al.*, 1979).

In order to elucidate the secretory pathway of the agglutination substances in relation to the site of the sex pheromone action in inducing sexual agglutinability, we used four kinds of temperature-sensitive secretory (*sec*) mutants which are defective in the export of the secretory proteins from ribosomes to the endoplasmic reticulum (ER), from the ER to the Golgi bodies, from the Golgi bodies to secretory vesicles, or from secretory vesicles to the cell surface (Novick, Ferro & Schekman, 1981; Ferro-Novick *et al.*, 1984). When

Fig. 4.8. Schematic representation of induction by opposite mating type sex pheromone of sexual agglutinability (cf. Fig. 4.7). $\mathbf{a}^i$ and α^i cells; inducible cells which produce no agglutination substances in the absence of sex pheromone of the opposite mating type (cf. Fig. 4.5). $\mathbf{a}^c$ and α^c cells; constitutive cells which produce the respective agglutination substance constitutively in the absence of the sex pheromone of the opposite mating type.

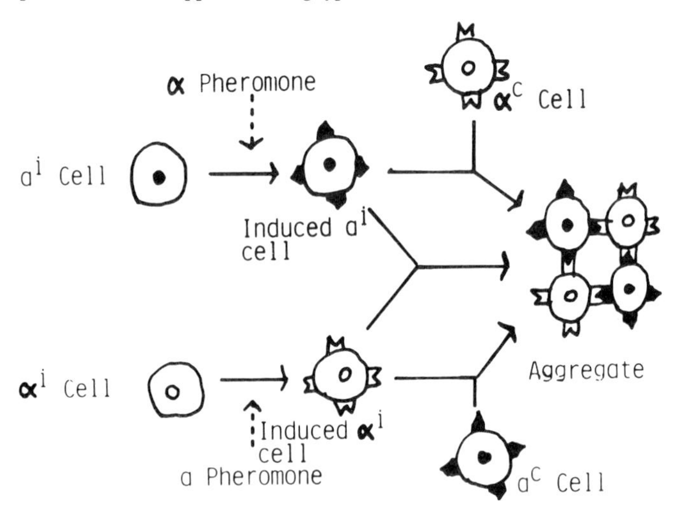

Fig. 4.9. Primary structure of the sex pheromones in *S. cerevisiae*.

a-Pheromone	H-Tyr-Ile-Ile-Lys-Gly-Val-Phe-Trp-Ala-Asx-Pro-OH
	H-Tyr-Ile-Ile-Lys-Gly-Leu-Phe-Trp-Ala-Asx-Pro-OH
α-Pheromone	H-Trp-His-Trp-Leu-Gln-Leu-Lys-Pro-Gly-Gln-Pro-Met-Tyr-OH

the glycerol-grown temperature-sensitive *sec* mutants (all **a** mating type) were treated with sex pheromone of the opposite mating type α at the restrictive temperature, no induction of sexual agglutinability occurred. This suggested an absence of secretion of the agglutination substances into the cell wall (Tohoyama & Yanagishima, 1985). When treated at the permissive temperature, all the *sec* mutants became sexually agglutinable. The *sec* mutant cells, treated at the restrictive temperature, were shown to contain the active **a** agglutination substance in the cytoplasm. The following experiment demonstrates that induction has in fact occurred but that secretion is inhibited at the restrictive temperature. We could detected only low levels of the **a** agglutination substance in the cytoplasm of a mutant defective in the export from ribosomes to the ER. This result suggests that the precursor molecules at this stage have little activity as the **a** agglutination substance. When the incubation temperature is lowered to below the restrictive level after addition of cycloheximide, which inhibits pheromone-induced production of the agglutination substance, cellular sexual agglutinability appeared and the amount of the agglutination substance in the cytoplasm decreased. This indicates that the cytoplasmic agglutination substances, which had accumulated during pheromone treatment at the restrictive temperature, were then secreted into the cell wall (Tohoyama & Yanagishima, 1985). The same kinds of result were also obtained with *sec* mutant α cells and the **a** pheromone (Tohoyama & Yanagishima, 1985, 1987). It can be concluded that the **a** and α agglutination substances are secreted through yeast secretory pathway (i.e. from the ER through the Golgi bodies and secretory vesicles to the cell surface) and that the sex pheromone of the opposite mating type acts at the site of synthesis of the precursor molecules of these substances. Indications that protein and RNA syntheses are required for pheromonal induction of the sexual agglutinability (Shimoda *et al.*, 1976; Tohoyama *et al.*, 1979) support the above conclusion.

During the early stage of shmooing in **a** cells, which are undergoing enlargement without DNA synthesis and at the same time are secreting under the influence of α pheromone, acid phosphatase is restricted to the cell tip (Field & Schekman, 1980). Since acid phosphatase and **a** and α agglutination substances are exported through the same pathway, secretion of the substances is thought to be restricted to cell tip region under the influence of sex pheromone of the opposite mating type, giving preferential binding capacity at cell tip. This must lead to preferential binding at cell tip between shmooing **a** and α cells. Indeed, we have observed **a**–α cell pair formation at cell tips when **a** and α cells, which had been detached initially on agar film, underwent the mating reaction (H. Miyata, K. Tachibana & N. Yanagishima, unpublished results; A. Shikata, Y. Shikata & N. Yanagishima, unpublished results).

Genetic control of sexual agglutinability

The mating type of *S. cerevisiae* is controlled by the mating type locus (*MAT*) which has regulatory action on mating-type-specific genes responsible for mating-type-specific functions. The *MAT*α allele has at least the two complementation groups, *MAT*α*1* and *MAT*α*2*, and the *MAT***a** allele one functional unit *MAT***a***1* (Strathern, Hicks & Herskowitz, 1981) (Fig. 4.10). α agglutination substance is not produced in *matα1* mutant cells (Tohoyama & Yanagishima, 1982). *matα2* mutant cells show no sexual agglutinability, but we can still detect α agglutination substance in their cell wall fraction after selective inactivation of **a** agglutination substance (Tohoyama & Yanagishima, 1982). These results indicate that both **a** and α agglutination substances are produced simultaneously in *matα2* cells, resulting in formation of an inactive complementary molecular complex. We could not detect any agglutination substances in *MAT***a**/*MATα* diploid cells. All of these results indicate that the production of **a** and α agglutination substances is controlled by the mating type locus through the mechanisms indicated by the α1–α2 hypothesis (Strathern *et al.*, 1981) (Fig. 4.10). The production of α agglutination substance is stimulated by the *MATα1* product and the production of **a** agglutination substance is inhibited by the *MATα2* product (Figs. 4.10 and 4.11).

Fig. 4.10. The α1–α2 hypothesis for the mating type control by the mating type locus. Straight arrows, stimulation of gene expression; T bars, inhibition of gene expression; wavy arrows, gene expression. **a**-sg, α-sg and h-sg; **a**-mating-type-, α-mating type- and haploid-specific genes unlinked to the mating type locus (*MAT*). **a**1-p, α1-p and α2-p; *MAT***a***1*, *MATα1* and *MATα2* products regulating expression of mating-type- and haploid-specific genes. In our case, *AGα1* is thought to be an α-sg, of which expression is stimulated by *MATα1* product. (Based on Strathern *et al.*, 1981.)

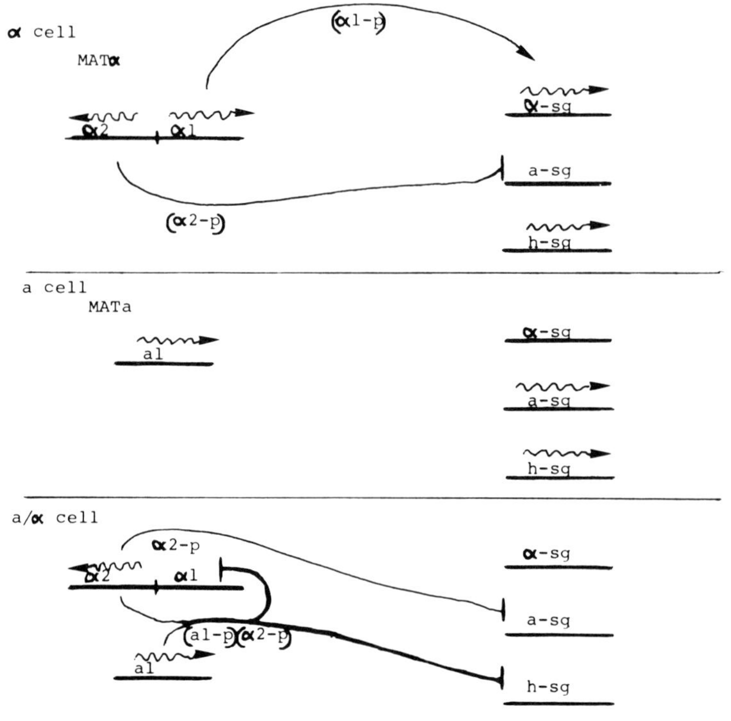

We have characterised the α- mating-type-specific gene *AGα1* which controls the production of α agglutination substance in α cells (Suzuki & Yanagishima, 1985*a*, *b* 1986*a*; Suzuki, 1986). *agα1* mutant α cells produce no detectable amount of the α agglutination substance and show no sexual agglutinability, but *agα1* mutant **a** cells produce the **a** agglutination substance normally and show high **a** agglutinability. The action of the *AGα1* gene is dose-dependent, suggesting that this gene is the structural gene of the α agglutination substance. The fact that *agα1* mutant α cells show normal sexual activities such as α pheromone production, sensitivity to **a** pheromone and mating ability except sexual agglutinability strongly supports the above suggestion. The *AGα1* gene is located near the centromere on chromosome X.

The expression of genes directly responsible for sexual agglutinability is

Fig. 4.11. Schematic representation of the hypothesis for genetic regulation of sexual agglutinability in inducible strains (cf. Fig. 10).

MATα1 product stimulates expression of the α-mating-type-specific agglutination substance gene, *ASα*. The *AGα1* gene described in text is probably the *ASα* gene. *MAT***a***1* product gives no effect on expression of the *AS***a** gene (**a**-mating-type-specific agglutination substance gene). RSE (regulator of sex expression) inhibits expression of the mating-type-specific genes *ASα* and *AS***a** during vegetative growth. This inhibition is removed by a mutation at this locus or the sex pheromone of the opposite mating type, resulting in high expression of sex-specific functions including sexual agglutinability. Ph (pheromone)-receptor determined mating-type specificity of pheromone actions. Many other genes are thought to affect expression of mating-type-specific genes, but as there is little information about them they are excluded from this scheme.

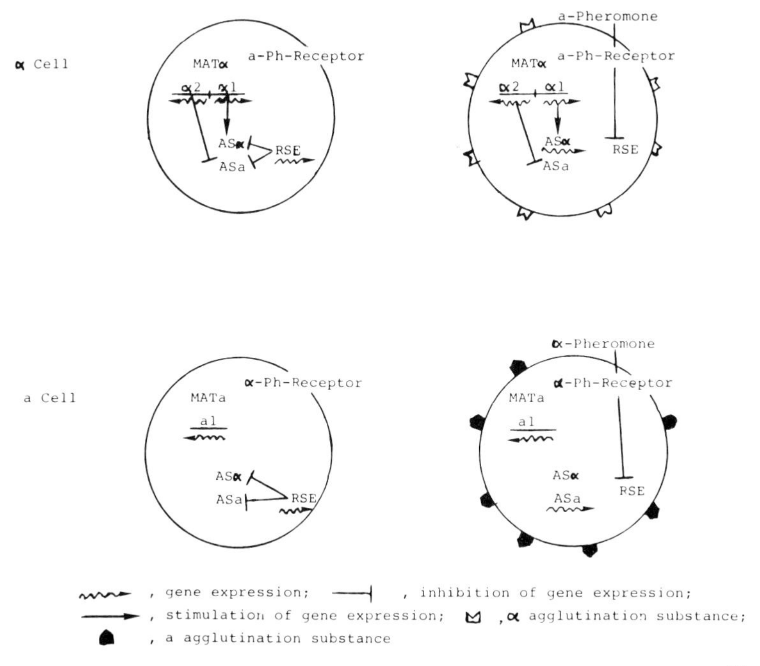

controlled both genetically and physiologically. There are two types of strains which affect sexual agglutinability; one is constitutive and the other inducible. The former strains produce a considerable amount of the respective sexual agglutination substance constitutively without the sex pheromone of the opposite mating type, but the latter strains produce the substance only in response to the pheromone (Sakai & Yanagishima, 1972) (Figs. 4.7 and 4.8). The genetic control system relating to the above phenomena is highly complex and has yet to be resolved. We can bring about the inducible state physiologically by changing culture conditions even in genetically constitutive strains.

We have recently found a possible way for analysing the genetic system regulating sex expression in relation to sex pheromone (Suzuki & Yanagishima, 1986*b*; Suzuki, 1986). A mating-type-non-specific gene regulates the expression of sexual activities such as sexual agglutinability, production of sex pheromone, sensitivity to the opposite mating type pheromone, retardation of vegetative growth and the formation of large pear-shaped cells (shmooing). A mutation of this gene causes a high level of expression of the above sexual activities during vegetative growth. In the wild-type cells, a high level of expression of the sexual activities is brought about only in response to the sex pheromone of the opposite mating type. It is notable that the degree of constitutive sexual agglutinability in the mutant is not only very high, but is also not enhanced by the sex pheromone of the opposite mating type. Hence, the gene is thought to inhibit the expression of sexual activities during vegetative growth. The inhibition is removed by the mutation of this gene or by the action of the sex pheromone of the opposite mating type. Although the gene is mating-type-non-specific, the action of the sex pheromones is mating-type-specific in that the sex pheromones act only on cells of the opposite mating type. The mating type specificity is probably brought about by the action of a mating-type-specific receptor for the **a** or α pheromone, whose production is under the control of the mating type locus. Based on the above results, we propose a genetic control system which relates to pheromonal regulation of the production of the sexual agglutination substance, as shown in Fig. 4.11.

Sexual agglutination and zygote formation

Role of sexual agglutination in the mating reaction

The formation of conjugating cells and zygotes (Fig. 4.3) is restricted almost exclusively to sexual cell aggregates in mixed liquid cultures of **a** and α cells, thus leading to the notion that sexual agglutination is a prerequisite for zygote formation (Kawanabe *et al.*, 1979). The α-mating-type-specific agglutinability-defective *agα1* mutant allowed us to study the role played by sexual agglutination in zygote formation (Suzuki & Yanagishima, 1985*b*, 1986*a*; Suzuki, 1986). This mutant α strain retains all sexual activities except agglutinability as mentioned previously. We investigated the effect of the *agα1* mutation on zygote formation in relation to culture conditions.

If the only role of sexual agglutination in the mating reaction is to bring about cell to cell contact between opposite mating type cells, the *agα1* mutant should readily mate with a wild type **a** strain. By establishing a mating condition which ensures cell to cell contact. On dry agar plates the *agα1* mutant mates at about the same frequency as wild type *α* cells. The mutant cells mated only at low frequency on wet agar plates and hardly at all in liquid media. The wild type (*AGα1*) *α* cells mated at the same frequency on both dry and wet agar plates and in liquid media. These results indicate that the fundamental action of sexual agglutination is to ensure contact between **a** and *α* cells.

However, simple contact alone between **a** and *α* cells is insufficient to cause a high frequency of mating. Packing of **a** and *α* cell mixtures by centrifugation did not improve mating efficiency in the *agα1* mutant, thus suggesting that sexual agglutination has a role other than merely bringing about the formation of simple cell masses where cell to cell contact is ensured. Even in wild type *α* cells packing by centrifugation lowers mating efficiency, although here the efficiency is much higher than in the mutant. In packed cell masses nutritional supply and regulation of the cell cycle by sex pheromones may become critical.

In connection with the above results, it is interesting to note that the sexually agglutinable mixtures of wild type **a** and *α* cells form spreading rough pellets but non-agglutinable mixtures of *agα1* mutant *α* cells and wild type **a** cells form compact dense cell pellets when centrifuged (Suzuki & Yanagishima, 1985*b*; Suzuki, 1986). Sexual agglutinability is retained after a short period at 100 °C but the cells are killed. When a mixture of *agα1* mutant *α* cells and wild type **a** cells was supplemented with heat-killed wild type *α* cells, the mixture formed spreading rough pellets when centrifuged. It is notable that the *agα1* mutant forms zygotes much more frequently in such spreading pellets than in compact dense pellets formed without addition of heat-killed wild type *α* cells. The addition of heat-killed wild type *α* cells is far more marked in the *agα1* mutant than in the wild type (Suzuki & Yanagishima, 1985*b*; Suzuki, 1986). These results strongly support the notion that sexual agglutination not only brings about cell to cell contact between **a** and *α* cells, but also gives rise to multicellular structure suitable for zygote formation.

The possibility that sexual agglutination directly causes physiological stimulation in cells of the opposite mating type receives scant support from experimental results obtained so far (Suzuki, 1986) but cannot be ruled out completely.

Sexual cell–cell interactions in sexual aggregates

To learn more about the process of zygote formation in sexual cell aggregates, we observed the mating process by time-lapse photomicroscopy, following gentle spreading of mixed suspensions of **a** and *α* cells on the surface of a nutritional agar film. The **a** and *α* cells formed sexual aggregates on the agar film but were smaller. Hence, we could trace the mating process in relation to cell division of individual cells both in and outside the aggregates (Miyata, Tachibana & Yanagishima, 1984; A. Shikata, Y. Shikata & N. Yanagishima, unpublished results). Under our experimental conditions, the increase in number of budding cells slowed down at about 45 min after inoculation of the **a** and *α* cells on to agar film. No similar arrest of budding was detected in single cultures of **a** or *α* cells. About 35 min after the arrest of budding, number of budding cells began to increase again. The **a**–α cell pairs (conjugating cells) began to appear at the end of the budding-arrest-period, a little before the initiation of the second increase in budding. These events can be explained by the facts that the **a** and α pheromones cause arrest of the cell cycle at the G_1 phase in the opposite mating type, (probably at the step controlled by the *CDC28* gene) and only cells in this condition can undergo the mating reaction (Hartwell, 1973; Reid & Hartwell, 1977).

a Cells not only become less sensitive to the α pheromone (Chan, 1977), but also increasingly secrete a substance which inactivates the α pheromone (Barrier protein) during growth in the presence of the α pheromone (Hicks & Herskowitz, 1976). Although the biological significance of this behaviour is far from clear, it may have a role in regulating the α pheromone level in mating mixtures, thus preventing the death of unmated cells in mating mixtures. The above phenomena seem to be responsible for the second increase in the number of budding cells in our mating system.

When **a** and α cells at different stages in the cell cycle are forming a mating pair, the cell at the more advanced stage in the cycle is arrested at G_1 until the other of the opposite mating type also reaches G_1 (Miyata *et al.*, 1984). The latter then undergoes the mating reaction without any arrest at G_1. Hence, the G_1 phase, but not necessarily arrested, appears essential, for **a** and α cells to undergo the mating reaction.

In relation to these findings, we have proposed that the α pheromone regulates mating ability of **a** cells not only by arresting the cell cycle at G_1 but also by an unknown action which causes maturation (or gametogenesis) of **a** cells (Miyata *et al.*, 1984; Tachibana, Miyata & Yanagishima, 1985). This notion came from the following two facts. Even when **a** and α cells at G_1 undergo the mating reaction, **a**–α mating pairs appear only after 60 min lag period (Miyata *et al.*, 1984). Pretreatment with the α pheromone of **a** cells, in a condition where the pheromone has no effect on cell division, causes significant enhancement of zygote formation; that is, **a**–α cell pairing is initiated earlier in the α-pheromone-pretreated **a** cells than in untreated or **a**-pheromone-pretreated **a** cells (Tachibana, *et al.*, 1985). These results suggest that the α pheromone causes the cell maturation necessary for the initiation of zygote formation. This activity of the α pheromone may operate even in cells which are not at G_1 because **a** or α cells can undergo mating reaction without becoming arrested at G_1. Sexual interactions in cell aggregates in relation to mating-type-specific substances are summarised in Fig. 4.12.

Phylogenetic significance of sexual interactions

Sexual agglutination is restricted to opposite mating type cells belonging to the same species: no complementary molecular complexes having any obvious biological activity are formed between agglutination substances extracted from different species by the autoclave method (Yamaguchi *et al.*, 1984*a*; Yamaguchi, 1986). Since sexual agglutination is thought to be the first step in the mating reaction in ascosporogenous yeasts, the species specificity mentioned above is an obvious mechanism of sexual isolation. Changes in specificity of sexual agglutination may be one of the important stages in species differentiation.

In spite of the high species specificity in sexual cell–cell recognition (agglutination), it has been found that the agglutination substances have common features which extend beyond the species boundaries. Burke, Mendonça-Previato & Ballou (1980) classified opposite mating types of three yeasts into **a** and α mating types corresponding to **a** and α mating types in *S. cerevisiae*, based on the stability of agglutination substances to heat, reducing agents and response of one of the three yeasts, *S. kluyveri* to the *S. cerevisiae* α pheromone (McCullough & Herskowitz, 1979). We have succeeded in classifying opposite mating types of ten yeasts into **a** and α mating types as follows. Sexual agglutinability, or biological activity of the solubilised agglutination substances, of one of the opposite mating types in a species is more sensitive to 2-mercaptoethanol treatment (5% at 28 °C) and more resistant to acetic acid treatment (3% at 100 °C) than that of the other mating type (Yamaguchi *et al.*, 1984*a*; Yamaguchi, 1986). The former mating type is thought to be **a** and the latter mating type α, because in *S. cerevisiae*

Fig. 4.12. Sexual interactions in sexual cell aggregates through sex-specific substances.

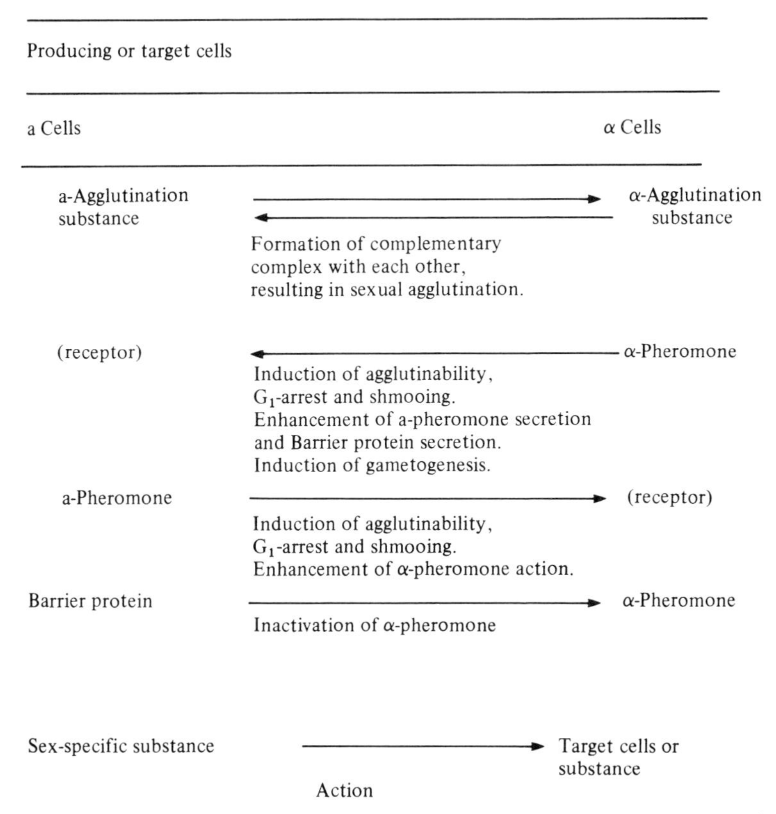

the **a** mating type agglutination substance is more resistant to acetic acid and more sensitive to 2-mercaptoethanol than the α mating type.

We have found sex pheromones which induce sexual agglutinability and/or formation of large pear-shaped cells in the opposite mating type cells in the genera *Saccharomyces*, *Hansenula* and *Pichia* (Sakai & Yanagishima, 1972; Yanagishima *et al.*, 1976; Yanagishima & Fujimura, 1981; Fujimura *et al.*, 1983; Yanagishima, 1986; Hisatomi, Yanagishima & Banno, 1986). α Pheromones were purified and their amino acid sequences were determined in three *Saccharomyces* yeasts (Fig. 4.13) (Sakurai *et al.*, 1976, 1984, 1986). These pheromones show the pheromonal action on **a** cells of all four species of *Saccharomyces* tested, and the *S. cerevisiae* α pheromone induces sexual agglutinability in **a** cells of *H. wingei* and *H. anomala*. In contrast to these α pheromones, the **a** pheromones of these yeasts act only on α cells of the same species. The mating type, determined on the basis of the chemical nature of the agglutination substances, coincides completely with the mating type as determined by pheromone production and responsiveness (Yamaguchi *et al.*, 1984).

Based on the above results, we conclude that ascosporogenous yeasts have two complementary mating types corresponding to the **a** and α mating types in *S. cerevisiae*. These findings may give clues for further studies on mechanisms of species differentiation in ascosporogenous yeasts.

Fig. 4.13. Primary structure of α pheromones in *S. cerevisiae*, *S. kluyveri* and *S. exiguus*.

S. cerevisiae	H-Trp-His-Trp-Leu-Gln-Leu-Lys-Pro-Gly-Gln-Pro-Met-Tyr-OH
S. kluyveri	H-Trp-His-Trp-Leu-Ser-Phe-Ser-Lys-Gly-Glu-Pro-Met-Tyr-OH
S. exiguus	H-Trp-His-Trp-Leu-Arg-Leu-Ser-Tyr-Gly-Gln-Pro-Ile-Tyr-OH

Acknowledgements

I express my sincere thanks to present and former members of my laboratory who performed many of the experiments described in this article. Thanks are extended to Drs K. Yoshida, K. Suzuki and H. Tohoyama for critically reading the manuscript. I wish to thank Dr M. Osumi for the electron micro-photographs and Ms T. Noto for help in preparation of the manuscript. The author's research described in this article was partly supported by Grants-in-Aid from the Ministry of Education, Science and Culture, Japan.

References

Betz, R., Manney, T. R. & Duntze, W. (1981). Hormonal control of gametogenesis in the yeast *Saccharomyces cerevisiae*. *Gamete Res.*, **4**, 571–84

Burke, D., Mendonça-Previato, L. & Ballou, C. E. (1980). Cell–cell recognition in yeast: Purification of *Hansenula wingei* 21–cell sexual agglutination factor and comparison of the factors from three genera. *Proc. Natl. Acad. Sci. USA*, **77**, 318–22

Chan, R. K. (1977). Recovery of *Saccharomyces cerevisiae* mating-type **a** cells from G_1 arrest by α factor. *J. Bact.*, **130**, 766–74

Doi, S. & Yoshimura, M. (1977). Temperature-sensitive loss of sexual agglutinability in *Saccharomyces cerevisiae*. *Arch. Microbiol.*, **114**, 287–8

Duntze, W., MacKay, V. & Manney, T. R. (1970). *Saccharomyces cerevisiae:* A diffusible sex factor. *Science*, **168**, 1472–3

Ferro-Novick, S., Novick, P., Field, C. & Schekman, R. (1984). Yeast secretory mutants that block the formation of active cell surface enzymes. *J. Cell Biol.*, **98**, 35–43

Field, C. & Schekman, R. (1980). Localized secretion of acid phosphatase reflects the pattern of cell surface growth in *Saccharomyces cerevisiae*. *J. Cell Biol.*, **86**, 123–8

Fujimura, M., Ishikawa, S., Fujimura, H., Yanagishima, N. & Sakurai, A. (1983). Sex pheromones in *Hansenula* and *Pichia* yeasts. *Proc. 48th Ann. Meeting Bot. Soc. Japan*, p. 195 (in Japanese)

Hagiya, M. (1980). Studies on the mechanism of sexual agglutination in the yeast *Saccharomyces cerevisiae*. *PhD thesis Nagoya University, Nagoya*

Hagiya, M., Yoshida, K. & Yanagishima, N. (1977). The release of sex-specific substances responsible for sexual agglutination from haphloid cells of *Saccharomyces cerevisiae*. *Exp. Cell Res.*, **104**, 263–72

Hartwell, T. H. (1973). Synchronization of haploid yeast cell cycles, a prelude to conjugation. *Exp. Cell Res.*, **76**, 111–7

Herskowitz, I. & Oshima, Y. (1981). Control of cell type in *Saccharomyces cerevisiae*: Mating type and mating type interconversion. In *Molecular Biology of the Yeast* Saccharomyces. *Life cycle and Inheritance*, ed. J. N. Strathern, E. W. Jones & J. R. Broach, pp. 181–209. Cold Spring Harbor: Cold Spring Harbor Laboratory

Hicks, J. B. & Hershowitz, I. (1976). Evidence for a new diffusible element of mating pheromones in yeast. *Nature (Lond.)*, **260**, 246–8

Hisatomi, T., Yanagishima, N. & Banno, I. (1986). Induction of heterothallic strains and their genetic and physiological characterization in a homothallic strain of the yeast *Saccharomyces exiguus*. *Curr. Genetics*, **10**, 887–92

Kawanabe, Y., Yoshida, K. & Yanagishima, N. (1979). Sexual cell agglutination in relation to the formation of zygotes in *Saccharomyces cerevisiae*. *Plant & Cell Physiol.*, **20**, 423–33

Levi, J. D. (1956). Mating reaction in yeast. *Nature (Lond.)*, **177**, 753–4

McCullough, J. & Herskowitz, I. (1979). Mating pheromones of *Saccharomyces kluyveri*: Pheromone interactions between *Saccharomyces kluyveri* and *Saccharomyces cerevisiae*. *J. Bact.*, **138**, 146–54

Miyata, H., Tachibana, K. & Yanagishima, N. (1984). Prelude for conjugation of the yeast *Saccharomyces cerevisiae*. *Proc. 49th Ann. Meeting Bot. Soc. Japan*, p. 22 (in Japanese)

Novick, P., Ferro, S. & Schekman, R. (1981). Order of events in the yeast secretory pathway. *Cell*, **25**, 461–9

Osumi, M., Shimoda, C. & Yanagishima, N. (1974). Mating reaction in *Saccharomyces cerevisiae* V. Changes in the fine structure during the mating reaction. *Arch. Microbiol.*, **97**, 27–38

Reid, B. J. & Hartwell, L. H. (1977). Regulation of mating in the cell cycle of *Saccharomyces cerevisiae*. *J. Cell Biol.*, **75**, 355–65

Sakai, K. & Yanagishima, N. (1972). Mating reaction in *Saccharomyces serevisiae* II. Hormonal regulation of agglutinability of **a** type cells. *Arch. Mikrobiol.*, **84**, 191–8

Sakurai, A., Sakamoto, M., Kitada, H., Esumi, Y., Takahashi, N., Fujimura, H., Yanagishima, N. & Banno, I. (1984). Isolation and amino acid sequence of a mating pheromone produced by mating type α cells of *Saccharomyces kluyveri*. *FEBS Lett.*, **166**, 339–42

Sakurai, A., Tamura, S., Yanagishima, N. & Shimoda, C. (1977). Structure of a peptidyl factor, α substance-1A inducing sexual agglutinability in *Saccharomyces cerevisiae*. *Agri. & Biol. Chem.*, **41**, 395–8

Sakurai, A., Tamura, S., Yanagishima, N. & Shimoda, C. (1976). Structure of the peptidyl factor inducing sexual agglutination in *Saccharomyces cerevisiae*. *Agric. & Biol. Chem.*, **41**, 1057–8

Sakurai, A., Tanaka, H., Esumi, Y., Takahashi, N., Hisatomi, T., Yanagishima, N. & Banno, I. (1986). Isolation and amino acid sequence of a mating pheromone produced by mating type α cells of *Saccharomycs exiguus*. *FEBS Lett.*, **203**, 285–8

Shimoda, C., Kitano, S. & Yanagishima, N. (1975). Mating reaction in *Saccharomyces cerevisiae* VII. Effect of proteolytic enzymes on sexual agglutinability and isolation of crude sex-specific substances responsible for sexual cell recognition. *Antonie van Leeuwenhoek. J. Microbiol. & Serol.*, **45**, 513–19

Shimoda, C. & Yanagishima, N. (1975). Mating reaction in *Saccharomyces cerevisiae* VIII. Mating-type-specific substances responsible for sexual cell agglutination. *Antonie van Leeuwenhoek. J. Microbiol. & Serol.*, **41**, 521–32

Shimoda, C., Yanagishima, N., Sakurai, A. & Tamura, S. (1976). Mating reaction in *Saccharomyces cerevisiae* IX. Regulation of sexual cell agglutinability of **a** type cells by a sex factor produced by α types cells. *Arch. Microbiol.*, **108**, 27–33

Shimoda, C., Yanagishima, N., Sakurai, A. & Tamura, S. (1978). Induction of sexual agglutinability of **a** mating type cells as the primary action of the peptidyl sex factor from α mating type cells in *Saccharomyces cerevisiae*. *Plant & Cell Physiol.*, **19**, 513–17

Sprague, G. F., Jr., Blair, L. C. & Thorner, J. (1983). Cell interactions and regulation of cell type in the yeast *Saccharomyces cerevisiae*. *Ann. Rev. Microbiol.*, **37**, 623–60

Stöltzler, D., Kiltz, H.-H. & Duntze, W. (1976). Primary structure of α factor peptides from *Saccharomyces cerevisiae*. *Eur. J. Biochem.*, **69**, 397–400

Strathern, J. N., Hicks, J. & Herskowitz, I. (1981). Control of cell type in years by the mating type locus, the $\alpha1$–$\alpha2$ hypothesis. *J. Mol. Biol.*, **157**, 357–72

Suzuki, K. (1986). Genetic and physiological studies on sexual cell–cell recognition in yeast. *PhD thesis, Nagoya University, Nagoya*

Suzuki, K. & Yanagishima, N. (1985*a*). An α-mating-type-specific mutation causing specific defect in sexual agglutinability in the yeast *Saccharomyces cerevisiae*. *Curr. Genetics*, **9**, 185–9

Suzuki, K. & Yanagishima, N. (1985*b*). Genetic and physiological characterization of an α-mating-type-specific agglutinability-defective mutant. *Proc. 50th Ann. Meeting Bot. Soc. Japan*, p. 227 (in Japanese)
Suzuki, K. & Yanagishima, N. (1986*a*). Genetic characterization of an α-specific gene responsible for sexual agglutinability in *Saccharomyces cerevisiae*: Mapping and gene dose effect. *Curr. Genetics*, **10**, 353–7
Suzuki, K. & Yanagishima, N. (1986*b*). A mutation leading to constitutive expression of high sexual activities in the yeast *Saccharomyces cerevisiae. Plant & Cell Physiol.*, **27**, 523–31
Tachibana, K., Miyata, H. & Yanagishima, N. (1985). Role of sex pheromones in the prelude for conjugation in the yeast *Saccharomyces cerevisiae. Proc. 50th Ann. Meeting Bot. Soc. Japan*, p. 247 (in Japanese)
Thorner, J. (1981). Pheromonal regulation of development in *Saccharomyces cerevisiae*. In *Molecular Biology of the yeast* Saccharomyces. *Life cycle and Inheritance*. Ed. J. N. Strathern, E. W. Jones, J. R. Broach, pp. 143–80. Cold Spring Harbor: Cold Spring Harbor Laboratory
Tohoyama, H., Hagiya, M., Yoshida, K. & Yanagishima, N. (1979). Regulation of the production of the agglutination substances responsible for sexual agglutination in *Saccharomyces cerevisiae*: Changes associated with conjugation and temperature shift. *Mol. & Gen. Genetics*, **174**, 269–80
Tohoyama, H. & Yanagishima, N. (1982). Control of the production of the sexual agglutination substances by the mating type locus in *Saccaromyces cerevisiae*: Simultaneous expression of specific genes for **a** and α agglutination substances in *matα2* mutant cells. *Mol. & Gen. Genetics*, **196**, 322–7
Tohoyama, H. & Yanagishima, N. (1985). The sexual agglutination substance is secreted through the yeast secretory pathway in *Saccharomyces cerevisiae. Mol. & Gen. Genetics*, **201**, 446–9
Tohoyama, H. & Yanagishima, N. (1987). Site of pheromone action and secretion pathway of a sexual agglutination substance during its induction by pheromone **a** in α cells of *Saccharomyces cerevisiae. Curr. Genetics*, **12**, 271–5
Wilkinson, L. E. & Pringle J. R. (1974). Transient G_1 arrest of *S. cerevisiae* cells of mating type α by **a** factor produced by cells of mating type **a**. *Exp. Cell Research*, **89**, 175–87
Yamaguchi, M. (1986). Sexual agglutination in ascosporogenous yeasts. Molecular basis and comparative physiology. *PhD thesis, Nagoya University, Nagoya*
Yamaguchi, M., Yoshida, K. & Yanagishima, N. (1982). Isolation and partial characterization of cytoplasmic α agglutination substance in the yeast *Saccharomyces cerevisiae. FEBS Lett.*, **192**, 125–9
Yamaguchi, M., Yoshida, K. & Yanagishima, N. (1984*a*). Mating-type differentiation in ascosporogenous yeasts on the basis of mating-type-specific substances responsible for sexual cell–cell recognition. *Mol. & Gen. Genetics*, **194**, 24–30
Yamaguchi, M., Yoshida, K. & Yanagishima, N. (1984*b*). Isolation and biochemical and biological characterization of an **a**-mating-type-specific glycoprotein responsible for sexual agglutination from the cytoplasm of **a**-cells in the yeast *Saccharomyces cerevisiae. Arch. Microbiol.*, **40,** 113–19

Yanagishima, N. (1978). Sexual cell agglutination in *Saccharomyces cerevisiae*: Sexual cell recognition and its regulation. *Bot. Mag. Tokyo Special Issue*, **1**, 61–81

Yanagishima, N. (1984). Mating systems and sexual interaction in yeast. In *Cellular Interactions, Encyclopedia of Plant Physiology New Series 17*, ed. H.-F. Linskens & J. Heslop-Harrison, pp. 402–23. Berlin: Springer-Verlag

Yanagishima, N. (1986). Sexual differentiation and interactions in yeasts. *Microbiol. Sci.*, **3**, 45–9

Yanagishima, N. & Fujimura, H. (1981). Sex pheromones of the yeast *Hansenula wingei* and their relationship to sex pheromones in *Saccharomyces cerevisiae* and *Saccharomyces kluyveri*. *Arch. Microbiol*, **129**, 281–4

Yanagishima, N. & Nakagawa, Y. (1980). Mutants inducible for sexual agglutinability in *Saccharomyces cerevisiae*. *Mol. & Gen. Genetics*, **178**, 241–51

Yanagishima, N., Yoshida, K., Hamada, K. Hagiya, M., Kawanabe, Y., Sakurai, A. & Tamura, S. (1976). Regulation of sexual agglutinability in *Saccharomyces cerevisiae* of **a** and α types by sex-specific factors produced by their respective opposite mating types. *Plant & Cell Physiol.*, **17**, 439–50

Yanagishima, N., Yoshida, K., Hagiya, M., Kawanabe, Y., Shimoda, C., Sakurai, A., Tamura, S. & Osumi, M. (1977). Sexual cell agglutination in *Saccharomyces cerevisiae*. In *Growth and Differentiation in Microorganisms*, ed. T. Ishikawa, Y. Maruyama & H. Matsumiya, pp. 193–208. Tokyo: University of Tokyo Press

Yanagishima, N. & Yoshida, K. (1981). Sexual interactions in *Saccharomyces cerevisiae* with special reference to the regulation of sexual agglutinability. In *Sexual Interactions in Eukaryotic Microbes*, ed. D. H. O'Day & P. A. Horgen, pp. 261–95. New York: Academic Press

Yoshida, K., Hagiya, M. & Yanagishima, N. (1976). Isolation and purification of sexual agglutination substance of mating type **a** cells in *Saccharomyces cerevisiae*. *Biochem. & Biophys. Research Comm.*, **71**, 1085–94

Yoshida, K. & Yanagishima, N. (1978). Intra- and intergeneric mating behaviour of ascosporogenous yeasts. I. Quantitative analysis of sexual agglutination. *Plant & Cell Physiol.*, **19**, 1519–33

5

An O-glycosylated agglutinin induced by α factor in *Saccharomyces cerevisiae*

M. Watzele and W. Tanner

Institut für Botanik der Universität Regensburg,
Universitätsstr. 31,
8400 Regensburg, West Germany

Abstract

A number of cell surface glycoproteins can be specifically and totally released from intact cells of *Saccharomyces cerevisiae* with 0.5% mercaptoethanol leaving the cells completely viable. Among these proteins is one with an M_r of 22 kD which is synthesised only in haploid *a* cells treated with the peptide mating pheromone α factor. This protein can be radiolabelled *in vivo* with [2-^{3}H]-mannose, [^{14}C]-phenylalanine, and [^{35}S]-sulphate. Its synthesis, as well as its export to the cell surface, is not inhibited by tunicamycin. β-Elimination released almost all radioactivity from the [2-^{3}H]-mannose labelled protein, 36% of its radioactivity being recovered subsequently as mannose, and 43% as a dimannoside. Evidence has been obtained that this 22 kD O-glycosylated protein is a mating type specific *a* cell agglutinin. Material released from two α factor induced *a* cells is sufficient to prevent agglutination of one *a* with one α cell (both pretreated with the corresponding mating pheromone). The inhibitory activity obtained from α factor treated *a* cells is thirty-fold higher than from control *a* cells.

Introduction

Species and mating type specific cell agglutination has been extensively studied in the yeast *Hansenula wingei* and *Saccharomyces cerevisiae* (Yanagishima, 1984). A number of cell surface molecules, which are thought to be involved in the mating type specific agglutination reaction (Yanagishima, 1984 and this symposium) have been characterised. In *S. cerevisiae* a significant increase in cell agglutinability is brought about by treating haploid heterothallic cells with the mating peptide pheromones α and *a* factors (Betz *et al.*, 1981; Terrance & Lipke, 1981; Thorner, 1981). Until recently, no changes in particular cell surface molecules resulting from treatment with

Table 5.1. *Agglutination of Mata and Matα cells*

	a_i, α_i*	a_i, α_{ni}	a_{ni}, α_i	a_{ni}, α_{ni}	
Time†	2 min	11 min	9 min	15 min	−Cycloheximide
Size	large	large	large	small	
Time	2 min	22 min	20 min	∞	+Cycloheximide
Size	large	very small	very small	no	

* a_l and α_i are pheromone induced cells, a_{ni} and α_{ni} are non-induced cells.
† Time elapsed until the appearance of the first agglutinates. 50 μl× 2180 Mat*a* cells (1.5 × 10^8 cells per ml) were mixed with 150 μl agglutination buffer, 0.1 M Na-PO_4 pH 6.0, and 50 μl × 2180 Matα cells (1.5 × 10^8 cells per ml) in flat bottom microtitre plates (15 × 17 mm wells). Agglutinates forming during gentle rotary shaking (110 strokes of 1.5 cm per min) can easily be seen with the naked eye. Where indicated Mat*a* cells were treated with α factor (6 μM), and Matα cells with *a* factor (20 units per ml) for 2 h in a medium containing 2% bactopeptone, 1% yeast extract, and 2% glucose prior to the assay. In the second experiment the assay was carried out in the presence of 10 μg/ml cycloheximide.

the mating factor have been reported. Here we describe the synthesis of an O-glycosylated agglutinin induced in *a* cells of *S. cerevisiae* by the α factor. Part of these results have been published in more detail elsewhere (Orlean *et al.*, 1986).

Results and discussion

The extent of mating type specific cell agglutination in *S. cerevisiae* depends in part on the strains used partly on the culture, media and partly on temperature (Yanagishima *et al.*, 1976; Doi & Yoshimura, 1978). With certain strains a high level of agglutination occurs even without prior treatment of the cells with the *a* and α factors (Yanagishima *et al.*, 1976). This is not the case with *S. cerevisiae* ×2180 grown at 30 °C in defined medium. When such haploid cells of both mating types are mixed in wells of a microtitre plate and shaken gently, after 15 min small aggregates can be observed with the naked eye (Table 5.1, fifth column). When both cell types are pre-treated with *a* and α factors, respectively, large aggregates are formed within 2 min. 15 minutes may be sufficient time for one mating type to induce agglutinin production in the opposite type (Thorner, 1981). To exclude this possibility the *a* and α cells were mixed in the presence of cycloheximide (Table 5.1). Under this regime agglutinates do not form, whereas cells pretreated with the mating factors agglutinate to the same extent and at the same rate as in the absence of cycloheximide (Table 5.1, second column). If only one mating type is induced with the corresponding mating factor, intermediate results are obtained (Table 5.1, columns 3 and 4). It has to be pointed out, however, that agglutination in these latter cases also takes place in the presence of cycloheximide. If a direct interaction between *a* and *α* agglutinin causes cells to stick together (as suggested by Yamaguchi *et al.* (1984)), the latter observation indicates that a small amount of agglutinin is probably present constitutively on each cell type.

The above results strongly suggest that a cell surface component in strain ×2180 of *S. cerevisiae* should dramatically increase following pretreatment

Fig. 5.1. The release of [2-^{3}H]mannose labelled cell surface glycoproteins from *S. cerevisiae* (×2180-1A) by mercaptoethanol. Cells were grown as described in Orlean *et al.* (1986), radiolabelled for 90 min, and extracted with 0.5% mercaptoethanol for 2 h at room temperature. Extracts were submitted to NaDodSO$_4$/PAGE and glycoproteins were visualised by fluorography. In the presence of *α* factor (12 μM) an additional glycoprotein gp 22 (arrow) is visible. For further details see Orlean *et al.* (1986).

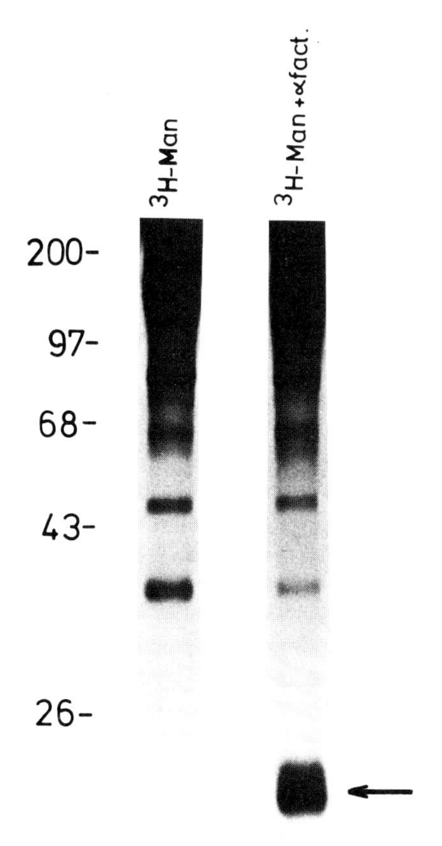

Fig. 5.2. Scheme showing mating factor induced agglutinins and agglutination in *S. cerevisiae*.

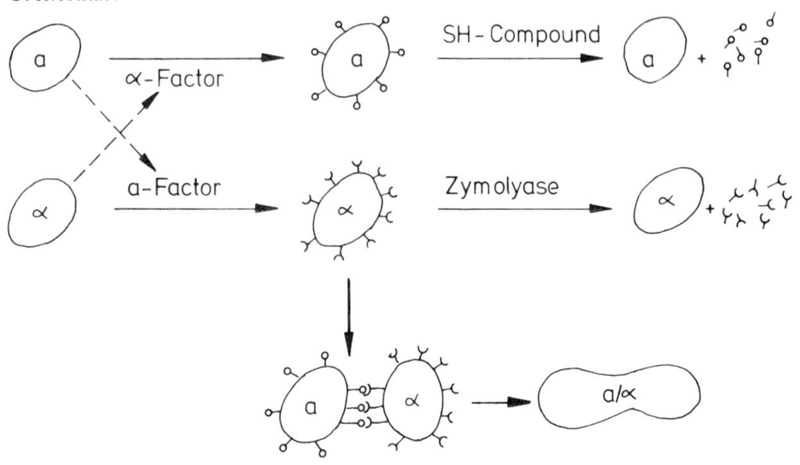

Fig. 5.3. Mild periodate oxidation of *a* agglutinin. The Gp 22 purified by NaDodSO$_4$/PAGE and by HPLC gelfiltration was treated with 10 mM NaIPO$_4$ in 20 mM phosphate buffer pH 6 for 4 h at 4 °C. The control was incubated for the same time in phosphate buffer. Subsequently the biological activity was tested as described in Orlean *et al.* (1986).

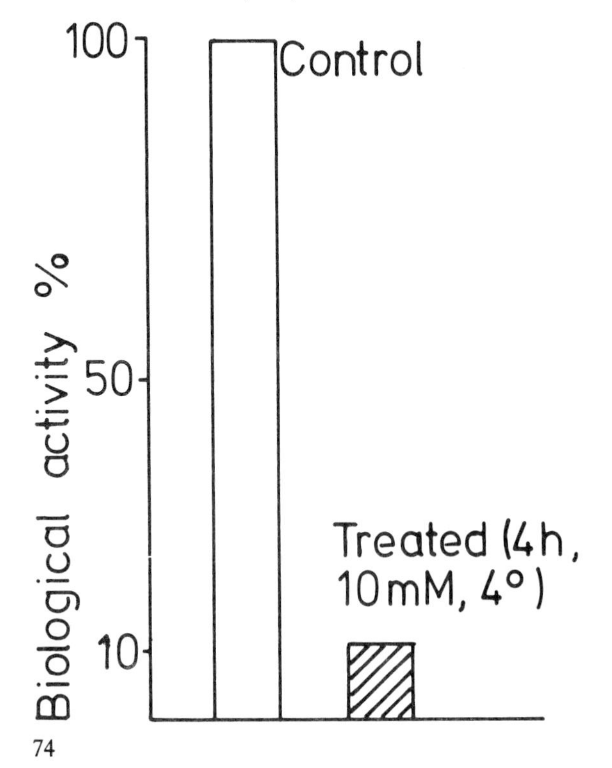

with mating factors. When haploid *a* cells were radiolabelled with ^{3}H-mannose in the presence of α factor a mannoprotein was observed which was absent from control cells (Orlean *et al.*, 1986). A number of cell surface glycoproteins could be released by treatment with 70 mM mercaptoethanol. When separated by SDS PAGE this produced the ^{3}H-mannoprotein pattern shown in Fig. 5.1. A 22 kD protein (arrow) was present only when the *a* cells were treated with α factor. Cells treated in this way remained fully viable, but lost their ability to agglutinate with induced α cells (Fig. 5.2). No similar protein was released from *a* factor induced α cells, nor did these cells lose their agglutination ability following the same treatment with mercapthoethanol. The mercaptoethanol extract from the *a* cells and the 22 kD protein are inhibitory when added in the agglutination assay to α cells before the addition of competent *a* cells (Orlean *et al.*, 1986). Again, this kind of inhibition was not observed with the mercaptoethanol extract from induced α cells. When α cells, pretreated with *a* factor, were incubated with small amounts of zymolyase (insufficient to release protoplasts) an activity inhibitory to agglutination was released (Fig. 5.2); *a* factor pretreated α cells showed a ten-fold increase in inhibitory activity compared with control cells.

Since the ^{3}H-mannose radioactivity in the *a* agglutinin (Fig. 5.1) could be released by treatment with weak bases and the biosynthesis of this 22 kD protein was not inhibited by tunicamycin, it was concluded to comprise an O-glycosylated mannoprotein (Orlean *et al.*, 1986). To see whether or not the carbohydrate moities in the molecule (estimated to amount to about 29% of its MW) are necessary for its biological function, the 22 kD protein was oxidised with periodate under mild conditions. It then lost about 90% of its agglutination inhibiting activity (Fig. 5.3). Similar behaviour has been observed previously with the agglutinins with *S. kluyveri* and *H. wingei* (Yen & Ballou, 1974; Pierce & Ballou, 1983).

Acknowledgements

We would like to thank Dr Frans Klis for introducing the bioassay to our laboratory, and Karin Hauser, who carried out the experiments with α cells and zymolyase. The work reported has been supported by a grant from Deutsche Forschungsgeme inschaft (SFB 43).

References

Betz, R., Manney, T. R. & Duntze, W. (1981). Hormonal control of gametogenesis in the yeast *Saccharomyces cerevisiae. Gamete Res.*, **4**, 571–84

Doi, S. & Yoshimura, M. (1978). Temperature dependent conversion of sexual agglutinability in *Saccharomyces cerevisiae. Mol. Gen. Genet.*, **1**, 251–8

Orlean, P., Ammer, H., Watzele, M. & Tanner, W. (1986). Synthesis of an O-glycosylated cell surface protein induced in yeast by α factor. *Proc. Natl. Acad. Sci. USA*, **83**, 6263–6

Pierce, M. & Ballou, C. E. (1983). Cell–cell recognition in yeast. *J. Biol. Chem.*, **258**, 3576–82

Terrance, K. & Lipke, P. N. (1981). Sexual agglutination in *Saccharomyces cerevisiae*. *J. Bact.*, **148**, 889–96
Thorner, J. (1981). Pheromonal regulation of development in *Saccharomyces cerevisiae*. In: The Molecular biology of the yeast *Saccharomyces*. Life cycle and inheritance. J. N. Strathern, E. W. Jones and J. R. Broach eds. Cold Spring Harbor Laboratory, pp. 143–80
Yamaguchi, M., Yoshida, K. & Yanagishima, N. (1984). Isolation and biochemical and biological characterization of an a-mating-type-specific glycoprotein responsible for sexual agglutination from the cytoplasm of α-cells in the yeast, *Saccharomyces cerevisiae*. *Arch. Microbiol.*, **140**, 113–19
Yanagishima, N. (1984). Mating systems and sexual interactions in yeast. In: *Encyclopedia of Plant Physiology*, New Series, Vol. 17 Cellular Interactions. H. F. Linskens and J. Heslop-Harrison eds., Springer-Verlag, Berlin, pp. 402–23
Yanagishima, N., Yoshida, K., Hamada, K., Hagiya, M., Kawanabe, Y., Sakurai, A. & Tamura, S. (1976). Regulation of sexual agglutinability in *Saccharomyces cerevisiae* of α and α types by sex-specific factors produced by their respective opposite mating types. *Plant & Cell Physiol.*, **17**, 439–50
Yen, P. H. & Ballou, C. E. (1974). Partial characterization of the sexual agglutination factor from *Hansenula wingei* Y-2340 type 5 cells. *Biochem.*, **13**, 2428–37

6 Sexual recognition in *Paramecium*

Koichi Hiwatashi

Biological Institute,
Tohoku University
Sendai, Japan

Abstract

The first event in mating in *Paramecium* is the formation of large agglutinations of cells. Here individuals of opposite mating type stick to each other by the tips of the ventral cilia. Molecules involved in this specific cell contact are called the mating-type substances. The mating-type substances are proteins bound tightly to the ciliary membranes. Although isolation of the substances in pure form has so far been unsuccessful, small membrane vesicles which can induce mating agglutination and conjugation have been isolated. Genetic analyses utilising intersyngenic hybrids revealed three loci controlling the specificities of the mating-type substances. The genetic results suggest that sexual recognition in *Paramecium* is the reaction between monomer ligand molecules of one mating type and dimer receptor molecules of the complementary mating type.

Introduction

Taxonomically recognised species of ciliates, such as *Paramecium caudatum* are in fact groups of morphologically indistinguishable sibling species referred to as syngens (Sonneborn, 1957). In some ciliates like *Paramecium aurelia* and *Tetrahymena pyriformis*, the syngens have been given binominal species names because each has been found to be identifiable by isozyme techniques (Sonneborn, 1975; Nanny & MacCoy, 1976). Conjugation of ciliates occurs when cells of complementary mating types within a same syngen are brought together. This means that cells have a specific recognition mechanism which distinguish the cell of the complementary mating type from other cells. In *Paramecium*, this recognition mechanism resides in the initial agglutination reaction observed immediately after mixing cells of complementary mating types.

In this article, mechanisms of mating recognition in *Paramecium*, mostly in *Paramecium caudatum*, will be discussed and explanations of where in the cells the mating recognition occurs, what kind of molecules are involved in it and how those molecules are genetically controlled.

Specific and non-specific cell adhesion in conjugation of *Paramecium*

When sexually mature cells of complementary mating types of *Paramecium* in their stationary phase are mixed together, they instantaneously form large agglutinates (Fig. 6.1). This agglutination reaction is called the mating reaction. In the mating reaction, cell adhesion occurs strictly between cells of different mating types. They stick to each other by the tip of the cilia.

Cells of *Paramecium* have mating ability when they are in the phase of sexual maturity. To reach the phase of sexual maturity, cells have to pass through an immaturity phase. This is measured in terms of the number of cell divisions after the previous conjugation and differs between species (Sonneborn, 1957). Even in the phase of sexual maturity, cells express mating ability only when the cultures are in the stationary phase or in the condition known as relative hunger.

About one hour after mixing mating types, the mating-reaction agglutinates disintegrate and release pairs which are united along their anterior surfaces. This cell adhesion is called the holdfast union; the cell are united

Fig. 6.1. The mating reaction of *Paramecium caudatum*. *Left:* cells of a single mating type; *right:* one minute after mixing complementary mating types.

not by their cilia but at their naked cell surfaces. It is apparent that during the mating reaction cilia along the anterior portion of the cell are absorbed in the cell body. Another important difference between the holdfast and mating-reaction unions is the formation of selfing pairs (Fig. 6.2). Thus, cell adhesion at this stage is not mating-type specific. The non-specificity of the union at this stage extends even beyond the species level as shown by Miyake (1968) in his experiments on the chemical induction of conjugation. He succeeded in inducing pair formation in every combination between *P. aurelia*, *P. caudatum* and *P. multimicronucleatum*.

The third step of cell adhesion in conjugation occurs at the paroral region of the cell and is thus termed the paroral union. In paroral union pairs, absorption of cilia does not stop at the region of cell contact but extends to the posterior end of the cell. Cell adhesion is finally established in the region between holdfast and paroral regions thus completing the formation of conjugating pairs.

Ciliary membranes as the site of specific sexual recognition

As mentioned in the previous section, cilia are the organelles by which cells of complementary mating types stick together initially and form cell agglutinates. However, not all cilia have the mating reactivity. Only the cilia on the ventral surface of a cell are reactive (Hiwatashi, 1961; Cohen & Siegel, 1963). Thus, when detached cilia of one mating type are applied to intact reactive cells of the complementary mating type, they adhere only to cilia along the ventral surface.

In some ciliates, such as several species of *Blepharisma*, *Euplotes*, the cells excrete a kind of sexual pheromone called the gamone which induces conjugation in cells of the complementary mating type (Miyake, 1981; Luporini,

Fig. 6.2. Three steps of cell adhesion in the conjugation of *P. caudatum*. A, mating reaction; B, holdfast unions; C, paroral unions. The white and dotted cells indicate different mating types.

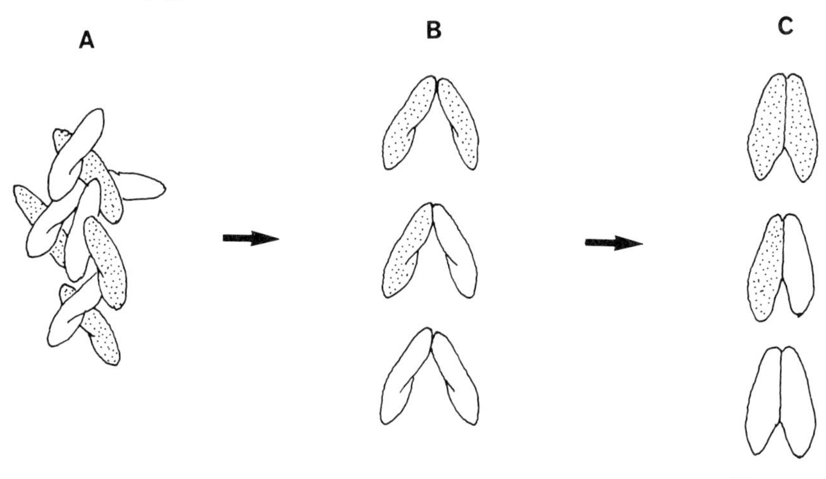

Miceli & Ortenzi, 1983; Heckmann & Kuhlmann, 1986). In *Paramecium*, no such soluble factors are involved in the initial mating reaction. This is clearly shown by the fact that detached cilia from mating reactive cells of complementary mating types form agglutinates when they are mixed after repeated washing (Takahashi, Takeuchi & Hiwatashi, 1974).

Since cilia were discovered to be the site of specific sexual recognition in *Paramecium*, investigators in several laboratories tried to extract the recognition molecules, i.e. the mating substances. However, all attempts at direct extraction of the mating substances from the reactive cilia have failed. In order to find the structural component bearing the mating substances, Watanabe (1977) fractionated mating reactive detached cilia and tested the mating reactivity of each fraction. The mating reactivity was always associated with the membrane fraction and not with axonemal or matrix components of the cilia.

Ciliary membranes of *Paramecium* are covered with a surface coat which can be observed under the electron microscope using ruthenium red staining (Wyroba & Przelecka, 1973; Watanabe, 1981). When detached cilia are treated with sialidase (neuraminidase), the surface coat is completely removed but the mating reactivity of the cilia is not destroyed (Watanabe, 1981; Kitamura & Hiwatashi, 1978). These results strongly suggest that the surface coat of cilia does not contain the mating substances but that these reside within the ciliary membrane itself.

Mating-reactive membrane vesicles and nature of the mating substances

When mating-reactive detached cilia are treated with a solution containing 2 M urea and 0.1 mM Na_2-EDTA, mating-reactive membrane vesicles with a diameter of 100 to 150 nm are obtained. These membrane vesicles not only induce large mating clumps among cells of the complementary mating type but also induce many selfing pairs (Kitamura & Hiwatashi, 1976) (Fig. 6.3). Another kind of membrane vesicles of 50 to 100 nm in diameter with mating reactivity is obtained when mating-reactive cilia are treated with 4 mM lithium diiodosalicylate (LIS) (Kitamura & Hiwatashi, 1980). LIS solutions are known to produce small vesicles of red cell ghosts when used in lower concentrations but higher concentrations dissolve the membranes completely (Marchesi & Andrews, 1971). Since it was found that treatment of cilia with LIS solution did not destroy the mating reactivity of the ciliary membranes, attempts were made to solubilise the mating-reactive membrane vesicles by LIS and then to reconstitute the vesicles from the soluble fraction. The mating-reactive membrane vesicles isolated with the urea–EDTA solution were treated with 9 mM LIS and the soluble fraction was dialysed to remove LIS completely. Membrane vesicles with a diameter of 50 to 100 nm were obtained from the dialysate. These reconstituted membrane vesicles had a strong conjugation-inducing activity but not an agglutination-inducing activity when applied to mating-reactive cells of the complementary mating type.

The same kind of membrane vesicles was obtained when membrane vesicles isolated with LIS were stored for more than 12 days at 4 °C. Though these membrane vesicles did not induce agglutination of cells of the complementary mating type, they induced agglutination of cilia on the ventral surface of the complementary type cells. Thus when the LIS membranes without agglutination-inducing activity were applied to mating-reactive cells of the complementary type, bundles of cilia sticking together at their tips were observed on the ventral surface of the cells. This suggests that the membrane vesicles with no activity to induce cell agglutination still hold the mating substances and their activity is high enough to induce conjugation, if not sufficient to induce cell agglutination.

When the urea-EDTA membrane vesicles and their LIS-soluble fraction, used for the reconstitution of membrane vesicles, were compared on a SDS polyacrylamide gel, a notable difference was observed. The LIS-soluble fraction lacks three predominant glycoprotein bands seen in the urea-EDTA membrane vesicles when stained with PAS (Kitamura & Hiwatashi, 1980). This result suggests that glycoproteins are not essential components of the mating substances. If this is the case, the chemical nature of the sexual cell contact in *Paramecium* must be different from that in yeast, *Chlamy-*

Fig. 6.3. Mating-reactive membrane vesicles from cilia of *P. caudatum* prepared by the urea-EDTA method. Negatively stained with uranyl acetate (courtesy of A. Kitamura).

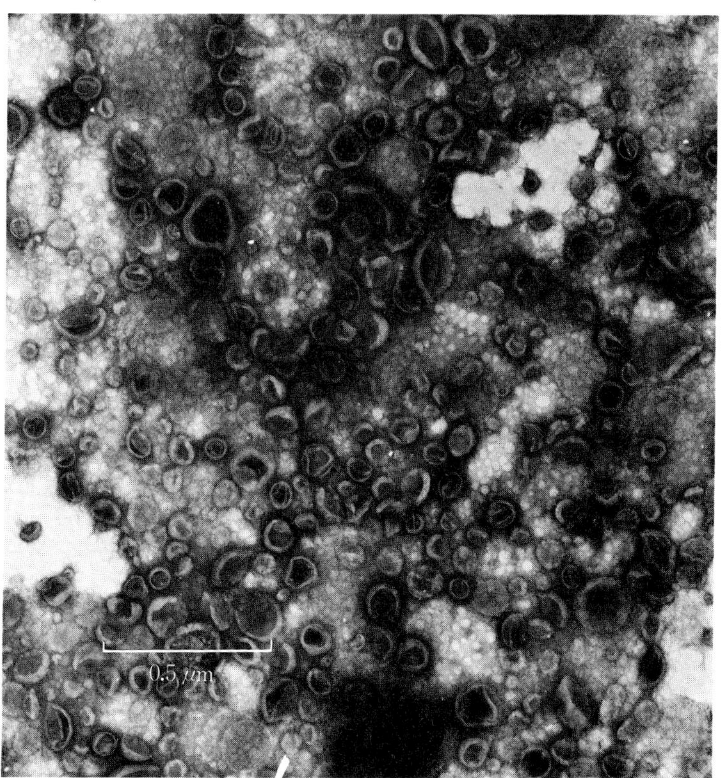

domonas and sea urchin, where glycoproteins are known to be involved in the reaction of gamete to gamete or sperm to eggs (Crandall, Lawrence & Saunders, 1974; Messina & Monroy, 1956; Wiese, 1974). To further explore this hypothesis, inactivation experiments with glycosidases were performed on mating-reactive cilia.

Mating-reactive detached cilia were treated with a mixture of α-mannosidase (110 μg/ml), α-L-fucosidase (10 μg/ml), β-galactosidase (110 μg/ml), β-glucosidase (230 μg/ml), and neuraminidase (230 μg/ml) for two hours at 30 °C and then were tested for mating reactivity. No decrease in mating reactivity was observed. The cilia showed strong agglutination-inducing activity as well as conjugation-inducing activity after the two hour incubation in the glycosidase mixture. By contrast, trypsin digestion for two hours destroyed the mating activity completely even at a concentration as low as 0.001%. It is well known that mating reactivity in *Paramecium* is easily destroyed by various kinds of proteases and the main components of the mating substances are proteins (Metz, 1954; Hiwatashi, 1969). The above results not only confirm that the main components of the mating substances are protein but also exclude the possible involvement of sugar residues in sexual recognition in *Paramecium*.

Expression of mating activity and the hydrophobic interactions on ciliary surfaces

As mentioned previously, cells of *Paramecium* express mating reactivity only when they are in the sexually mature phase of the clonal life cycle and in the stationary phase of culture. When they are mating-reactive, the reactivity, i.e. the ability to stick to cilia of the complementary mating type, appears only in the cilia on the ventral surface of the cell. A parallel to this restricted expression of mating reactivity has been found in the ability of cells to adhere to a hydrophobic surface (Kitamura, 1982). When mating-reactive cells are introduced into a polystyrene Petri dish for bacterial culture, they stick to the bottom by their ventral cilia. Cell in immature or log phases of growth never show this adhesion to the polystyrene surface. The same kind of adhesion has also been observed when mating-reactive cells are applied to paraffin or other hydrophobic surfaces. When the hydrophobic surface of a polystyrene Petri dish is treated with sulphuric acid and thus made hydrophilic, cells in any stage never stick to the surface. Thus, the adhesion involves a hydrophobic interaction between the ciliary surface and the artificial surface.

The question arises as to why only mating-reactive cells adhere to hydrophobic surfaces. One possible explanation may be that the mating substances are proteins of hydrophobic nature and the adhesion is brought about by the interaction between the mating substances and the hydrophobic artificial surface. This explanation, seems implausible however in the light of the following experiment. Kitamura (1982) tested cell adhesiveness to a polystyrene surface when the mating reactivity was completely inactivated with trypsin. He found not a loss but an increase in adhesiveness comparing with that of reactive cells without trypsin treatment. A more likely explanation is that there is an extension of free lipid surfaces from the ventral cilia where the mating substances are supposed to be carried (Fig. 6.4). As mentioned before, ciliary membrane surfaces of *Paramecium* are covered with a thick, polysaccharide surface coat which has no activity in mating agglutination (Watanabe, 1981). Thus, the problem arises as to how mating substances on ciliary membrane surfaces can interact during the mating reaction while the surfaces are covered with non-reactive materials. This problem can be explained if the mating substances and the outer lipid surfaces of the ciliary membranes are to some extent exposed free from the surface coat when cells enter the mating reactive phase. This explanation explicitly indicates that the mating reactive cells adhere to the hydrophobic artificial surface by the exposed free lipid surfaces of their ventral cilia. The increased adhesiveness of cells to a polystyrene surface by trypsin treatment can be interpreted as resulting from an increase in the free lipid surfaces of ciliary membranes. This is supported by the fact that the trypsin treatment removes the ciliary surface coat of *Paramecium* (Wyroba & Przelecka, 1973; Watanabe, 1981).

The ligand and receptor view of sexual recognition based on genetics of mating type

Mode of determination and inheritance of mating type in *Paramecium* is somewhat complicated. It differs between species and both genetic and epigenetic controls are involved. In general, however, one gene, designated mt^+, is important in the control of mating type differences. In species of *P. aurelia* complex and syngens of *P. caudatum*, when the mt^+ gene is active, cells can express 'even' (E) mating types and when the mt^+ is absent or inactive, cells show the complementary 'odd' (O) mating types (Sonneborn, 1937, 1974; Hiwatashi, 1968). In some species of the *P. aurelia* complex, genes at more than one locus are known to have the function like mt^+ (Taub, 1963; Byrne, 1973). However, the actions of mt^+ and other genes like mt^+ are unknown. Mating-type specificity in *Paramecium* is known to depend upon the specificity of mating substances (Metz, 1954). Is mt^+ the gene coding the E mating-type substances? If so, what is the gene coding the O substance? Butzel (1955) proposed the

Fig. 6.4. Schematic illustrating the exposure of free lipid surfaces of cilia and the appearance mating activity. A, cilia in log phase cells; B, mating-reactive cilia in stationary phase cells.

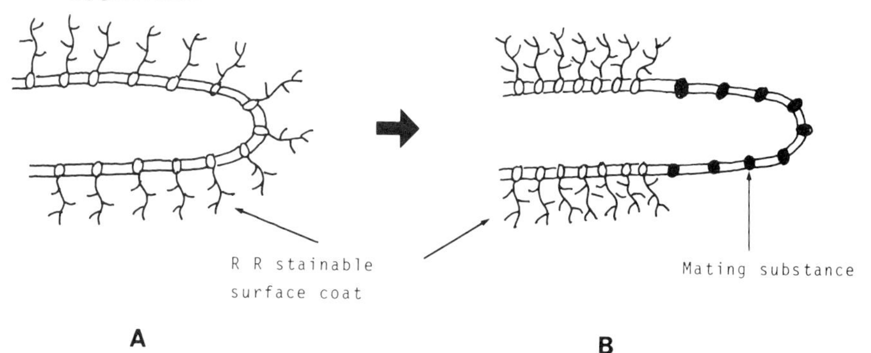

hypothesis that the O mating-type substance is the precursor of the E mating-type substance and the gene mt^+ controls the reaction converting the O substance to E substance. According to his hypothesis, there must be other genes controlling syngen-specificity of mating substances (Butzel, 1955).

A completely different picture of genetics of mating type in *Paramecium* has recently been shown by Tsukii & Hiwatashi (1983) using intersyngenic hybrids of *P. caudatum*. In *P. aurelia* and in *P. bursaria*, intersyngenic mating is known to be lethal or to produce sterile F_1 hybrids (Sonneborn, 1974; Jennings & Opitz, 1944; Haggard, 1974). In *P. caudatum*, however, intersyngenic hybrids produced by bypassing the barrier of the mating reaction are found to be highly viable and completely fertile (Tsukii & Hiwatashi, 1983). This situation makes it possible to analyse syngen specificity of mating types genetically in *P. caudatum*.

Two methods of obtaining intersyngenic hybrids were used. One is to mix four different mating types belonging to two different syngens. In this mixture, initial mating reaction occurs strictly between cells of complementary mating types within each syngen. The second step of reaction, the holdfast union formation, is, however, neither mating-type specific nor syngen specific as mentioned earlier. Thus, intersyngenic pairs are formed together with intrasyngenic cross- and self-pairs. The intersyngenic pairs were isolated from other kinds of pairs using the behavioural mutant *cnrB* (Takahashi, 1979) as the marker for both complementary mating types of one syngen. A high K^+ solution was used to distinguish intersyngenic pairs from other pairs. Pairs between wild type and the *cnrB* can be easily distinguished from pairs of wild type and of the mutant by their swimming behaviour when stimulated by the high K^+ solution (Takahashi, 1979). The other method of obtaining intersyngenic pairs is that of chemical induction of conjugation which induces conjugation directly without the prior occurrence of the mating reaction (Miyake, 1968). The use of the behavioural marker for isolating intersyngenic pairs from other pairs is essentially the same as the first method.

Extensive cross breeding analyses were performed using various clones of intersyngenic hybrids and, based on the results of those crosses, the three-gene model for the determination of syngenic specificity of mating type was established (Table 6.1) (Tsukii & Hiwatashi, 1983).

The three gene loci are *Mt* which controls E mating types, and *MA* and *MB* the interaction of which controls O mating types. Each locus has co-dominant multiple alleles, each of which controls the specificity of each syngen. Thus, Mt^1 controls the E mating type of syngen 1 and Mt^3 controls that of syngen 3, superscripts of gene symbols designating the number of syngens. Since Mt^1 and Mt^3 are co-dominant, the heterozygote between them expresses dual mating types of E^1 and E^3. To know whether a single cell of the dual E type clone expresses dual mating types, tester cells (cells of standard clones for the mating type test) of one mating type were fed with carmine particles and those of the other with India ink. When white cells of a dual mating type clone were mixed with red and black cells together, single white cells reacted with red and black cells at the same time. This shows that clones of dual mating types are not a mixture of cells of two mating types but consist of cells expressing two mating types.

Table 6.1. *Various mating types and their genotypes based on the three gene model (Tsukii & Hiwatashi, 1983)*

	Locus		
Mating type	Mt	MA	MB
E^1	Mt^1/Mt^1 or Mt^1/mt	$-/-$	$-/-$
O^1	mt/mt	$MA^1/-$	$MB^1/-$
E^3	Mt^3/Mt^3 or Mt^3/mt	$-/-$	$-/-$
O^3	mt/mt	$MA^3/-$	$MB^3/-$
E^1E^3	Mt^1/Mt^3	$-/-$	$-/-$
O^1O^3	mt/mt	MA^1/MA^3	MB^1/MB^3
O^0	mt/mt	MA^1/MA^1	MB^3/MB^3
O^0	mt/mt	MA^3/MA^3	MB^1/MB^1

–, Any allele of the locus O^0, mating-typeless.

On the other hand, for the expression of O types, the *Mt* locus must be the recessive homozygote (*mt*/*mt*) because *Mt* is epistatic to *MA* and *MB*. In addition to the above requirement, at least one allele each at the *MA* and *MB* loci must have a common syngenic specificity. Thus, when one of the two loci is heterozygous for the syngen specificity and the other homozygous, e.g. MA^1/MA^3; MB^1/MB^1, a single O type (in the above sample, mating type O^1) is expressed. Dual O types are expressed only in the double heterozygotes. An inevitable consequence of the above rule is that cells cannot express mating type when one of the two loci is homozygous for one syngen and the other is homozygous for another syngen. This actually occurred and the clones of double homozygote of different syngens became mating-typeless clones. Though the mating-typeless clone expresses no mating type, it is different from the clone in the sexually immature period. The mating-typeless clone can be induced to conjugate by conjugation-inducing chemicals, whereas cells in the immature period cannot be induced to conjugate by any means.

What are the functions of these three genes? The simplest hypothesis may be that *Mt* codes for the E mating-type substance and, *MA* and *MB* for the O type substance. If we assume that the E mating-type substance is a monomer protein coded by *Mt* and the O type substance is a dimer of two protein subunits coded by *MA* and *MB*, specificity of mating reaction can be explained as binding of the E substance as the ligand to the O substance as the receptor (Fig. 6.5). By this hypothesis, expression of a single O type in the intersyngenic hybrid of single heterozygote can be explained as shown in Fig. 6.6. The double heterozygote of *MA* and *MB* loci may produce four different receptor molecules but only two among them have corresponding ligands and thus, expresses dual O types (Fig. 6.7). If both *MA* and *MB* are homozygous but they control different syngen specificities, the receptor molecules formed would have no corresponding ligand in the existing E types and thus, the clone would become the mating-typeless (Fig. 6.8). Though the hypothesis, tentatively called the ligand and receptor hypothesis

Fig. 6.5. A schematic illustration of the ligand and receptor hypothesis of sexual recognition in *Paramecium*. The ligand is a monomer molecule controlled by *Mt* and the receptor is a dimer molecule consisting of two subunits controlled by *MA* and *MB*. The upper and the lower figures show the difference of syngens.

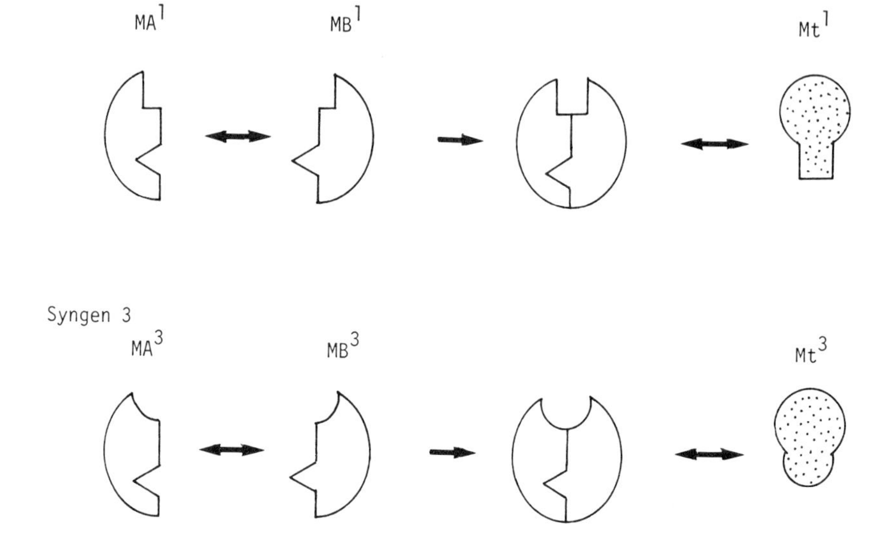

Fig. 6.6. The ligand and receptor hypothesis (abbreviations as in Fig. 6.5). When *MA* is homozygous and *MB* heterozygous (e.g. *mt*/*mt*; MA^1/MA^1; MB^1/MB^3), only one of the two receptor formed has the corresponding ligand and thus, the hybrid shows a single O type.

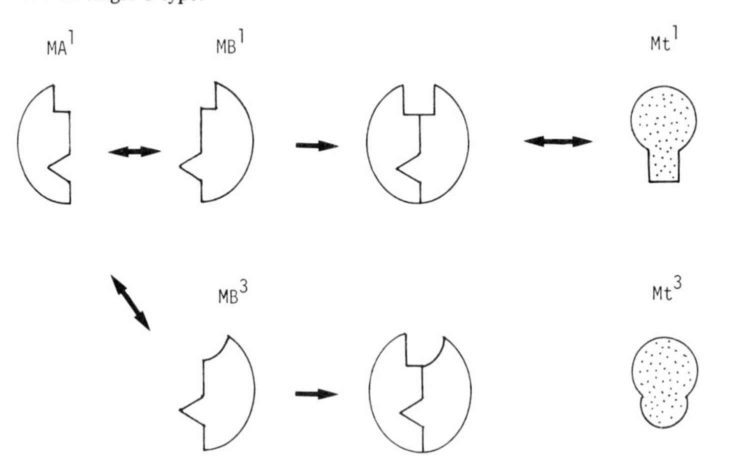

of mating recognition, can explain all the results of intersyngenic crosses in *Paramecium caudatum*, we have no direct evidence showing that *Mt* codes for the E type substance and, *MA* and *MB* for the O type substance. Reconstitution of mating-reactive membrane vesicles from the soluble fractions of two mating-typeless clones of opposite genotypes by the technique reported by Kitamura & Hiwatashi (1980), would provide a means of verifying the above hypothesis.

Based on his studies of the actions of cell-excreted mating signals, the so-called gamones, of *Blepharisma*, Miyake (1981) proposed the gamone receptor hypothesis for preconjugant interaction in ciliates. According to this hypothesis, a basic process of preconjugant interaction is the reaction between the gamone of one mating type and the receptor on the complementary mating type. Each mating type has its own gamone and the receptor(s) for the gamone(s) of the complementary mating type(s). Binding of gamone to its corresponding receptor activates the cell and brings about conjugation. Though the word gamone originally indicated preconjugant signal pheromone excreted from cells, Miyake (1974, 1981) extended the concept of gamone to cover the mating substances of *Paramecium* and called the latter 'cell-bound gamone'. Thus, according to the gamone receptor hypothesis, cells of each mating type in *Paramecium* should have the mating substance (the cell-bound gamone) of their own mating type and the receptor for the mating substance of the complementary mating type.

In his pioneer works on mating interaction in *Paramecium*, Metz (1948, 1954) showed two possible mechanisms for mating-substance interactions (Fig. 6.9). In the first model, the mating reaction is due to the reaction between a single pair of complementary substances, each locating on each of the complementary mating types (Fig. 6.9(a)). In the second model, an inducing substance *I* on one mating type combines with a reacting substance *R* on the other mating type and, at the same time, *I'* on the latter combines with *R'* on the former (Fig. 6.9(b)). The ligand and receptor hypothesis set out in this paper is in the same scheme as Metz's first model, and Miyake's gamone receptor hypothesis is in the same scheme as Metz's second model. Proof the validity of the models now awaits biochemical evidence. However, the gamone receptor hypothesis does not adequately explain the mating-typeless clone produced by a simple replacement of alleles at one locus with those controlling a different syngen. Without assuming that the *MA* and/or the *MB* control specificity of the whole set of respective gamone and receptors, the mating-typeless clone would not appear because the mating-typeless clone can neither activate cells of the complementary mating type nor be activated by the latter.

Conclusion and perspectives

Sexual recognition in *Paramecium* depends upon the interaction of mating substances on the ciliary surfaces of cells of complementary mating types. Molecules involved in the recognition, the mating substances, are proteins tightly bound to the ciliary membranes. These proteins are the key molecules in the isolation of sibling species or syngens in *Paramecium*. However, every attempt to isolate the mating substances has so far been unsuccessful.

Fig. 6.7. The ligand and receptor hypothesis (abbreviations as in Fig. 6.5). When *MA* and *MB* are doubly heterozygous (e.g. *mt/mt; MA^1/MA^3; MB^1/MB^3*), four different kinds of receptors are formed. Two of them have corresponding ligands and thus, the hybrid shows a dual mating type.

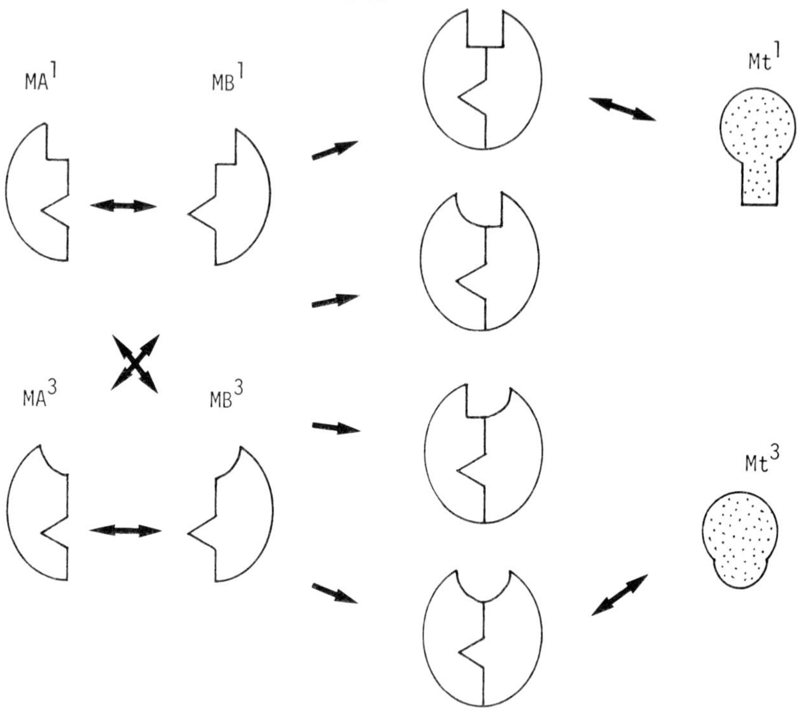

Fig. 6.8. The ligand and receptor hypothesis (abbreviations as in Fig. 6.5). When *MA* and *MB* are both homozygous but have different syngen specificities (e.g. *mt/mt; MA^1/MA^1; MB^3/MB^3*), the receptor formed has no corresponding ligand and thus, the hybrid becomes mating-typeless.

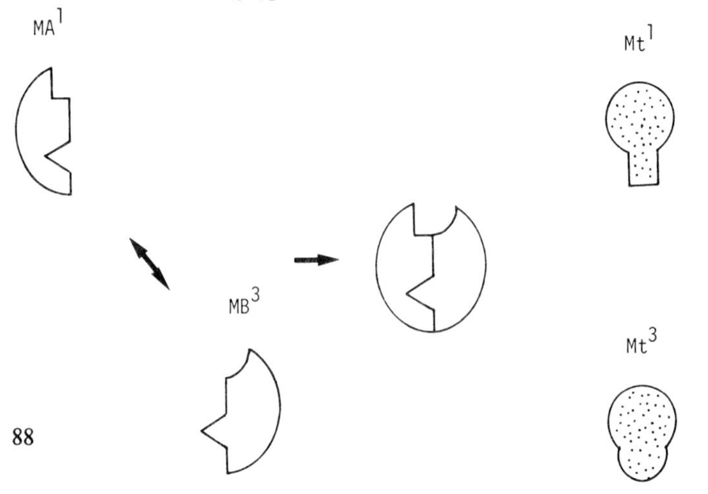

On the other hand, genetical studies of mating type and syngen specificities using intersyngenic hybrids in *P. caudatum* reveals genes most probably coding for the mating substances. Based on the three-gene model for the determination of specificity of mating type, the ligand and receptor hypothesis of mating recognition in *Paramecium* has been proposed. Crucial to testing the hypothesis is the isolation of the mating substances. Though the various methods so far attempted have failed, detergent solubilisation and fractionation of membrane-bound proteins may yet prove feasible. Methods for assaying mating activity need to be improved since the mating agglutination of living cells so far employed as an assay seems insufficiently sensitive for detection of small amounts of mating substances. Mating-type-specific inactivation of ciliary movement, which was observed when LIS membrane vesicles with conjugation-inducing activity but lacking agglutination-inducing activity was applied to cells of the complementary mating type, may be one of the most sensitive and easiest methods for the assay. This method, however, is still limited to living cells and thus cannot be used for assaying toxic substances and specific inhibitors.

Use of immunological methods for identifying the mating substances, could be much improved: all mating-blocking antibodies so far produced lack mating-type specificity (Hiwatashi & Takahashi, 1967; Barnett & Steers, 1980).

Fig. 6.9. Metz's (1954) scheme of two possible interaction mechanisms of mating reaction in *Paramecium*. Each series of arrows represents the main activation chain in one conjugant. a, simultaneous activation of conjugants by a single pair of surface substances (A, a); b, simultaneous activation of conjugants by interaction of two pairs of surface substances. I and I′, inducers; R and R′ reactors.

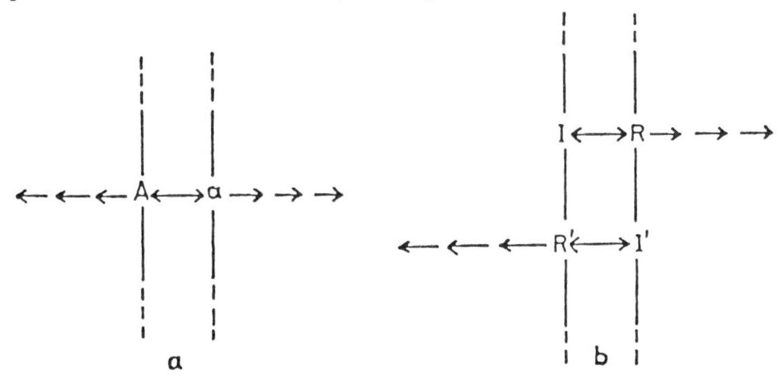

References

Barnett, A. & Steers, E., Jr (1980). Antibodies blocking the mating reaction in *Paramecium multimicronucleatum* syngen 2. *J. Protozool.*, **27**, 103–7

Butzel, H. M., Jr (1955). Mating type mutations in variety 1 of *Paramecium aurelia*, and their bearing upon the problem of mating type determination. *Genetics*, **40**, 321–30

Byrne, B. C. (1973). Mutational analysis of mating type inheritance in syngen 4 of *Paramecium aurelia. Genetics*, **74**, 63–80

Cohen, L. W. & Siegel, R. W. (1963). The mating-type substances of *Paramecium bursaria*. *Genet. Res.*, **4**, 143–50

Crandall, M., Lawrence, L. M. & Saunders, R. M. (1974). Molecular complementarity of yeast glycoprotein mating factors. *Proc. Natl. Acad. Sci. USA*, **71**, 26—9

Haggard, B. W. (1974). Interspecies crosses in *Paramecium aurelia* (syngen 4 by syngen 8). *J. Protozool.*, **21**, 152–9

Heckmann, K. & Kuhlmann, H. W. (1986). Mating types and mating inducing substances in *Euplotes octocarinatus*. *J. Exp. Zool.*, **237**, 87–96

Hiwatashi, K. (1961). Locality of mating reactivity on the surface of *Paramecium caudatum*. *Science Reports of the Tohoku University, 4th Series (Biology)*, **27**, 93–9

Hiwatashi, K. (1968). Determination and inheritance of mating type in *Paramecium caudatum*. *Genetics*, **58**, 373–86

Hiwatashi, K. (1969). Paramecium. In *Fertilization*, vol. 2, ed. C. B. Metz & A. Monroy, pp. 255–93. New York. Academic Press

Hiwatashi, K. & Takahashi, M. (1967). Inhibition of mating reaction by antisera without ciliary immobilization in *Paramecium*. *Science Reports of the Tohoku University, 4th Series (Biology)*, **33**, 281–90

Jennings, H. S. & Opitz, P. (1944). Genetics of *Paramecium bursaria*. IV. A fourth variety from Russia. Lethal crosses with American variety. *Genetics*, **29**, 576–83

Kitamura, A. (1982). Attachment of *Paramecium* to polystyrene surfaces: A model system for the analysis of sexual cell recognition and nuclear activation. *J. Cell Sci.*, **58**, 185–99

Kitamura, A. & Hiwatashi (1976). Mating-reactive membrane vesicles from cilia of *Paramecium caudatum*. *J. Cell Biol.*, **69**, 736–40

Kitamura, A. & Hiwatashi, K. (1978). Are sugar residues involved in the specific cell recognition of mating in *Paramecium*? *J. Exp. Zool.*, **203**, 99–108

Kitamura, A. & Hiwatashi, K. (1980). Reconstitution of mating active membrane vesicles in *Paramecium*. *Exp. Cell Res.*, **125**, 486–9

Luporini, P., Miceli, C. & Ortenzi, C. (1983). Evidences that the ciliate *Euplotes raikovi* releases mating inducing factors (gamones). *J. Exp. Zool.*, **226**, 1–9

Marchesi, V. T. & Andrews, E. P. (1971). Glycoproteins: Isolation of cell membranes with lithium diiodosalicylate. *Science*, **174**, 1247–8

Messina, L. & Monroy, A. (1956). Evidence for the inhomogeneity of the jelly coat of sea urchin egg. *Publ. di Stazione Zool. Napoli*, **28**, 266–8

Metz, C. B. (1948). The nature and mode of action of the mating type substances. *Am. Naturalists*, **82**, 85–95

Metz, C. M. (1954). Mating substances and the physiology of fertilization in ciliates. In *Sex in Microorganisms*, ed. D. W. Wenrich, pp. 284–334. Washington, D.C. American Association for the Advancement of Science

Miyake, A. (1968). Induction of conjugation by chemical agents in *Paramecium*. *J. Exp. Zool.*, **167**, 359–80

Miyake, A. (1974). Cell interaction in conjugation of ciliates. *Current Topics in Microbiol. & Immunol.*, **64**, 49–77

Miyake, A. (1981). Physiology and biochemistry of conjugation in ciliates.

In *Biochemistry and Physiology of Protozoa*, 2nd ed., vol. 4, ed. M. Levandowsky and S. H. Hutner, pp. 125–98. New York: Academic Press

Nanney, D. L. & MacCoy, J. W. (1976). Characterization of the species of *Tetrahymena pyriformis* complex. *Trans. Am. Microscopical Soc.*, **95**, 664–82

Sonneborn, T. M. (1937). Sex, sex inheritance and sex determination in *Paramecium aurelia. Proc. Nat. Acad. Sci. USA*, **23**, 378–85

Sonneborn, T. M. (1957). Breeding systems, reproductive methods, and species problems in Protozoa. In *The Species Problem*, ed. E. Mayr, pp. 155–324. Washington, D.C. American Association for the Advancement of Science

Sonneborn, T. M. (1974). Paramecium aurelia. In *Handbook of Genetics*, vol. 2, ed. R. C. King, pp. 469–594. New York & London: Plenum Press

Sonneborn, T. M. (1975). The *Paramecium aurelia* complex of fourteen sibling species. *Trans. Amer. Microscopical Soc.*, **94**, 155–78

Takahashi, M. (1979). Behavioral mutants in *Paramecium caudatum. Genetics*, **91**, 393–401

Takahashi, M., Takeuchi, N. & Hiwatashi, K. (1974). Mating agglutination of cilia detached from complementary mating types of *Paramecium. Exp. Cell Res.*, **87**, 415–17

Taub, S. R. (1963). The genetic control of mating type differentiation in *Paramecium. Genetics*, **48**, 815–34

Tsukii, Y. & Hiwatashi, K. (1983). Genes controlling mating-type specificity in *Paramecium caudatum*: three loci revealed by intersyngenic crosses. *Genetics*, **104**, 41–62

Watanabe, T. (1977). Ciliary membranes and mating substances in *Paramecium caudatum. J. Protozool.*, **24**, 426–9

Watanabe, T. (1981). Electron microscopy of cell surface of *Paramecium caudatum* stained with ruthenium red. *Tssue & Cell*, **13**, 1–7

Wiese, L. (1974). Nature of sex specific glycoprotein agglutinins in *Chlamydomonas. Ann. NY Acad. Sci*, **234**, 383–95

Wyroba, E. & Przelecka, A. (1973). Studies on the surface coat of *Paramecium aurelia*. I. Ruthenium red staining and enzyme treatment. *Zeitschrift für Zellforschung*, **143**, 343–53

7 Sexual agglutination in *Chlamydomonas eugametos*

H. van den Ende, F. M. Klis, A. Musgrave and D. Stegwee

Department of Plant Physiology,
University of Amsterdam

Abstract

As a system for instigating cell–cell recognition in plants, the study of sexual reproduction in the unicellular green alga Chlamydomonas is proving to be highly successful. The two species most commonly used for these investigations are *C reinhardtii* and *C eugametos*; they are sexually incompatible and only distantly related to each other (Lemieux and Lemieux, 1985), and rather obvious differences exist between them. In this review of our most recent work on *C. eugametos*, some such differences will be described. Nonetheless, the two morphologically similar species, being faced with the same challenges in that sexual reproduction have maintained or co-evolved comparable molecular strategies. The two lines of research on these two species *Chlamydomonas* are to be regarded as complementing each other for the differences tend to be quantitative rather than qualitative.

Introduction

C. eugametos is heterothallic and isogamous. Clones of mating type *plus* (mt^+) or *minus* (mt^-) are usually cultivated agar in Petri dishes on a minimal medium (Mesland, 1976). When full-grown, cultures are flooded with water, the cells produce flagella and appear to be sexually competent, i.e. gametes. This indicates that the cells have been subject to nutrient stress (see below). Between 80 and 100% of them will fuse when mixed with an excess of gametes of the opposite mating type. In liquid cultures, gametes also form during the last stage of the exponential growth phase (Tomson *et al.*, 1985).

What happens when we mix suspensions of *C. eugametos* gametes of different mating types? First, a practically instantaneous clumping of the cells takes place, due to a specific adhesiveness of the flagella. This process is called 'sexual agglutination'. The cells stick to each other and are dancing or twitching in a characteristic way. Gradually, they reassociate to form pairs of cells, which cling together with both flagella, and orient themselves in such a way that they can fuse by their anterior surface region. This comprises a specialised zone of the plasma membrane, called the 'mating structure', or fusion papilla which during sexual agglutination protrudes outwards and penetrates the cell wall (Goodenough & Weiss, 1975). Two cells, fused via their papillae, form a tandem of two cells, connected by a plasma bridge, referred to as '*vis-à-vis pairs*'. which have a life-time of several hours. The flagella of such a pair then lose their adhesive properties (Musgrave *et al.*, 1985). Using only the original mt^+ flagella, they swim free of the clumps of other still agglutinating cells. Dependent on the light conditions, they fuse completely and form a zygospore.

Sexual agglutination has a number of interesting features. From scanning electron micrographs, as well as from light microscopy using Nomarski optics, it is evident that during sexual agglutination, particularly the flagellar tips become associated. Also, the flagella have a tendency to align along their whole length, which seems to be essential to bring about the required mutual orientation of the conjugating cells (Mesland, 1976). A second remarkable characteristic is the high degree of specificity displayed by gamete flagella. Those of the same mating type have no mutual affinity whatsoever, while those of different mating type associate effectively. Sexual agglutination has a remarkable dynamic nature. Cells, originally agglutinating *en masse* with many potential partners reassort to yield cell pairs, so efficiently that zygote yields regularly approach 100%. A fourth property feature worth mentioning is that the capacity for sexual agglutination is well regulated, so that flagellar adhesiveness disappears soon after a plasma bridge has been established between two conjugating cells. Evidence that this is also dependent on external factors, such as light and nutrients will be presented below. Finally, sexual agglutination has a signalling function, which triggers subsequent events leading to cell fusion. Thus in *C. eugametos* the protrusion of the fusion papilla through the cell wall must be preceded by local hydrolysis of the cell wall, which otherwise is highly resistant to enzymic attack, and remains present until the zygote has produced a thick, ornamented, secondary cell wall. In *C. reinhardtii*, the cell wall is released completely during sexual agglutination, due to the action of an autolytic enzyme (Claes, 1971). This results in naked gametes which fuse rapidly and completely, without the intermediary *vis-à-vis* pair stage. Since neither cell wall lysis nor the protrusion of the fusion papilla are seen before agglutination, it is generally assumed that in this species both processes are elicited by the interaction of the flagella. Thus, sexual agglutination not only serves to orientate the mating partners for fusion, but also induces the formation of structures by which they fuse.

Agglutinins, their composition and properties

The key to a molecular understanding of sexual agglutination is the identification of the molecules at the flagellar surface which mediate flagellar adhesiveness ('agglutinins'). Because of their biological activity, their presence can be assayed *in vitro* (Musgrave *et al.*, 1981) Both the mt^+ and mt^- agglutinins of *C. eugametos* and *C. reinhardtii* have been identified, partially characterised and visualised. The agglutinins of *C. eugametos* can be extracted from the flagellar surface or whole cells by extraction, using the detergent Triton X-100 or the chaotropic agent guanidine thiocyanate. The agglutinins can also be solubilized by a short period of sonication (Homan *et al.*, 1980) while the *mt*$^+$ agglutinin is brought into solution by a pH- shock (Klis *et al.*, 1985). The agglutinins of *C. reinhardtii* can be extracted quantitatively from living cells with millimolar concentrations of EDTA (Adair *et al.*, 1982). These data suggest that the agglutinins are extrinsic membrane proteins, possibly anchored to the flagellar membrane by an intrinsic membrane protein.

Samson *et al.* (1987) developed a simple three-step procedure for large scale purification of the sexual agglutinins of *C. eugametos*. First, the extract was fractionated by gel filtration. Second, the biologically active fractions were pooled and subjected to anion exchange chromatography. The material was finally purified to homogeneity by high performance gel filtration. The purity was confirmed by gel electrophoresis in SDS and electron microscopy. Saito and Matsuda (1984) introduced a two-step procedure for the isolation of the agglutinins from *C. reinhardtii*. After gel filtration, the biologically active material was further purified by high performance hydroxyapatite chromatography.

The agglutinins of *C. eugametos* appeared to be high-molecular weight glycoconjugates, the carbohydrates content (w/w) being 60 and 50%, for the *mt*$^+$ and *mt*$^-$ agglutinins, respectively. The molecular masses were determined from the Stokes radii and sedimentation coefficients, and were found to be 1.2 and 1.3 MDa (Samson *et al.*, 1987). More conventional methods for determining the molecular mass, such as gel filtration led to erroneously high estimates, due to the highly asymmetric structure of the molecules (see below). The agglutinins of *C. reinhardtii* are similar in mass and carbohydrate content to those of *C. eugametos* (Collin-Osdoby & Adair, 1985).

The amino acid compositions of the agglutinins of *C. eugametos* and *C. reinhardtii* show a high degree of similarity. The major amino acids are hydroxyproline, serine and glycine. The only obvious difference is the presence of phosphoserine in the *mt*$^+$ agglutinin of *C. eugametos* (Samson *et al.*, 1987). The sugar components of the *C. eugametos* agglutinins also have much in common. The major monosaccharides are arabinose and galactose. O-methylated sugars, typical of many other membrane-associated glycoproteins in *C. eugametos* (Schuring *et al.*, 1987), are absent in both agglutinins. The only striking difference between the two mating types is the presence of N-acetylglucosamine in the *mt*$^+$ agglutinin.

The relatively high content of hydroxyproline, serine, arabinose and galactose mark the sexual agglutinins as typical plant proteoglycans, similar to those present in the extracellular matrices of higher plants and green algae (Clarke *et al.*, 1979) where they have also been implicated in cell adhesion and recognition (Fincher *et al.*, 1983). It is therefore likely that the alkali-stable linkages between hydroxyproline and galactose/arabinose, found in these compounds, also occur in the agglutinins. Similarly, the high serine/threonine content suggests the presence of O-glycosidic linkages. This is vindicated by the observation that mild alkali treatment, leading to beta-elimination of the O-glycosidic linkages, destroys the biological activity of the agglutinins and results in the release of mainly arabinose and galactose

(Lens *et al.*, 1983; Samson *et al.*, 1987). The presence of hydroxyproline might help explain the linear configuration of the agglutinins as demonstrated by electron microscopy (see below), since in many other glycoproteins containing hydroxyproline, a left-handed polyproline II helix is found that is thought to confer fibrous properties on the protein (Homer & Roberts, 1979; van Holst & Varner, 1984).

The sensitivity of the biological activity of isolated agglutinins towards chemical and enzymic attack has been studied in detail by Collin-Osdoby and Adair (1985) for *C. reinhardtii* and by Samson *et al.* (1987) for *C. eugametos*. In all four agglutinins studied, mild periodate treatment destroys the biological activity, indicating that the carbohydrate part is necessary for biological activity, either indirectly by contributing to the stability of the overall structure of the molecule or directly, because certain carbohydrate side chains are involved in recognition *per se*. Samson *et al.* (1987) contributed evidence for the latter possibility. First, they showed in *C. eugametos* that the mt^+ agglutinin but not the mt^- agglutinin is bound by Con A Sepharose, indicating that mt^+ agglutinin possesses terminal mannose/glucose residues. Further they demonstrated that they were differentially susceptible to glycosidase treatment. The mt^+ agglutinin was sensitive towards alpha-mannosidase but resistent to alpha-galactosidase, whereas the mt^- agglutinin showed the reverse pattern. This implies that terminal mannose residues are essential for biological activity in the mt^+ agglutinin and terminal galactose residues essential for the biological activity of the mt^- agglutinin. The involvement of terminal mannose residues in the biological activity is further supported by the observation of Wiese and Mayer (1982) that alpha-mannosidase can inactivate live mt^+ gametes of *C. eugametos* but not mt^- gametes. In this connection, the presence of N-acetyl glucosamine in the mt^+ agglutinin is relevant. Its presence suggests the occurrence of N-glycosidic side chains linked to asparagine in the molecule, which are often characterised by having terminal mannose residues (Lennarz, 1980) and the synthesis of which can be specifically inhibited by tunicamycin. In fact, Wiese and Mayer (1982) demonstrated that only the synthesis of mt^+ agglutinin was affected by tunicamycin. Taken together, these data form strong evidence that in mt^+ agglutinin of *C. eugametos* terminal alpha-glycosidically linked mannose residues of N-linked carbohydrate side chains are directly involved in recognition and contribute to the specificity of the agglutination reaction.

Alpha-galactosidase specifically inactivates the mt^- agglutinin of *C. eugametos*, pointing to a role for terminal alpha-linked galactose residues. Other lines of evidence support this idea. Wiese and Mayer (1982) showed that the agglutinability of mt^- gametes, but not mt^+ gametes, was affected by alpha-galactosidase. Lens *et al.* (1982, 1983) demonstrated that an antiserum raised against the mt^- agglutinin blocked its activity and specifically bound a disaccharide containing galactose and/or arabinose (galactose : arabinose $= 1:1$) from a mixture of oligosaccharides released by beta-elimination. These data suggest that in the mt^- agglutinin, terminal alpha-linked galactose residues are involved in the biological activity of the molecule and co-determine the specificity of the sexual agglutination reaction.

Due to their large size, the agglutinins can be visualised by a variety of methods. In the study of the *C. reinhardtii* agglutinins, the quick-freeze, deep-etch technique of Heuser (1983) was used. In this species, both agglutinins appeared to be highly asymmetric molecules (Goodenough *et al.*, 1985). They exhibited morphologically distinct regions, a terminal head, a fairly rigid shaft, a flexible domain and a terminal hook. In addition, the mt^- agglutinin displayed a 'shepherds crook' conformation near the head region. In *C. eugametos*, the mt^+ agglutinin was visualised by rotary shadowing (Klis *et al.*, 1985) and was shown to resemble strongly the *C. reinhardtii* agglutinins, being a long molecule, with a terminal head domain and a flexible region. While this technique failed to give consistent images of the mt^- agglutinin, a negative staining technique, described by Smith and Seegan (1984) was used very successfully to visualise both agglutinins of *C. eugametos*. Mt^+ agglutinin preparations were found to contain long structures with an average length of 286 nm and a globular head at one end of the molecule. The mt^- agglutinin was also found to be a long molecule, but unlike the mt^+ agglutinin appeared not to have a globular head (K. Crabbendam, pers. communication).

Goodenough *et al.* (1985) have produced electron micrographs of flagella, showing that the *C. reinhardtii* agglutinins are oriented more or less perpendicular to the membrane, with the globular heads pointing outward. This orientation in itself immediately suggests that the distal heads may contain the binding site. The authors have argued that this is the case because inactivated agglutinins invariably possess abnormal heads. Likewise, we have found that preparations of mt^+ agglutinins of *C. eugametos*, that were inactivated by freeze-thawing, were aggregated via their globular heads, perhaps sterically blocking the binding site.

The relative ease by which the agglutinins can be extracted from gametes suggests that the agglutinins are peripheral membrane proteins, attached to the membrane via an intrinsic membrane protein (termed: anchor protein). A. Musgrave (pers. communication) has further shown that the mt^- agglutinin of *C. eugametos* cannot be labelled by phenyl thiocyanate, a compound that specifically binds to hydrophobic domains typical of intrinsic membrane proteins (Kempf *et al.*, 1981). Two lines of evidence point to the existence of the putative anchor protein. When mt^- isoagglutinins (membrane vesicles shed into the culture fluid, and enriched in agglutinin, Förster *et al.*, 1956) are sonicated, they lose their ability to bind to the flagella of mt^+ gametes, and the agglutinin is found in the supernatant (Homan *et al.*, 1982). The sonicated vesicles can be reactivated by incubating them overnight in the supernatant. This reactivation is inhibited by prior treatment of the inactivated vesicles with trypsin, implying that the agglutinin molecule is indeed held to the membrane by another (glyco) protein. The existence of an anchor protein could also explain the different susceptibility between gametes and isolated agglutinin towards proteolytic enzymes. Whereas live mt^- gametes of *C. reinhardtii* lose their agglutinability completely within thirty min in the presence of 0.25 mg/ml trypsin, isolated agglutinin is unaffected by this concentration (Matsuda *et al.*, 1982). Similarly, Wiese and Mayer (1982) showed that chymotrypsin inactivated live mt^+ gametes of *C. reinhardtii*, whereas isolated mt^+ agglutinin was resistant (Collin-Osdoby and Adair, 1985).

Do agglutinins of different mating type react with each other (homophilic interaction) or has each agglutinin its own receptor (heterophilic interaction)? Until now, we have not succeeded in demonstrating any form of interaction between sexual agglutinins *in vitro*, although isolated flagella may interact with each other *in vitro* (Köhle *et al.*, 1980; Goodenough, 1986). First, when agglutinins of both mating types are mixed, their adhesive properties can be independently measured and are the same as when measured without mixing. Secondly, when mixed agglutinins of *C. eugametos* are centrifuged over a sucrose gradient, they sediment with the same velocity as when applied separately and there is no indication of a faster moving complex (M.R. Samson, pers. communication). This might mean that the recognition system operating in *Chlamydomonas* consists of two ligand-receptor combinations in which the agglutinins function as ligand, presenting their terminal mannose or galactose residues towards an unknown receptor. However, a more trivial explanation might be that, as the inactivation experiments have been carried out at low agglutinin concentrations – of the order of 10^{-10} M – the ligand affinity-constant was not high enough to obtain a detectable complex formation between the sexual agglutinins.

Dynamics of flagellar agglutination

The question is, how, in a random clump of agglutinating cells, an efficient sorting out can take place, which results in the orderly formation of cell pairs. By labelling one of the two mating types of *C. eugametos* with fluorescein, it was found that within the first minute of mixing the gametes, the mating types agglutinate more or less randomly, but then rapidly rearrange themselves so that the gametes at the outer surface of the clumps are mt^+ and those within mt^- (Musgrave *et al.*, 1985). In such cell clumps the flagella of all cells are oriented around the mt^- cell bodies (Fig. 7.1). This is in contrast to the orientation of flagella in *vis-à-vis* pairs, where the flagella have de-agglutinated and the motile mt^+ flagella are held around the mt^+ cell body (Lewin, 1952). On close visual inspection, it appeared that all cell pairs fuse with their flagella oriented around the mt^- cell bodies, but when the agglutinating power of the flagella has sufficiently diminished for the mt^+ flagella to dissociate and to regain their swimming activity, they become oriented in the typical *vis-à-vis* pair position.

Another feature of agglutinating cells, mentioned earlier, is that the adhering flagella are fully aligned but always have their tips associated. To explain this preference to tip-to-tip binding, various authors have postulated that occupied agglutination sites have the tendency to accumulate at the flagellar tips, a process called tipping (e.g. Goodenough and Jurivich, 1978). As is depicted in Fig. 7.1, the two features, the preferred orientation of the flagella in pairs of agglutinating cells, and the transport of the sites of contact to the flagellar tips, would largely explain the behaviour of agglutinating cells, and make it easier to understand the way cells find a partner cell in a random clump, and become properly oriented towards it in order to be able to bring their fusion papillae into close contact.

The idea of accumulating agglutination sites at the flagellar tips as a function of sexual agglutination, requires demonstration of the actual lateral mobility of the agglutinins in the plane of the flagellar membrane. Recently, a redistribution of agglutinins, resulting in strong tipping, was shown, using a monoclonal antibody directed against the mt^- agglutinin, which blocked the activity of isolated agglutinin and of intact mt^- gametes specifically, and therefore probably has its epitope at, or near the binding site (W. L. Homan, pers. communication).

The prime function of tipping is considered to be the aligning of the flagella in a tip-to-tip fashion, as depicted in Fig. 7.1, but other functions, like, the generation of the signal by which cell fusion is initiated, cannot

Fig. 7.1. Diagram illustrating the importance of tipping and flagellar orientation for papillar contact. The papillae just protrude through the flagellar ridge. A: the most obvious form of flagellar alignment, with all flagella projecting forward, does not bring the papillae into juxtaposition; B: reorientation of the flagella around the mt^- cell body, that occurs during agglutination, brings the papillae in contact; C: transport of bound agglutinins to the flagellar tips results in sorting out and realignment. Tip-to base contacts will eventually be lost as illustrated on the left. Tip-to-tip (and base-to-base) contacts are stable and the whole flagellar surface can adhere, which naturally brings the papillae into juxtaposition; D: the orientation of flagella around the mt^- cell bodies encloses the mt^- cells into the middle of large clumps because the flagella can only make secondary contacts with other gametes on the mt^- side of agglutinating pairs (from Musgrave, A. (1987). Sexual agglutination in *Chlamydomonas eugametos*. In *Algal Development*, eds. W. Wiessner, D. G. Robinson and R. C. Starr, pp. 83–9. Berlin, Heidelberg: Springer-Verlag).

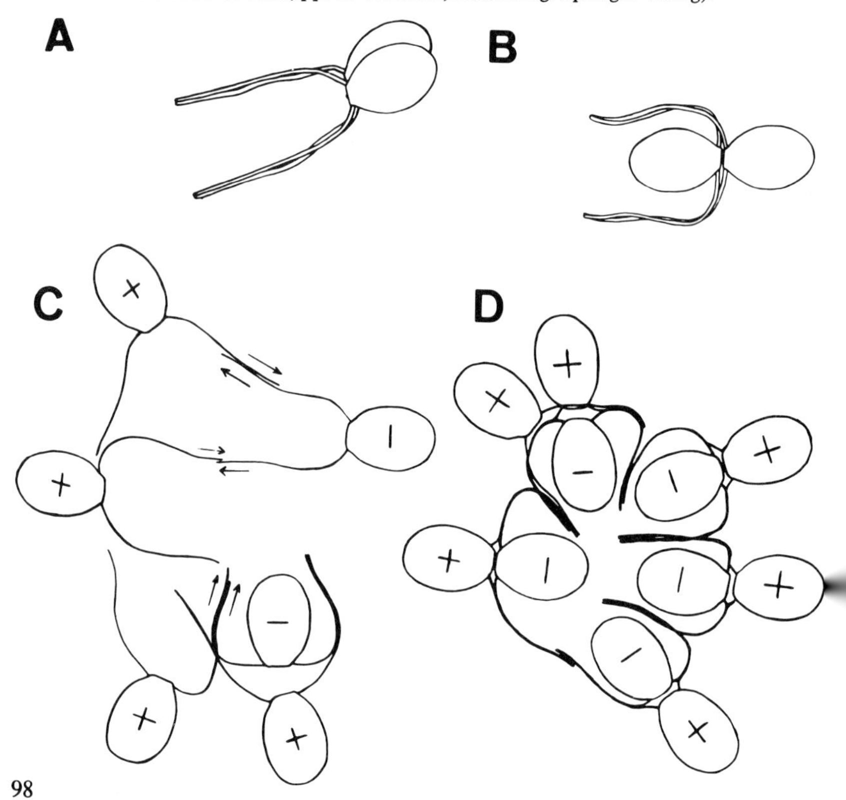

be excluded (cf. Goodenough *et al.*, 1980). It remains difficult to distinguish effects that are causally related from those that simply occur in chronological sequence. Another example of this uncertainty is the correlation that has been demonstrated between tipping and 'flagellar tip activation' (Mesland *et al.*, 1980). The latter phenomenon is an ultrastructural change in the flagellar tip, including the extension of the outer doublets of the flagellar axoneme and the accumulation of electron dense material. The connection between these two processes remains unclear.

Control of agglutinability

Several authors have pointed out that agglutinins at the flagellar surface are subject to a turnover, whose rate is considerably increased by the agglutination process (Snell and Moore, 1980; Pijst *et al.*, 1983). It is logical to assume that the density of agglutinin molecules at the flagellar surface is maintained by the supply from an intracellular pool. This notion was strengthened by the fact that compared with the flagella, the cell body of gametes contain a considerable amount of agglutinin, which is mainly localised at the plasma membrane (Pijst *et al.*, 1983). However, from subsequent studies it appeared that there is a functional membrane barrier in the transition zone at the base of each flagellum of *C. eugametos* gametes, which for example is demonstrated by the fact that several glycoconjugates are restricted to the flagellar membrane, just as others are restricted to the cell body membrane (Musgrave *et al.*, 1986). This makes it unlikely that agglutinin molecules present at the surface of the cell body can diffuse or be transported to the flagellar membrane. Apparently, the transport of agglutinin molecules from the cell body to the flagella takes place via some unknown route, so that the transition zone is bypassed. A possible explanation for the fact that flagellar adhesiveness disappears rapidly after cell fusion, is that the formation of the plasma bridge signals the arrest of agglutinin supply to the flagella, where its level would drop due to a continuous high turnover rate. The data in Table 7.1 seem to be in accord with this idea: *vis-à-vis* pairs contain as much agglutinin in their cell bodies as unmated gametes, while, as expected, the flagella are devoid of agglutinin. So the agglutinability of flagella might be controlled by regulating the agglutinin density at the surface. The following results, however, show that this is not the only means by which the cell can control the agglutinability of gamete flagella.

Once gametes have differentiated, and agglutinins have been incorporated into their flagellar membrane, one might expect them to be sexually competent, but this need not be the case. In some strains of *C. eugametos*, light is an absolute requirement for agglutinability. This particular property is not sex-linked. If such strains are kept in the dark, they always remain non-agglutinable, but after a short illumination, maximum agglutinability is achieved. On returning to the dark, they become non-agglutinable after 30 min. This activity can be turned on and off for at least four cycles in the absence of protein synthesis, which indicates that the mechanism involves the modification/demodification of the same agglutinins rather than their turnover in the flagellar membrane (Kooyman *et al.*, 1986).

Table 7.1. *Agglutinin present on the flagellar and cell-body surfaces of gametes and* vis-à-vis *pairs of* Chlamydomonas eugametos.

Gametes and *vis-à-vis* pairs were deflagellated and the cell bodies separated from the flagella by differential centrifugation. Both fractions were extracted in 1% Triton X-100 and the concentration of agglutinin determined via the charcoal assay, described by Musgrave *et al.* (1981). The concentration is expressed as the agglutination titer of a 1 ml extract of 0.1 ml packed cell bodies, or their flagella. Since half the volume of *vis-à-vis* pairs is mt^+ and the other half mt^-, twice as much material (0.2 ml packed volume) was extracted to make the results comparable with those from free gametes. The preparation of *vis-à-vis* pairs also contained 5% non-fused gametes (from Musgrave *et al.*, 1985).

Material extracted		Agglutination titre
Gametes		
mt^+	Flagella	2^2
	Cell bodies	2^5
mt^-	Flagella	2^3
	Cell bodies	2^6
Vis-à-vis pairs		
mt^+	Flagella	inactive
	Cell bodies	2^5
mt^-	Flagella	2^0
	Cell bodies	2^6

Gametogenesis

A number of studies point to a nutritional control of gametogenesis in *Chlamydomonas reinhardtii*. Sager and Granick (1954) confirmed older findings that flooding agar plate cultures with distilled water yielded suspensions of motile cells capable of forming zygotes. Pair formation generally required light, but the older the cultures, the less strict was this requirement. The inference was made that light stimulated the intracellular depletion of nutrients essential for vegetative growth. In older cultures these compounds would already be near their critical level so that less light would be required to induce gametogenesis. A systematic study revealed that shorter light requirements for gametic activity were found only in nitrogen-deficient cultures, even to the extreme case of no light requirement at all. The authors claimed reversibility of gametogenesis by the addition of nitrogen salts, although they only showed inactivation of gametes but no resumption of cell division. Kates and Jones (1964) showed that synchronous cultures gradually developed inducibility of gametogenesis by nitrogen depletion; quantitative gametogenesis was obtained only at the end of the linear growth phase. Although the results presented suggest an absolute light requirement for gametogenesis, in their discussion the authors mention the fact that gamete induction can take place in the dark, provided that the first 6 h in N-free medium are spent in the light, thus confirming the results of Sager and Granick (1954). They also showed that nitrogen salts inhibited gamete activity; resumption of cell divisions 24 h after the addition of nitrogen (i.e. 9–12 h after gamete inactivation) indicated true reversibility of gamete differentiation. Martin & Goodenough (1975) demonstrated that agar plate cells suspended in N-containing medium exhibited gametic activity, depending on the age of the culture. The first activity was detectable at day 5 and 100% mating efficiency was reached at day 9 of one particular set of cultures. At day 6 the agar medium contained less than 2% of the original ammonium. Upon reinoculation with vegetative cells, this medium yielded gametes within one day. These observations once more stress the importance of nitrogen depletion as a trigger for gametogenesis.

As was stated before, a light requirement for gamete differentiation can be explained in terms of metabolic depletion of residual intracellular nitrogen (Sager and Granick, 1954). Martin and Goodenough (1975) made a point of this by showing that gamete induction can be rendered completely light-independent by providing the cells with acetate, which is utilised by *C. reinhardtii* for heterotrophic growth. We might add here that since in all these studies either potassium or ammonium nitrate was used as a source of nitrogen, the photocontrol of nitrate reductase activity in *C. reinhardtii* as described by Azuara and Aparicio (1983) could well play a role in the observed effect of light upon gametogenesis.

As far as other species are concerned, Bernstein & Jahn (1955) working *inter alia* with *C. eugametos*, essentially confirmed the results of Sager & Granick (1954). Trainor (1975) reported the need of a reduced level of nitrogen in initiating gametogenesis in cultures of *C. eugametos*. However, he pointed out that 10 mg/l N may still be present in the medium without affecting mating competence. Tomson *et al.* (1985) confirmed this observation by experiments in which they obtained spontaneous gametogenesis in asynchronous liquid cultures using 1% Kates & Jones (1964) medium, except for KNO_3 which was of the original strength. Gametogenesis occurred at the end of the linear growth phase when the medium still contained 86% of the original nitrogen source. The authors suggested that any nutrient stress could be the cause of gametogenesis, since for example the phosphate content had dropped to 3% of the original level.

Gametogenesis goes hand in hand with biochemical and morphological alterations. Jones *et al.* (1968) showed that net protein synthesis ceases, although there is a persistent protein turnover. Martin & Goodenough (1975) reported an extensive breakdown of RNA during gametogenesis to compensate for a persisting synthesis of DNA. Most of the chloroplastic and cytoplasmic ribosomes disappeared. However, as was stressed by these authors, not all cellular changes occurring during gametic induction need be strictly related to gametogenesis *per se*, but may be simply stress responses. Essential features would be: 1) the loss of the capacity to perform cell division; 2) the formation of the 'mating structure'; 3) acquisition of flagellar agglutinability. Tomson *et al.* (1985) reported the synthesis of extractable agglutinin concomitant with the acquisition of agglutinability and mating competence. For the present it would therefore seem appropriate to focus attention on the triggering of agglutinin synthesis, on the localisation of this process and on the incorporation of agglutinin in the flagellar membrane, features truly characteristic of gamete differentiation.

References

Adair, W. S., Monk, B. C., Cohen, R. Hwang, C. & Goodenough, U. W. (1982) Sexual agglutinins from the *Chlamydomonas* flagellar membrane. Partial purification and characterization. *J. Biol. Chem.*, **257**, 4593–602

Azuara, M. P. & Aparicio, P. J. (1983). *In vivo* blue-light activation of *Chlamydomonas reinhardtii* nitrate reductase. *Plant Physiol.*, **71**, 286–90

Bernstein, E. & Jahn, T. L. (1955). Certain aspects of the sexuality of two species of Chlamydomonas. *J. Protozool.*, **2**, 81–5

Claes, H. (1971). Autolyse der Zellwand bei den Gameten von *Chlamydomonas reinhardtii*. *Arch. Microbiol.*, **78**, 180–8

Clarke A. E., Anderson, R. L. and Stone, B. A. (1979). Form and function of arabinogalactans and arabinogalactan-proteins. *Phytochemistry*, **18**, 521–40

Collin-Osdoby, P. & Adair, W. S. (1985). Characterization of the purified Chlamydomonas minus agglutinin. *J. Cell Biol.*, **101**, 1144–52

Fincher G. B., Stone, B. A. & Clarke, A. E. (1983). Arabinogalactans: structure, biosynthesis and function. *Ann. Rev. Plant Physiol.*, **34**, 47–70

Förster, H., Wiese, L. & Braunitzer, G. (1956). Über das agglutinierend wirkende Gynogamon von *Chlamydomonas eugametos*. *Zeitschrift für Naturforschung*, **11b**, 315–7

Goodenough, U. W. & Weiss R. L. (1975). Gametic differentiation in *Chlamydomonas reinhardtii*. *J. Cell Biol.*, **67**, 623–37

Goodenough, U. W., Adair, W. S., Caligor, E., Forest, C. L., Hoffman, J. L., Mesland, D. A. M. & Spath, S. (1980). Membrane–membrane and membrane–ligand interactions in *Chlamydomonas* mating. In *Membrane–Membrane Interactions*, ed. N. B. Gilula, pp. 131–52. New York: Raven Press

Goodenough, U. W. & Jurivich, D. (1978). Tipping and mating structure activation induced in *Chlamydomonas* gametes by flagellar membrane antisera. *J. Cell Biol.*, **79**, 680–91

Goodenough, U. W. (1986). Experimental analysis of the adhesion reaction between isolated *Chlamydomonas* flagella. *Exp. Cell Res.*, **166**, 237–46

Goodenough, U. W., Adair, W. S., Collin-Osdoby, P. & Heuser, J. (1985). Structure of the *Chlamydomonas* agglutinin and related flagellar surface proteins *in vitro* and in situ. *J. Cell Biol.*, **101**, 924–41

Heuser, J. E. (1983). Procedure for freeze-drying molecules on mica flakes. *J. Mol. Biol.*, **169**, 155–95

Holst, G. J. van, & Varner, J. E. (1984). Reinforced polyproline II conformation in a hydroxyproline-rich cell wall glycoprotein from carrot root. *Plant Physiol.*, **74**, 247–51

Homan, W. L., Musgrave, A., Molenaar, E. M. & van den Ende, H. (1980). Isolation of monovalent sexual binding components from *Chlamydomonas eugametos* flagellar membranes. *Arch. Microbiol.*, **128**, 120–5

Homan, W. L., Hodenpijl, P. G., Musgrave, A. & van den Ende, H. (1982). Reconstitution of biological activity in isoagglutinins from *Chlamydomonas eugametos*. *Planta*, **155**, 529–35

Homer, R. B. & Roberts, K. (1979). Protein conformation in plant cell walls. Circular dichroism reveals a polyproline II structure. *Planta*, **146**, 217–22

Jones, R. F., Kates, J. R. & Keller, S. J. (1968). Protein turnover and macromolecular synthesis during growth and gametic differentiation in *Chlamydomonas reinhardtii*. *Biochim. Biophys. Acta*, **157**, 589–98

Kates, J. R. & Jones, R. F. (1964). The control of gametic differentiation in liquid cultures of *Chlamydomonas*. *J. Cell. & comparative Physiol.*, **63**, 157–64

Kempf, C., Brock, C., Sigrist, H., Tanner, M. J. A. & Zahler, P. (1981). Interaction of phenylisothiocyanate with human erythrocyte band 3 protein. II Topology of phenylisothiocyanate binding sites and influence of p-sulfophenylisothiocyanate on phenylisothiocyanate modification. *Biochim. et Biophys. Acta*, **641**, 88–98

Klis, F. M., Samson, M. R., Touw, E., Musgrave, A. & Van den Ende, H. (1985). Sexual agglutination in the unicellular green alga *Chlamydomonas eugametos*. Identification and properties of the mating type plus agglutination factor. *Plant Physiol.*, **159**, 740–46

Köhle, D., Lang, W. & Kauss, H. (1980). Agglutination and glycosyltransferase activity of isolated gametic flagella from *Chlamydomonas reinhardtii*. *Arch. Microbiol.*, **127**, 239–43

Kooyman, R., Elzenga, T. J. M., de Wildt, P., Musgrave, A., Schuring, F. & van den Ende, H. (1987). Light dependence of sexual agglutinability in *Chlamydomonas eugametos*. *Planta*, **169**, 370–8.

Lemieux B. & Lemieux, C. (1985). Extensive sequence rearrangements in the chloroplast genomes of the green algae *Chlamydomonas eugametos* and *Chlamydomonas reinhardtii*. *Curr. Genetics*, **10**, 213–9

Lennarz, W. J. (1980). *The Biochemistry of Glycoproteins and Proteoglycans.* Plenum Press, New York

Lens, P. F., Olofsen, F., Nederbragt, A., Musgrave, A. & van den Ende, H. (1982). An antiserum against a glycoprotein functional in flagellar adhesion between *Chlamydomonas eugametos* gametes. *Arch. Microbiol.*, **131**, 241–6

Lens, P. F., Olofsen, F., van Egmond, P., Musgrave, A. & van den Ende, H. (1983). Isolation and partial characterization of an antigenic determinant from flagellar glycoproteins of *Chlamydomonas eugametos*. *Arch. Microbiol.*, **135**, 311–8

Lewin, R. F. (1952). Studies on the flagella of algae. I. General observations on *Chlamydomonas moewusii* Gerloff. *Biol. Bull.*, **102**,74–9

Martin, N. C. & Goodenough, U. W. (1975). Gametic differentiation in *Chlamydomonas reinhardtii* I. Production of gametes and their fine structure. *J. Cell Biol.*, **67**, 587–605

Matsuda, Y., Sakamoto, K., Kiuchi, N., Mizuochi, T., Tsubo, Y. & Kobata, A. (1982). Two tunicamycin-sensitive components involved in agglutination and fusion of *Chlamydomonas reinhardtii* gametes. *Arch. of Microbiol.*, **131**, 87–90

Mesland, D. A. M. (1976). Mating in *Chlamydomonas eugametos*. A scanning electron microscopic study. *Arch. of Microbiol.*, **109**, 31–5

Mesland, D. A. M., Hoffman, J. L., Caligor, E. & Goodenough, U. W.

(1980). Flagellar tip activation stimulated by membrane adhesions in *Chlamydomonas reinhardtii*. *J. Cell Biol.*, **84**, 599–617

Musgrave, A., van Eijk, E., te Welscher, R., Broekman, R., Lens, P. F., Homan, W. L. & van den Ende, H. (1981). *Planta*, **153**, 362–9

Musgrave, A., de Wildt, P., Schuring, F., Crabbendam, K. & van den Ende, H. (1985). Sexual agglutination in *Chlamydomonas eugametos* before and after cell fusion. *Planta*, **166**, 234–43

Musgrave, A., de Wildt, P., van Etten, I., Pijst, H., Scholma, C., Kooyman, R., Homan, W. L. & van den Ende, H. (1986). Evidence for a functional membrane barrier in the transition zone between the flagellum and cell body of *Chlamydomonas eugametos* gametes. *Planta*, **167**, 544–53

Pijst, H. L. A., Zilver, R. J., Musgrave, A. & van den Ende, H. (1983). Agglutination factor in the cell body of *Chlamydomonas eugametos*. *Planta*, **158**, 403–9

Sager, R. & Granick, S. . (1954). Nutritional control of sexuality in *Chlamydomonas reinhardtii*. *J. Gen. Phys.*, **37**, 729–42

Saito, T. & Matsuda, Y. (1984). Sexual agglutination of mating-type minus gametes in *Chlamydomonas reinhardtii*. II Purification and characterization of minus agglutinin and comparison with plus agglutinin. *Arch. Microbiol.*, **139**, 95–9

Samson, M. R., Klis, F. M., Homan, W. L., van Egmond, P., Musgrave, A. & van den Ende, H. (1987). Composition and properties of the sexual agglutinins of the flagellated green alga *Chlamydomonas eugametos*. *Planta*, **170**, 314–21

Schuring, F., Smeenk, J. W., Homan, W. L., Musgrave, A. & van den Ende, H. . (1987). Occurrence of O-methylated sugars in surface glycoconjugates in *Chlamydomonas eugametos*. *Planta*, **170**, 322–7

Smith, C. A. & Seegan, G. W. (1984). The platelet sheet: an unusual negative staining method for transmission electron microscopy of biological molecules. *J. Ultrastructural Res.*, **89**, 111–22

Snell, W. J. & Moore, W. S. (1980). Aggregation dependent turnover of flagellar adhesion molecules in *Chlamydomonas* gametes. *J. Cell Biol.*, **84**, 203–10

Tomson, A. M., Demets, R., Bakker, N. P. M., Stegwee, D. & van den Ende, H. (1985). Gametogenesis in liquid cultures of *Chlamydomonas eugametos*. J. Gen. Microbiol., **131**, 1553–60

Trainor, F. R. (1975). Is a reduced level of nitrogen essential for *Chlamydomonas eugametos* mating in nature? *Phycologia*, **14**, 167–70

Wiese, L. & Mayer, R. A. (1982). Unilateral tunicamycin sensitivity of gametogenesis in dioecious isogamous *Chlamydomonas* species. *Gamete Res.*, **5**, 1–9

PART III

Recognition in multicellular organisms

8 Cell type specific glycoproteins involved in cell differentiation and movement in *Dictyostelium discoideum* slugs

Elizabeth Smith and Keith L. Williams

School of Biological Sciences,
Macquarie University,
Sydney, NSW, 2109, Australia

Abstract

A monoclonal antibody, MUD50, defines a surface epitope found only on the prespore subclass of differentiated cells in the *D. discoideum* slug. It also reveals a distinctive molecular lattice in the extracellular matrix. Different proteins carrying this epitope, while largely cell-type specific, are found not only at the cell surface, but also as internal proteins.

The *D. discoideum* slug consists of approximately 100,000 cells which migrate in a cohesive fashion through an extracellular matrix of their own making. The role of MUD50 proteins in slug movement is discussed in relation to a model integrating slug movement with the maintenance of the prestalk–prespore pattern.

During development in multicellular organisms, differentiation results in the formation of an organised array of cell types or tissues. In order to achieve this, messages must pass between cells so that they can undergo appropriate differentiation, and so that the correct pattern is produced and maintained. These events are believed to involve cell surface receptors. Some of these may receive diffusible signals, while others may interact with molecules on the surfaces of adjacent cells (Chisholm *et al.*, 1984). In addition, there is increasing evidence for the interaction of cells with signals present in extracellular matrices (ECMs). A wide range of diverse materials comprise ECMs in different organisms. These range from the cell walls of plants, to the extracellular slime of plants and microorganisms, to the matrices of mammalian tissues.

Use of monoclonal antibodies in the study of development

Monoclonal antibodies (McAbs) are used extensively in the study of development processes to reveal the complexity of the changes occurring at the molecular level. If studies are performed with a well-characterised system, one can quickly progress from histochemical studies to gene isolation, and use mutants in further functional characterisation. A good example of this is *Drosophila melanogaster*, and several groups are currently investigating the changes in expression of cell surface and other antigens during *D. melanogaster* development (e.g. Wilcox & Leptin, 1985; Bedian *et al.*, 1986). By observation of the functional effects of mutations, the developmental role of the wild type protein may be unequivocally established (Beachy, 1985).

Using a similar approach we are investigating the role of cell surface antigens in the formation and maintenance of pattern in the multicellular stage of the cellular slime mould, *Dictyostelium discoideum*. In this relatively simple eukaryote a considerable amount is known about cell–cell signalling during aggregation (Chisholm *et al.*, 1984) and attention is now being focussed on the multicellular slug stage. We have identified, using McAbs, a complex group of developmentally regulated cell surface and related proteins which will be discussed in detail later in the paper.

Nature of epitopes recognised by McAbs

In a number of studies with McAbs it has been found that, when screened on Western blots, some McAbs recognise a set of proteins or glycoproteins, while others recognise only a single molecule. This raises two questions: first, the nature of the epitope recognised by the McAb and, second, whether the epitope itself has functional significance.

It is often assumed that a McAb which binds to a single protein is probably recognising a peptide sequence, while one that binds to several molecules is probably interacting with a common oligosaccharide modification. Two examples suffice to indicate that this generalisation should be treated with caution. In the first case McAbs were raised against the glycoproteins which self-assemble to form the two-dimensional crystalline array of the cell wall of the green alga, *Chlamydomonas reinhardii* (Smith, Roberts, Hutchings & Galfre, 1984). These McAbs were found to have properties consistent with their binding to oligosaccharide rather than peptide determinants. One of a group with overlapping specificities had a very limited distribution, despite the fact that the epitope is almost certainly carabohydrate. Similar results have been reported by Bertholdt *et al.* (1985) using McAbs to contact sites A glycoprotein of *D. discoideum*, and by Van den Ende (1986) with McAbs to the sexual agglutinin of *C. engametos* (this volume).

The second example is that of McAbs which recognise ubiquitin, a remarkable peptide of 76 amino acids which is widespread among eukaryotes and extremely highly conserved (Hart, 1986). This peptide is unusual in that a specific isopeptidase catalyses the formation of an isopeptide bond between the C-terminal end of the ubiquitin molecule and the ε-NH_2 of lysine in another protein, whether nuclear, cytoplasmic or cell surface. For this reason McAbs directed against peptide determinants on ubiquitin detect several proteins (Siegelman *et al.*, 1986).

Thus it seems that a McAb binding to a particular carbohydrate determinant on a glycoprotein may give a very simple pattern, while one recognising a peptide sequence may bind to several proteins. Such phenomena should alert us to consider carefully the chemical nature of antigenic determinants recognised by McAbs.

Chemical nature of signals

Although the precise molecular structure of signal molecules is not yet fully understood, there are some cases in which oligosaccharides or peptides with functional significance have been characterised.

An oligosaccharide is involved in the initiation of nodulation in various plants by the nitrogen-fixing *Rhizobium* bacterium, which produces exopolysaccharides (Djordjevic *et al.*, 1986). The oligosaccharide repeat units have been purified and their structures identified. The use of non-mucoid mutants of *Rhizobium* has allowed confirmation of the role of these carbohydrates.

An example of a system in which a peptide sequence is known to be involved in function is fibronectin. A tetrapeptide of this glycoprotein has been shown to be involved in some way in attachment of cells to the substratum (Pierschbacher & Ruoslahti, 1984). Fibronectin is an ECM and cell surface glycoprotein which has a role in the cell migrations observed during morphogenesis (Hay, 1983). Therefore, study of the molecular interactions between fibronectin and cells should provide some insights into how cells move with respect to ECMs and other cells. These processes must involve recognition, signal transduction and adhesion. An understanding of these molecular events may also lead to clarification of mechanisms of differentiation.

Dictyostelium discoideum – a simple multicellular organism

It seems probable that many of the basic molecular mechanisms of morphogenesis, and perhaps even the molecules themselves, are fundamentally similar in all multicellular organisms. However, in complex organisms it becomes difficult to study individual molecular components in isolation from the whole.

We have chosen the cellular slime mould *Dictyostelium discoideum* as a very simple multicellular organism, but one with a number of features that distinguish it as an excellent system for the study of development. Firstly, *D. discoideum* has a short well-defined asexual life-cycle with easily identified developmental stages (Fig. 8.1). The most intensively studied portion of the life cycle is aggregation, which occurs after the single-celled amoebae have consumed all available bacteria. Periodic pulses of cAMP are emitted from emerging aggregation centres and surrounding amoebae respond by moving towards the source of cAMP and also by relaying the cAMP pulses (Chisholm *et al.*, 1984). During this process end-to-end contacts are formed, resulting in streams of cells. Work on the molecular basis for cell–cell contact at this stage of development has been pioneered in Gerisch's laboratory, and one of the molecules involved has been identified and characterised as a cell-surface glycoprotein designated contact sites A (Bertholdt *et al.*, 1985) or gp80 (Siu, 1986).

Our studies, however, focus on a later stage of the *D. discoideum* life cycle, when the multicellular organism has been formed by aggregation of individual cells. After the formation of a tip, which is often considered to be an 'organiser' region, the slime mould can proceed along alternative deve-

Fig. 8.1. Life cycle of the cellular slime mould, *Dictyostelium discoideum*.

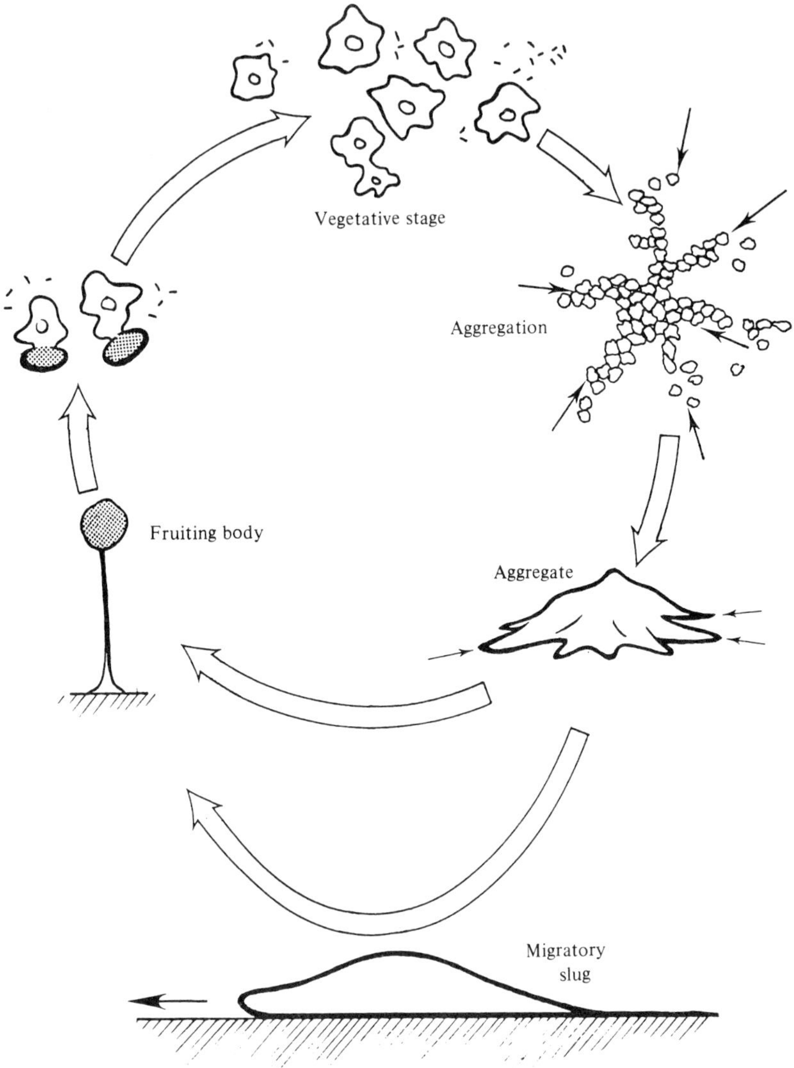

Vegetative stage

Aggregation

Aggregate

Fruiting body

Migratory slug

lopmental pathways (Rubin & Robertson, 1975). Either it can form a fruiting body immediately, or, given suitable conditions of light and humidity, it can be distracted to form a migrating 'slug' before going on to form a fruiting body (Fig. 8.1). The slug stage is of particular interest with regard to the role of cell–cell and cell–ECM interactions in differentiation and pattern formation. At least two cell types, prespore cells, which will go on to form the spores in the mature fruiting body, and prestalk cells, destined to become the dead vacuolated cells enclosed in cellulose-containing cell walls, are present in the slug. The latter will form the stalk which will support the spore head. The two kinds of cells are maintained in a pattern within the slug, with the prestalk cells occupying the front 10–20% and the prespore cells in the rear, intermingled with some anterior-like cells (MacWilliams & David, 1984). In this work the spatial pattern was considered to be stable, but, recently, mechanisms integrating internal cell movement with maintenance of pattern have been considered (Williams *et al.*, 1986).

The slug is surrounded by an ECM, the slime sheath, composed of cellulose fibrils in a glycoprotein matrix (Hohl & Jehli, 1973; Freeze & Loomis, 1977; Smith & Williams, 1979). This material, which is continuously synthesised and through which the mass of slug cells moves, is left behind like a collapsed sausage skin. The discarded sheath material can easily be collected and provides a source of relatively pure ECM for use in biochemical and immunohistochemical studies.

Thus *D. discoideum* has several characteristics which facilitate developmental studies, in particular the study of cell–cell and cell–ECM interactions. An added advantage is the availability of numerous well-characterised mutants and, since *D. discoideum* strains can easily be maintained in cold storage, it is possible to work with any appropriate mutant (Newell, 1982; Welker & Williams, 1982). Since the genetic system does not depend on the complete asexual cycle (i.e. slug and fruiting body formation), developmental mutants can easily be analysed genetically (Williams & Newell, 1976).

Experimental approach to developmental studies with D. discoideum

The approach we have taken is to use McAbs to identify and study developmentally regulated, cell-type specific and cell surface antigens of *D. discoideum* (Fig. 8.2). In looking for molecules involved in signal reception, McAbs are of great significance because of their unique specificity and sensitivity. They can be used for affinity-purification of molecules which are present only in minute amounts. This may be especially relevant when one is dealing with biologically important oligosaccharides, since relatively large amounts are required for structural analysis, and oligosacchardies are much more difficult than peptides to purify by conventional means. We routinely use a variety of methods to screen for McAbs recognising developmentally regulated antigens, especially those on the cell surface.

Screening monoclonal antibodies

(i) Western blots

Initial screening for positive hybridoma supernatants involves the dot-immunobinding assay (Smith, Roberts, Butcher & Galfre, 1984). The proteins recognised by McAbs are identified on Western blots of extracts of whole slug cells or slime trails after SDS–PAGE separation of the proteins (Laemmli, 1970; Grant, Welker & Williams, 1985).

(ii) Flow cytometry

McAbs are screened for binding to the cell surface using a flow cytometer (Voet *et al.*, 1984). *D. discoideum* slugs are disaggregated and the single cells incubated with McAb and fluoresent-labelled second antibody before being analysed. Using flow cytometry it is simple to determine if a particular

Fig. 8.2. Different stages of the life cycle are identified using the letters A to E. Part I: Cells taken at each of these times were subjected to reducing SDS-PAGE and Western blotting, followed by probing the blots with MUD50. Parts II, III and IV refer to the migratory slug stage.

Part II: slug cells were disaggregated, stained with FITC-coupled MUD1 and sorted on an EPICS V flow cytometer (see III). The sorted cells were subjected to SDS-PAGE under reducing conditions, Western blotted and stained with MUD50. Lane 1 – prestalk cells, Lane 2 – prespore cells, Lane 3 – unsorted slug cells, Lane 4 – slime sheath.

Part III: FACS analysis of disaggregated slug cells stained with MUD50. Two populations of cells are detected with the prespore cells labelled and the prestalk cells, which are larger, unlabelled.

Part IV: Underside of slime trail left by migrating slug stained with MUD50 and FITC-coupled anti-mouse IgG. U-shaped pattern left by shuffling slug (*upper photograph*), elliptical pattern left by hurdling slug (*lower photograph*).

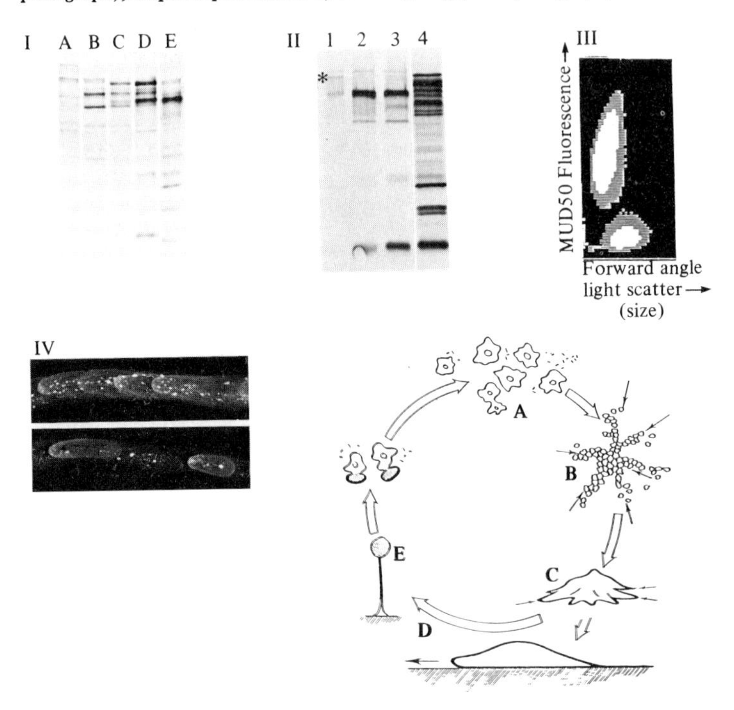

McAb recognises a cell-surface determinant, and whether or not the binding distinguishes subpopulations of cells.

(iii) Immunohistochemistry

McAbs are further characterised by immunohistochemistry on frozen sections of slugs (Gregg *et al.*, 1982) as well as on the slime trails left behind by migrating slugs (Vardy *et al.*, 1986).

Using the above techniques it is often possible to distinguish between McAbs recognising internal proteins, cell surface proteins and ECM proteins, or a combination of these. McAbs can then be used in conjunction with mutants which are in some way defective in migration, pattern formation, or other aspects of development, to investigate the role of the molecules recognised.

A group of McAbs recognising prespore-specific proteins

A prespore specific McAb, MUD1, recognising only a single band on Western blots, i.e. the cell surface molecule designated PsA, has been previously described (Krefft *et al.*, 1984). Most of the work to be discussed in this paper focuses on the McAb, MUD50, which recognises a different epitope from that recognised by MUD1 on the prespore-specific surface antigen PsA. It also binds to a number of other proteins in slug cells and the ECM (Grant & Williams, 1983; Grant *et al.*, 1985). MUD50 recognises complex groups of proteins which are developmentally regulated and vary greatly during the period from initiation of starvation of vegetative amoebae through to fruiting body formation, with about 12 bands visible at any one stage. The position and intensity of bands on Western blots is different at each time interval, so that the patterns at aggregation, slug and fruiting body stages are quite different (Fig. 8.2). Although several MUD50 positive proteins can be detected on Western blots of cells from the time of initiation of starvation, labelling of the cell surface with this McAb does not occur until tip formation in the aggregate. PsA becomes visible on Western blots shortly prior to tip formation (Fig. 8.2) and surface-labelling with MUD1 appears at the same time as MUD50 surface labelling, suggesting the possibility that surface-labelling by MUD50 is due to its binding to PsA. However, other surface proteins may also be recognised by MUD50.

Pure preparations of both major cell types, prespore and prestalk cells, can be obtained by sorting cells in a flow cytometer on the basis of their labelling with MUD1. These sorted cells can be subjected to analysis by SDS–PAGE and Western blotting. Probing Western blots of such sorted cells with MUD50 shows that the determinant recognised by this McAb is confined to prespore cells, with the exception of one faint band in prestalk cells (asterisk Fig. 8.2). Western blots of extracellular sheath material left behind by migrating slugs show a large number of MUD50 positive proteins present in the ECM, and most are different from those in slug cells (Fig. 8.2).

Studies on the mechanism of slug movement and its relevance to maintenance of pattern

Immunohistochemistry with MUD50 confirms the results obtained by flow cytometry, showing that prestalk cells at the front of the slug are essentially unstained, while most of those in the back portion are stained both on the surface and internally. Investigation of the staining of slime trails left behind by individual slugs whose progress had been followed in detail, has revealed interesting information about the interaction of cells with the slime sheath (Vardy *et al.*, 1986). The distribution of MUD50 positive material can be correlated with the way in which the slug has moved. Sometimes slugs are in contact with the substratum along their whole length with just the tip raised (shuffling), while at other times one or more arches are formed during movement (hurdling) (Vardy *et al.*, 1986).

Each of the types of movement described above, shuffling and hurdling, results in a different distribution of MUD50 staining in the trail (Vardy *et al.*, 1986). In the case of the shuffling slug a series of bright U-shaped areas are observed, whereas the hurdling slug leaves a series of individual elliptical areas of staining in the trail. Observations of the movement-related distribution of MUD50 positive material has led to proposals about slug movement and how this relates to maintenance of the pattern in the *D. discoideum* slug (Williams *et al.*, 1986). As previously discussed, the slug has two distinct classes of cells, prespore at the back and prestalk at the front, with a third possible cell type, the anterior-like cells scattered throughout the back part of the slug, but concentrated at the very back (Voet *et al.*, 1984). Staining of frozen longitudinal sections of slugs with MUD50 shows little or no labelling of the cells in the prestalk area at the front, or of the layer of cells on the underside of the slug adjacent to the slime sheath, or of the anterior-like cells in the rear (Williams *et al.*, 1984). This information, coupled with observations of the distribution of MUD50 staining in the slime trails, has led to a proposal as to how cells move relative to the slime sheath or ECM, and relative to each other (Williams *et al.*, 1986). In this model anterior prestalk cells, which are destined to die in the mature fruiting body, are seen as the 'engine'. Prespore cells form the 'cargo' – cells which will go on to produce the next generation of amoebae. The model integrates cell movement within the slug, due to locomotion, with the observed pattern of differentiated cells and provides a basis for further experimental study of slug locomotion (Williams *et al.*, 1986).

Time lapse video films of migrating slugs are being used in conjunction with McAb staining of specific cell and sheath components. This approach provides information which allows us to modify and refine our ideas on the relationships between maintenance of pattern and slug locomotion. For such studies mutants with altered proportions of stalk and spore cells are especially useful. From results obtained so far it seems that one, at least, of the MUD50-positive proteins may act as a 'glue' or 'traction protein' between cells and the sheath in contact with the substratum in the migrating slug (Vardy *et al.*, 1986).

The nature of the epitope recognised by MUD50

The question of whether the determinant recognised by MUD50 is carbohydrate or peptide in nature, and whether it has functional importance, remains to be addressed. We have recently examined *D. discoideum* strain DL118, which carries a mutation at the *modB* locus (Loomis *et al.*, 1985). This mutation results in a defect in carbohydrate processing of the aggregation cell contact molecule CsA (Murray *et al.*, 1984).

Strain DL118 does not contain the epitope recognised by MUD50 as determined by flow cytometry and Western blotting of slug cells. We have therefore concluded that the determinant recognised by MUD50 is carbohydrate in nature (Alexander *et al.*, submitted). Whether the appearance and disappearance of MUD50 positive proteins on Western blots over the developmental time course is due to synthesis and degradation of proteins or to modification of existing proteins remains open to question.

Future approaches to the study of this group of proteins involve characterisation of the epitope recognised by MUD50 and identification of the roles of individual MUD50 positive proteins. Since we are especially interested in cell surface antigens we are using the McAb MUD1, which is believed to recognise a peptide determinant on the cell surface glycoprotein, PsA (Krefft *et al.*, 1984) to purify this protein. The N-terminal amino acid sequence has been determined (Gooley, personal communication) and a DNA probe based on this sequence is being used to seek the gene for PsA.

It is becoming clear that such an approach to the determination of the function of a protein can be effective, especially, when peptide sequences in the functionally active domains are highly conserved. This has recently been demonstrated for the 70 kD family of highly conserved stress proteins (Chappell *et al.*, 1986). Antibodies to a particular amino acid sequence common to heat shock proteins were also found to bind to a protein with known enzymic activity, the uncoating ATPase which releases clathrin triskelions from bovine brain coated vesicles. In this case another property of the heat shock proteins, ATP binding activity, had led the researchers to propose a function for these proteins, but the high sequence homology of the peptides could equally have provided clues as to their function.

In cases where oligosaccharide substituents on glycoproteins are the reactive sites, the only approach would seem to be through use of cross-reactive McAbs. In laboratories all over the world a vast number of McAbs have been raised against tissues and individual proteins from a great diversity of organisms. Information about these needs to be more easily accessed so that it becomes feasible to test McAbs raised to a cell component or protein from one organism against equivalent or different material from another source. In this way information on similarities at the molecular level can be gained, whether the functional group is peptide or carbohydrate. However care should be taken in interpreting results since cross-reactions may arise through non-specific interactions (Ghosh & Campbell, 1986).

Using the McAbs available in our laboratory we have begun an investigation, using flow cytometry and Western blots, of the distribution of antigens among different species of slime moulds as well as among different wild-type strains of *D. discoideum* (Williams, Gooley & Bernstein, in preparation). Preliminary results show considerable variation in the distribution of antigens and it may prove possible to improve the classification of slime moulds by such analysis. Extensions of such studies using other species and a wider range of McAbs could well reveal important conserved molecules, especially on cell surfaces.

Acknowledgements

This research was supported by grants from Macquarie University, the Australian Research Grants Scheme to E.S. and K.L.W. and the N.H. & M.R.C. to K.L.W. We thank Michelle Thorpe for typing the manuscript and Phil. Vardy for help with the illustrations.

References

Beachy, P. (1985). *Drosophila* proteins pave the way to gene mutations. *Nature (Lond.)*, **317**, 18–19

Bedian, V., Oliver, C. E., McCoon, P. & Kauffman, S. A. (1986). A cell surface differentiation antigen of *Drosophila*. *Developmen. Biol.*, **115**, 105–18

Bertholdt, G., Stadler, J., Bozzaro, S., Fichtner, B. & Gerisch, G. (1985). Carbohydrate and other epitopes of the contact site A glycoprotein of *Dictyostelium discoideum* as characterized by monoclonal antibodies. *Cell Differentiation*, **16**, 187–202

Chappell, T. G., Welch, W. J., Schlossman, D. M., Palter, K. B., Schlesinger, M. J. & Rothman, J. E. (1986). Uncoating ATPase is a member of the 70 kilodalton family of stress proteins. *Cell*, **45**, 3–13

Chisholm, R. L., Fontana, D., Theibert, A., Lodish, H. F. & Devreotes, P. (1984). Development of *Dictyostelium discoideum*: chemotaxis, cell–cell adhesion and gene expression. In *Microbial Development*, Cold Spring Harbor Monograph 16, ed. R. Losick & L. Shapiro, pp. 219–54. Cold Spring Harbor, New York

Djordjevic, S. P., Chen, H., Batley, M., Redmond, J. W. & Rolfe, B. G. (1986). Restoration of nitrogen fixation ability of synthesis mutants of *Rhizobium* strains of NGR234 and *R. trifolii* by the addition of purified exopolysaccharide. *J. Bact.* (in press)

Freeze, H. & Loomis, W. F. (1977). Isolation and characterization of a component of the surface sheath of *Dictyostelium discoideum*. *J. Biol. Chem.*, **252**, 820–4

Ghosh & Campbell (1986). *Immunol. Today* (Aug.)

Grant, W. N., Welker, D. L. & Williams, K. L. (1985). A polymorphic, prespore-specific cell surface glycoprotein is present in the extracellular matrix of *Dictyostelium discoideum*. *Mol. & Cell. Biol.*, **5**, 2559–66

Grant, W. N. & Williams, K. L. (1983). Monoclonal antibody characterisation of slime sheath: the extracellular matrix of *Dictyostelium discoideum*. *EMBO J.*, **2**, 935–40

Gregg, J. H., Krefft, M., Haas-Kraus, A. & Williams, K. L. (1982). Antigenic differences detected between prespore cells of *Dictyostelium discoideum* and *Dictyostelium mucoroides* using monoclonal antibodies. *Exp. Cell Res.*, **142**, 229–33

Hart, G. W. (1986). Ubiquitination of cell-surface glycoproteins. *Trends in Biochem. Sci.*, **11**, 272

Hay, E. D. (1983). Cell and extracellular matrix: their organisation and mutual dependence. *Modern Cell Biol.*, **2**, 509–48

Hohl, H. R. & Jehli, J. (1973). The presence of cellulose microfibrils in the proteinaceous slime trace of *Dictyostelium discoideum*. *Arch. Microbiol.*, **92**, 179–87

Krefft, M., Voet, L., Gregg, J. H., Mairhofer, H. & Williams, K. L. (1984). Evidence that positional information is used to establish the prestalk-prespore pattern in *Dictyostelium discoideum* aggregates. *EMBO J.*, **3**, 201–6

Laemmli, U. K. (1970). Cleavage of structural proteins during the assembly of the head of bacteriophage T4. *Nature (Lond.)*, **227**, 680–5

Loomis, W. F., Wheeler, S. A., Springer, W. R. & Barondes, S. H. (1985). Adhesion mutants of *Dictyostelium discoideum* lacking the saccharide determinant recognized by two adhesion-blocking monoclonal antibodies. *Developmen. Biol.*, **109**, 111–17

MacWilliams, H. K. & David, C. N. (1984). Pattern formation in *Dictyostelium*. In: *Microbial Development*, Cold Spring Harbor Monograph 16, ed. R. Losick & L. Shapiro, pp. 255–74. Cold Spring Harbor, New York

Murray, B. A., Wheeler, S., Jongens, T. & Loomis, W. F. (1984). Mutations affecting a surface glycoprotein, gp80, of *Dictyostelium discoideum*. *Mol. & Cell. Biol.*, **4**, 514–19

Newell, P. C. (1982). Genetics. In *The development of* Dictyostelium *discoideum*, ed. W. F. Loomis, pp. 35–70. New York: Academic Press

Pierschbacher, M. D. & Ruoslahti, E. (1984). Cell attachment activity of fibronectin can be duplicated by small synthetic fragments of the molecule. *Nature (Lond.)*, **309**, 30–3

Rubin, J. & Robertson, A. (1975). The tip of *Dictyostelium discoideum* pseudoplasmodium as an organiser. *J. Embryol. & Exp. Morphol.*, **33**, 227–41

Siegelman, M., Bond, M. W., Gallatin, W. M., St. John, T., Smith, H. T., Fried, V. A. & Weissman, I. L. (1986). Cell surface molecule associated with lymphocyte homing is a ubiquinated branched-chain glycoprotein. *Science*, **231**, 823–9

Siu, C-H. (1986). This volume

Smith, E., Roberts, K., Butcher, G. W. & Galfre, G. (1984). Monoclonal antibody screening: Two methods using antigens immobilized on nitrocellulose. *Anal. Biochem.*, **138**, 119–24

Smith, E., Roberts, K., Hutchings, A. & Galfre, G. (1984). Monoclonal antibodies to the major structural glycoprotein of *Chlamydomonas* cell wall. *Planta*, **161**, 330–8

Smith, E. & Williams, K. L. (1979). Preparation of slime sheath from *Dictyostelium discoideum*. *FEMS Microbiol. Lett.*, **6**, 119–22

Van den Ende, H. (1986). This volume

Vardy, P. H., Fisher, L. R., Smith, E. & Williams, K. L. (1986). Traction proteins in the extracellular matrix of *Dictyostelium discoideum* slugs. *Nature (Lond.)*, **320**, 526–9

Voet, L., Krefft, M., Mairhofer, H. & Williams, K. L. (1984). An assay for pattern formation in *Dictyostelium discoideum* using monoclonal antibodies, flow cytometry, and subsequent data analysis. *Cytometry*, **5**, 26–33

Voet, L., Krefft, M., Bruderlein, M. & Williams, K. L. (1985). Flow cytometer study of anterior-like cells in *Dictyostelium discoideum*. *J. Cell Sci.*, **75**, 423–35

Welker, D. L. & Williams, K. L. (1982). A genetic map of *Dictyostelium discoideum* based on mitotic recombination. *Genetics*, **102**, 691–710
West, C. M. & Loomis, W. F. (1985). Absence of a carbohydrate modification does not affect the level or subcellular localization of three membrane glycoproteins in *modB* mutants of *Dictyostelium discoideum*. *J. Biol. Chem.*, **260**, 13803–9
Wilcox, M. & Leptin, M. (1985). Tissue-specific modulation of a set of related cell surface antigens in *Drosophila*. *Nature (Lond.)*, **316**, 351–4
Williams, K. L., Grant, W. N., Krefft, M., Voet, L. & Welker, D. L. (1984). Analysis of cell surface and extracellular matrix antigens in *D. discoideum* pattern formation. In *Molecular Biology of Development*, pp. 75–90. Alan R. Liss, Inc.
Williams, K. L. & Newell, P. C. (1976). A genetic study of aggregation in the cellular slime mould *Dictyostelium discoideum* using complementation analysis. *Genetics*, **82**, 287–307
Williams, K. L., Vardy, P. H. & Segel, L. A. (1986). Cell migrations during morphogenesis: some clues from the slug of *Dictyostelium discoideum*. *BioEssays*, **5**, 148–52

9

Mechanisms of cell–cell recognition and cell cohesion in *Dictyostelium discoideum* cells

Chi-Hung Siu, Lu Min Wong, Anthony Choi and Aesim Cho

Banting and Best Department of Medical Research,
Charles H. Best Institute, University of Toronto,
Toronto, Ontario, Canada M5G 1L6

Abstract

During development of the cellular slime mould *D. discoideum*, an increase in the expression of EDTA-sensitive and EDTA-resistant cell–cell binding sites can be observed at different stages. The molecular nature of the EDTA-resistant binding sites acquired at the aggregation stage was elucidated by monoclonal antibodies and a surface glycoprotein of M_r 80,000 (gp80) was identified. The primary structure of gp80 was deduced from its cDNA sequence. Regulation of gp80 synthesis was studied using these DNA probes. gp80-conjugated covaspheres were used to demonstrate that gp80 mediates cell–cell binding directly and undergoes homophilic interaction. Immuno-EM studies showed that gp80 is preferentially associated with contact areas and filopodia, which may play a vital role in the initial recognition process. Other phenomena, such as cell compaction and cell sorting, will also be discussed.

Introduction

Specific cell–cell recognition and cell cohesion are known to underlie many developmental processes. In recent years, much effort has been directed towards understanding the complexities of the cell surface, and plasma membrane components implicated in cell–cell binding have been purified from different sources. The cellular slime mould *Dictyostelium discoideum* is particularly suitable for such studies because it has a simple and well-defined life cycle (for reviews, see Bonner, 1967; Loomis, 1975, 1982), which is characterised by various types of cell–cell interactions. When nutrients become depleted, the unicellular amoebas begin to differentiate and transform from the non-social vegetative growth phase to the social developmental phase of their life cycle. About 4 to 5 hours after the initiation of development, cells begin to develop a sensitivity to pulses of cAMP, giving rise to chemotactic migration toward areas of higher cAMP concentrations. They also acquire an increase in mutual cohesiveness and form stable intercellular contacts, leading to the formation of multi-cellular entities called pseudoplasmodia or slugs. The slug phase eventually culminates in the formation of a fruiting body, consisting of spores and stalk cells.

Intercellular cohesiveness can be monitored by the roller tube assay developed by Gerisch (1961). Cells can be dissociated mechanically and then allowed to reassociate in a rotating medium. For cells which have developed cohesiveness, reassociation into large cell clumps takes only 30 min. Two types of contact sites have been identified. One type is sensitive to treatment with 1 to 2 mM EDTA, while the other is resistant in concentrations of EDTA up to 15 mM. When cells are assayed in 10 mM EDTA, a rapid increase in cell cohesiveness has been observed between 6 and 10 h of development, indicating that the acquisition of the EDTA-resistant contact sites is concomitant with the cell aggregation process (Rosen *et al.*, 1973).

Since this early observation, a number of cell surface molecules have been implicated in mediating the formation of these two types of contact site (Muller & Gerisch, 1978; Geltosky *et al.*, 1979; Steinemann & Parish, 1980; Saxe & Sussman, 1982; Brodie *et al.*, 1983; Chadwick & Garrod, 1983). It is hoped that the determination of the molecular nature of these contact sites will eventually lead to the unravelling of the different mechanisms by which cells undergo selective binding to other cells.

The EDTA-sensitive contact sites

The EDTA-sensitive cell–cell binding sites are expressed very early in the developmental cycle. Vegetative cells which have been grown by feeding on a bacterial lawn do not exhibit these binding sites. However, as soon as development is triggered by starvation, cells begin to acquire these sites on the surface and small aggregates can be observed at 1 to 2 h of development (Fig. 9.1). These aggregates are readily dissociated by the addition of 2 mM EDTA to the medium. The expression of these sites is not affected by cAMP. When the axenic strain AX2 is cultured in liquid medium, cells express these binding sites even in their vegetative growth stage. Expression of these binding sites are among the very early developmental events and they may be regulated in a similar manner to the early enzyme markers, such as α-mannosidase and N-acetyl-glucosaminidase, which are also expressed in axenically grown cells (Ashworth & Quance, 1972).

The exact physiological role of the EDTA-sensitive binding sites is not clear, but it is likely that they are responsible for the compacted morphology of aggregates observed at mid-development (Fig. 9.2(a)). When aggregation-stage cells are reassociated in the presence of 2 mM EDTA, they are still able to form large aggregates but they have lost the compacted morphology (Fig. 9.2(b)). In this respect, these binding sites are similar to the Ca^{2+}-dependent binding sites which mediate cell compaction in early mouse embryogenesis (Hyafil *et al.*, 1980; Damsky *et al.*, 1983; Shirayoshi *et al.*, 1983).

The molecular nature of the EDTA-sensitive binding sites is still not clear. A potential candidate is a surface glycoprotein of M_r 126,000 (Chadwick & Garrod, 1983). This glycoprotein is apparently capable of neutralising polyspecific Fab preparations that block the EDTA-sensitive binding sites. But further characterisation of this glycoprotein is needed to substantiate its role in cell cohesion.

The aggregation stage specific cell cohesion molecule

At the onset of the chemotactic migration period, cells begin to express EDTA-resistant binding sites. These sites were first defined serologically by Beug *et al.* (1973) and were called contact sites A. They prepared polyclonal antibodies against aggregation stage cells. After absorption with vegetative cells, Fab fragments of these antibodies were found to inhibit the EDTA-resistant contact sites. Subsequently, a membrane-associated glycoprotein of M_r 80,000 (gp80) was identified as the major component recognised by their antiserum and a fraction enriched in gp80 was able to neutralise the inhibitory effect of their Fab preparation (Muller & Gerisch, 1978). However, gp80 carries a highly immunogenic carbohydrate moiety, and a number of polyclonal and monoclonal antibodies raised against this protein

Fig. 9.1. Expression of EDTA-sensitive and EDTA-resistant cohesiveness during development. NC4 cells were collected at mid-exponential growth phase and resuspended in 17 mM phosphate buffer, pH 6.4, for development. The culture was shaken at 180 rpm at room temperature. Cell cohesion assay was carried out in either the presence (—) or absence (– – –) of 10 mM EDTA. Expression of EDTA-sensitive cohesiveness was also measured in AX2 cells, previously grown in an axenic liquid medium (----).

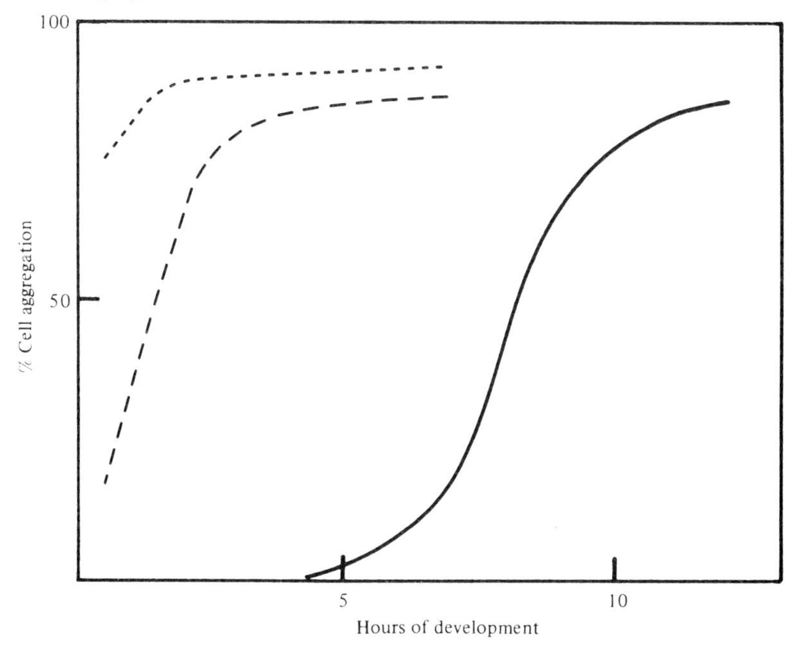

Fig. 9.2. Morphology of cell clumps derived from cells at 10 h of development. (a) Cells reassociated in the presence of 17 mM phosphate buffer, pH 6.4, and (b) cells reassociated in phosphate buffer plus 2 mM EDTA.

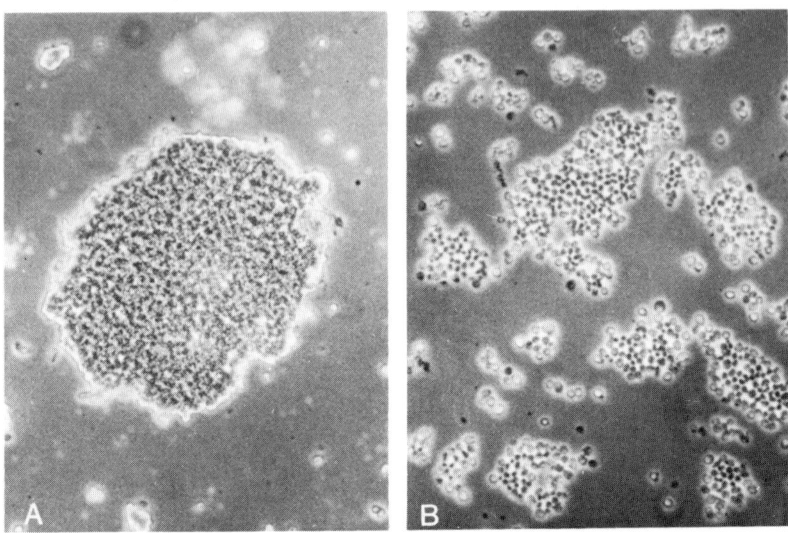

Fig. 9.3. Dosage effect of monoclonal antibody 80L5C4 on cell cohesion. The cell cohesian assay was carried out using 9 h cells in the presence of different concentrations of 80L5C4. Cells (2×10^7) cells/ml) were precoated with 80L5C4 and then diluted 1/10 in a buffer containing 0.2 mg/ml of goat Fab against mouse-IgG for reassociation, in the presence (●) or absence (■) of 10 mM EDTA. Alternatively, cells (2×10^6 cells/ml) were reassociated in the presence of 10 mM EDTA and different concentrations of Fab fragments derived from 80L5C4 (○).

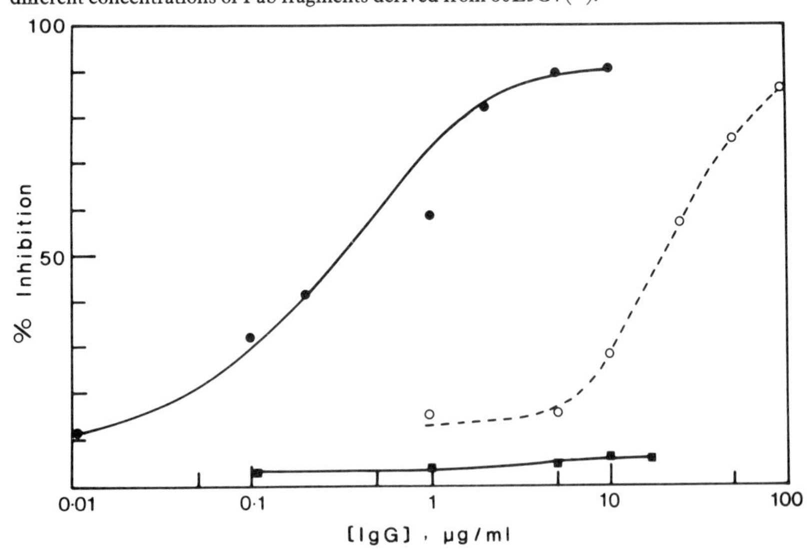

react with its carbohydrate and thus also show reactivity with other glycoproteins sharing similar carbohydrate structures (Ochiai *et al.*, 1982; Murray *et al.*, 1981; Springe & Barondes, 1983; Siu *et al.*, 1985). Although many of these antibodies block cell cohesion to some extent, for some time it has remained an enigma as to whether gp80 was directly involved in the formation of contact sites A or merely carried a common antigenic determinant shared by the *bona fide* contact site A glycoprotein.

Inhibition of cell cohesion by a monospecific monoclonal antibody

Recently, we succeeded in preparing a monoclonal antibody (80L5C4) which is monospecific for gp80 and is capable of blocking the EDTA-resistant binding sites (Siu *et al.*, 1985). Aggregation stage cells were precoated with the anti-gp80 monoclonal antibody 80L5C4 and Fab fragments of a goat anti-mouse-IgG antibody according to the method of Springer & Barondes (1983). These cells were dissociated mechanically and allowed to reassociate in the presence of 10 mM EDTA to assess the effect of the monoclonal antibody on the EDTA-resistant binding sites. Alternatively, cells were reassociated in the presence of 10 mM EDTA and Fab fragments derived from 80L5C4. In both cases, cell reassociation was effectively inhibited (Fig. 9.3). However, this monoclonal antibody has no effect on the EDTA-sensitive binding sites (Fig. 9.3). The inhibitory activity of 80L5C4 can be neutralised stoichiometrically by purified gp80 (Siu *et al.*, 1985). Since trypsin or pronase treatment of gp80 completely destroys its ability to neutralise 80L5C4, the 80L5C4 epitope is probably protein in nature (Siu *et al.*, 1985).

An interesting observation is that the inhibitory effect of monoclonal antibody 80L5C4 is restricted to the aggregation stage of development (Siu *et al.*, 1985). Cells in the post-aggregation stages become resistant to its cell dissociation effect. This suggests that gp80 has a unique, though transient, role in cell cohesion at the aggregation stage and other cohesion molecules must play a more active role than gp80 at the later stages.

Binding of gp80-conjugated covaspheres to cells

Although our antibody studies provide strong evidence that gp80 is involved in the formation of the EDTA-resistant binding sites, it is possible that the binding of antibody to gp80 may pose steric hinderance on an adjacent contact site A component, thus resulting in cell dissociation. To rule out this possibility, we conjugated purified gp80 to covaspheres and tested whether they would bind to aggregation stage cells. The binding of covaspheres to cells was analysed by laser-activated flow cytometry and the results are summarised in Table 9.1. The binding of gp80-covaspheres is about three times higher than the control. It is, therefore, evident that gp80 alone can mediate cell binding and that the binding is stage specific. The absence of gp80-covasphere binding to 0 h cells suggests that the receptor for gp80 is either absent or masked on vegetative cells.

Homophilic interaction of gp80

To determine whether gp80 interacts homophilically or heterophilically with surface components, 12 h cells were precoated with anti-gp80 monoclonal antibody and Fab fragments of an anti-mouse-IgG second antibody and then

Table 9.1. *Binding of gp80-conjugated covaspheres to NC4 cells*

Type of covaspheres	% Cell particles with covaspheres[a]	
	Single cells	Aggregates
gp80-covaspheres	27.2 ± 2.3%	70.0 ± 9.2%
BSA-covaspheres	10.7 ± 2.5%	24.0 ± 1.7%
80L5C4 IgG +gp80-covaspheres	13.0 ± 2.0%	nd[b]

Samples were analysed in an EPCIS V cell sorter. 10,000 particles were counted for each sample.

[a]Particles were separated into two categories, single cells and aggregates. Values represent the percentage of fluorescent particles to the total number of particles under each category. Results are expressed as mean ± SD of three independent determinations.

[b]nd = not determined. Under these conditions, about 97% of cells remained as single cells.

Fig. 9.4. Competition of [^{125}I]gp80 binding in the filter assay. 0.1 μg of gp80 was immobilised on nitrocellulose discs. 0.2 μg/ml of [^{125}I]gp80 was mixed with various concentrations of cold gp80 and then incubated with the nitrocellulose discs for 30 min at room temperature. Samples were washed extensively in buffer containing 0.05% octyl glucoside, before counting in a gamma counter.

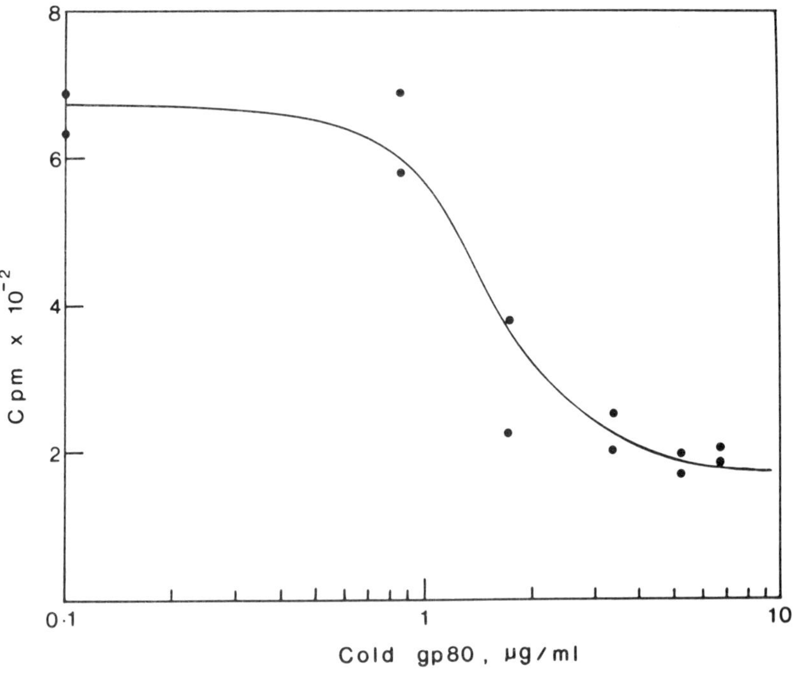

allowed to interact with gp80-covaspheres. Binding of gp80-covaspheres was reduced to the background level (Table 9.1), suggesting that gp80 binds homophilically to gp80 on the cell surface.

As a direct test of the ability of gp80 to bind homophilically to gp80, a filter binding assay was set up to quantitate the binding of [^{125}I]gp80 to gp80 bound on nitrocellulose. The binding of [^{125}I]gp80 to filters is dosage dependent and it can be competitively replaced by cold gp80, indicating that the binding is specific (Fig. 9.4). These results thus confirm the notion that gp80 is capable of undergoing homophilic interaction.

Primary structure of gp80

To investigate the detailed mechanism by which gp80 undergoes homophilic interaction, it is important to determine the primary and secondary structure of gp80. We have employed the recombinant DNA approach. cDNA of gp80 have been cloned into the expression vector λgt11 (Wong & Siu, 1986; Wong *et al.* submitted for publication). Figure 9.6 shows the primary structure of gp80 deduced from its cDNA sequence. The protein contains 514 amino acid residues, with a hydrophobic leader sequence of 19 amino acids. The leader sequence is cleaved off in the mature glycoprotein, leaving behind 495 amino acids.

The protein can be divided into three major domains. The carboxy-terminal domain consists of 18 hydrophobic amino acids. The hydropathy average of this domain is well above 1.5, which is characteristic of membrane spanning peptides. No other hydrophobic stretches in gp80 are long enough to form another membrane spanning region. Since gp80 is an integral protein (Stadler *et al.*, 1982), it must be anchored in the plasma membrane by the carboxy-terminal domain (see model in Fig. 9.7). Adjacent to the anchor domain is a hydrophilic region, located between amino acid positions 431 and 496 (Fig. 9.5). Within this region is a unique Pro-rich region (between positions 454 and 496), which contains two 8-amino acid repeats. Each repeat has 3 Pro residues alternating with either Thr or Ser. There are a total of 9 Pro-Thr/Ser dipeptides in this region. Many of the amino acids in this domain have a high probability for turns (Chou & Fasman, 1979), and this would probably produce a somewhat 'kinky' configuration. Therefore, this Pro-rich domain may provide a flexible extended stalk structure for the rest of the molecule (Fig. 9.6). The third domain consists of the amino-terminal 434 amino acids. The cell binding site of gp80 probably lies within this domain. The average hydropathic values are relatively low and this is characteristic of soluble proteins. Secondary structure predictions (Levitt, 1978) suggest that this domain contains many β-sheets and probably has a highly folded globular structure. Exactly how this domain is involved in homophilic interaction leading to cell–cell binding is not clear. The cDNA clones should allow us to make use of the molecular genetics approach to dissect this problem.

Biased distribution of gp80 on the cell surface

As a first step in understanding how cell–cell contacts are formed at the cellular level, the surface distribution of gp80 was localised using the monospecific monoclonal antibody 80L5C4 (Choi & Siu, 1984). Immunofluorescence microscopy showed that gp80 is preferentially localised at the two polar ends of the migrating cells (Fig. 9.7). These cells are often elongated and the surface has a large number of filopodial structures (De Chastellier & Ryter, 1980). It is of interest to note that bright fluorescence is often associated with cell protrusions, suggesting the occurrence of gp80 clusters on these membrane structures.

Cells were also labelled with protein A-conjugated colloidal gold for EM studies. In addition to those on the cell contact areas, gold particles were also found in approximately two-fold higher concentrations on the filopodial surfaces (Fig. 9.8). Filopodial structures are often found in abundance at regions where two cells are in close apposition. In many cell–cell contact zones, interdigitation of filopodia can be observed (Fig. 9.9) and the region is often characterised by clusters of gold particles. An examination of a large number of transmission and scanning electron micrographs leads to the conclusion that filopodia are involved in making the initial contacts between cells. It is conceivable that cells go through at least four stages in the formation of stable contacts as depicted in Fig. 9.10. In the first

Fig. 9.5. The primary structure of gp80 deduced from its cDNA sequence. The arrow under Ala at position 20 marks the beginning of the mature protein. Five potential N-glycosylation sites are boxed. The two identical repeats are underlined.

```
MET Lys Phe Leu Leu Val Leu Ile Ile Leu Tyr Asn Ile Leu Asn Ser Ala His Ser Ala   20
                                                                            →
Pro Thr Ile Thr Ala Val Ser Asn Gly Lys Phe Gly Val Pro Thr Tyr Ile Thr Ile Thr   40
Gly Thr Gly Phe Thr Gly Thr Pro Val Val Thr Ile Gly Gly Gln Thr Cys Asp Pro Val   60
Ile Val Ala Asn Thr Ala Ser Leu Gln Cys Gln Phe Ser Ala Gln Leu Ala Pro Gly Asn   80
Ser Asn Phe Asp Val Ile Val Lys Val Gly Gly Val Pro Ser Thr Gly Gly Asn Gly Leu  100
Phe Lys Tyr Thr Pro Pro Thr Leu Ser Thr Ile Phe Pro Asn Asn Gly Arg Ile Gly MET  120
Ile Leu Val Asp Gly Pro Ser Asn Ile Ser Gly Tyr Lys Leu Asn Val Asn Asp Ser Ile  140
Asn Ser Ala MET Leu Ser Val Thr Ala Asp Ser Val Ser Pro Thr Ile Tyr Phe Leu Val  160
Pro Asn Thr Ile Ala Gly Gly Leu Leu Asn Leu Glu Leu Ile Gln Pro Phe Gly Phe Ser  180
Thr Ile Val Thr Ser Lys Ser Val Phe Ser Pro Thr Ile Thr Ser Ile Thr Pro Leu Ala  200
Phe Asp Leu Thr Pro Thr Asn Val Thr Val Thr Gly Lys Tyr Phe Val Thr Thr Ala Ser  220
Val Thr MET Gly Ser His Ile Tyr Thr Gly Leu Thr Val Gln Asp Asp Gly Thr Asn Cys  240
His Val Ile Phe Thr Thr Arg Ser Val Tyr Glu Ser Ser Asn Thr Ile Thr Ala Lys Ala  260
Ser Thr Gly Val Asp MET Ile Tyr Leu Asp Asn Gln Gly Asn Gln Gln Pro Ile Thr Phe  280
Thr Tyr Asn Pro Pro Thr Ile Thr Ser Thr Lys Gln Val Asn Asp Ser Val Glu Ile Ser  300
Thr Thr Asn Thr Gly Thr Asp Phe Thr Gln Ile Ser Leu Thr MET Gly Thr Ser Ser Pro  320
Thr Asn Leu Val Ile Thr Gly Thr Asn Glu Lys Ile Val Ile Thr Leu Pro His Ala Leu  340
Pro Glu Gly Glu Ile Gln Phe Asn Leu Lys Ala Gly Ile Ser Asn Val Val Thr Ser Thr  360
Leu Leu Val Thr Pro Val Ile Asn Ser Val Thr Gln Ala Pro His Asn Gly Gly Ser Ile  380
Thr Ile Ser Gly Ile Phe Leu Asn Asn Ala His Val Ser Ile Val Val Asp Gln Asn Thr  400
Thr Asp Ile Val Cys Ala Pro Asp Ser Asn Gly Glu Ser Ile Ile Cys Pro Val Glu Ala  420
Gly Ser Gly Thr Ile Asn Leu Val Val Thr Asn Tyr Lys Asn Phe Ala Ser Asp Pro Thr  440
Ile Lys Thr Glu Ala Thr Thr Ser Thr Thr Tyr Thr Ile Pro Asp Thr Pro Thr Pro Thr  460
Asp Thr Ala Thr Pro Ser Pro Thr Pro Thr Glu Thr Ala Thr Pro Ser Pro Thr Pro Lys  480
Pro Thr Ser Thr Pro Glu Glu Thr Glu Ala Pro Ser Ser Ala Thr Thr Leu Ile Ser Pro  500
Leu Ser Leu Ile Val Ile Phe Ile Ser Phe Val Leu Leu Ile                          514
```

Fig. 9.6. Model of mature gp80, showing the three major domains of the protein.

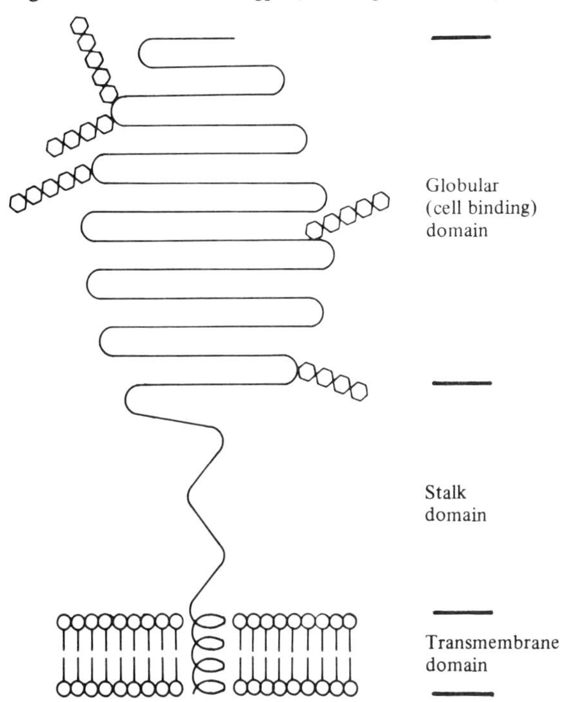

Fig. 9.7. Phase contrast and epifluorescence micrographs of NC4 cells stained indirectly with monoclonal antibody 80L5C4. 12 h cells were fixed with cold methanol (−20 °C) for 10 min, labelled with monoclonal antibody, followed by TRITC-conjugated goat anti-mouse-IgG antibody.

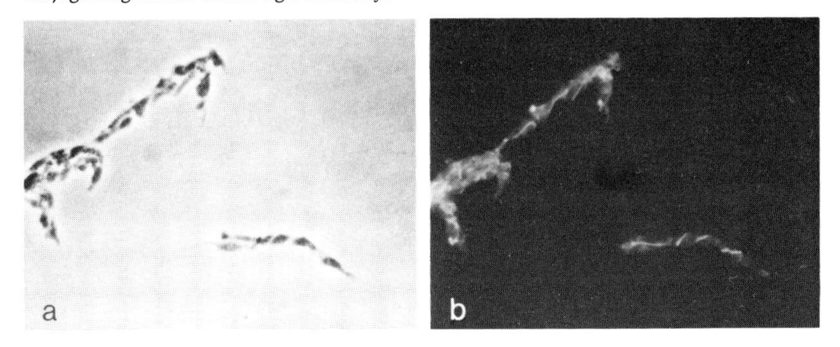

stage, aggregation-competent cells send out filopodia which establish initial contacts with adjacent cells. In the second stage, filopodia from both cells appear to proliferate and interdigitate with each other at the contact zone. Such intimate interactions among the filopodia may help to draw the two cells into closer contact. In the third stage, filopodia at the contact region are withdrawn by both cells, leading to the final stage whereupon stable contacts are formed between smooth membrane surfaces. Since gp80 is pre-

Fig. 9.8. Localisation of gp80 in contact zones and on filopodia of aggregating cells. Aggregation stage cells were labelled with 80L5C4 IgG and protein A-gold and then processed for TEM. (a) Clustering of gold particles at cell–cell contact zone; bar, 0.2 μm. (b) Clustering of gold particles on the surface of filopodia and pseudopod of two neighbouring cells; bar, 0.25 μm.

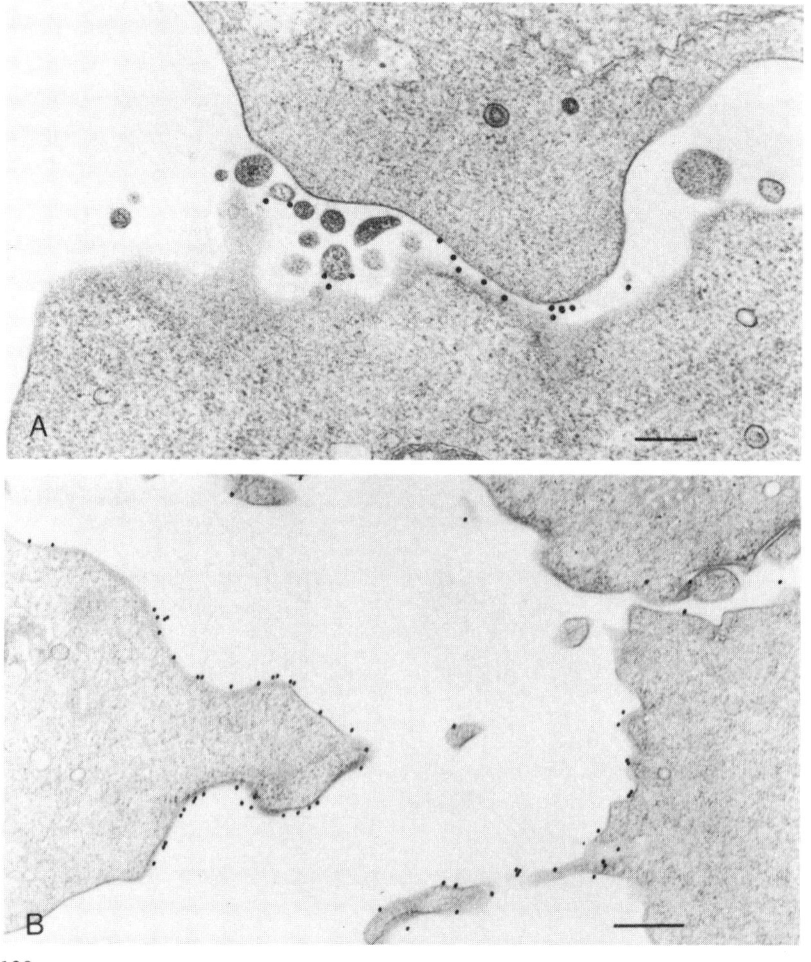

ferentially associated with filopodial structures, it is likely that gp80 plays a vital role in one or more of these stages.

Other cell cohesion molecules

As indicated previously, the action of gp80 in cell cohesion is restricted to the aggregation stage of *D. discoideum* development. Other molecules must take the place of gp80 and continue to maintain stable cell–cell contacts in the post-aggregation stages. We have studied another cell cohesion molecule with M_r = 150,000 (gp150). This molecule is maximally expressed

Fig. 9.9. Scanning electron micrographs of streaming cells. Cells were labelled sequentially with 80L5C4 IgG, rabbit anti-mouse-IgG antibody and protein A-gold. (a) Low magnification SEM micrograph showing interdigitation of filopodia between the polar ends of two cells. The arrow heads provide reference points for the higher magnification micrograph in (b). Bar, 5 μm. (b) Micrograph taken with the backscattering electron imaging mode in reverse polarity, showing the cell–cell contact zone. gp80 was visualised with protein A-gold which appeared as black particles. Bar, 1 μm.

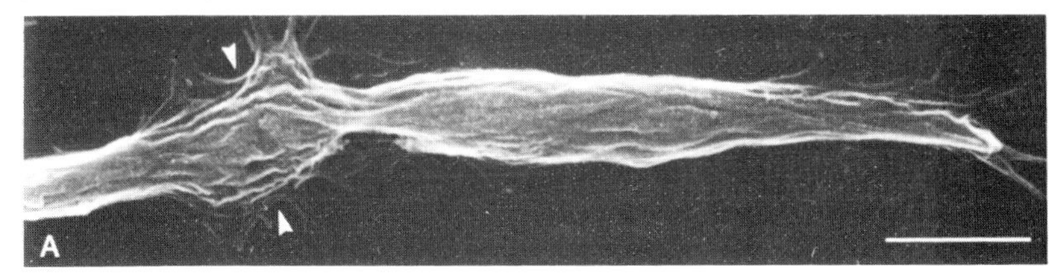

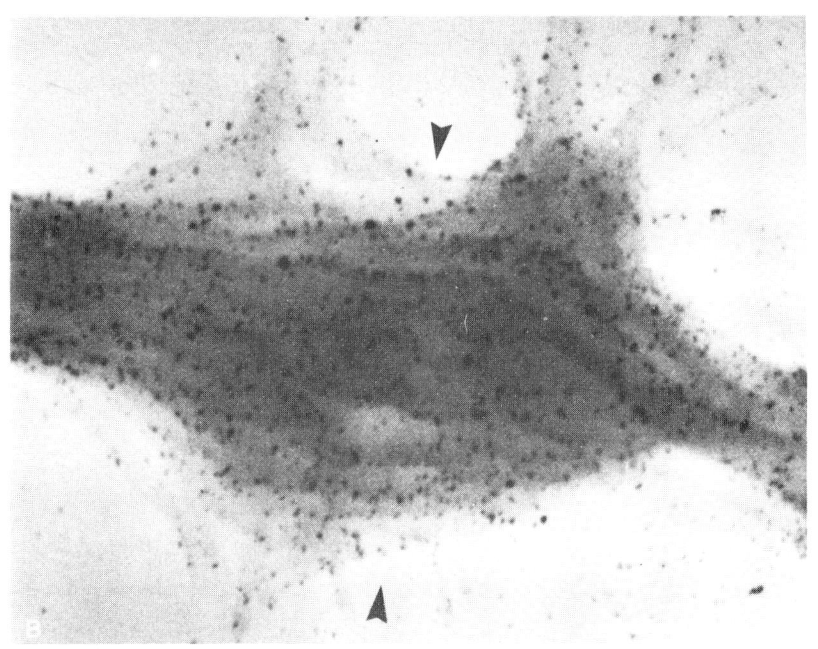

during the post-aggregation stage of development. It was first discovered as a major Con A-binding glycoprotein on the cell surface (Geltosky *et al.*, 1976). Subsequently, Geltosky *et al.* (1979) showed that unilavent antibodies directed against gp150 can block the EDTA-resistant cell binding sites, and

Fig. 9.10. Model of end-to-end contact formation between cells at the aggregation-stage of development.

Stage 1: Initial recognition and contact via filopodia

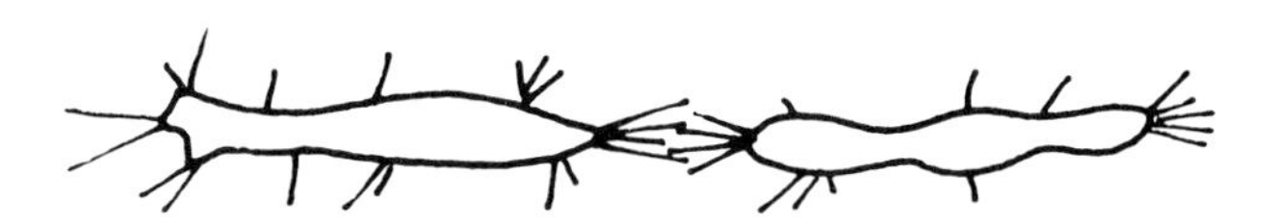

Stage 2: Interdigitation of filopodia

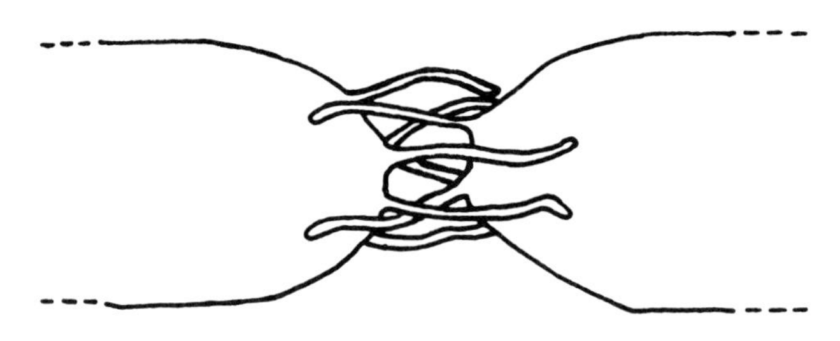

Stage 3: Retraction of filopodia

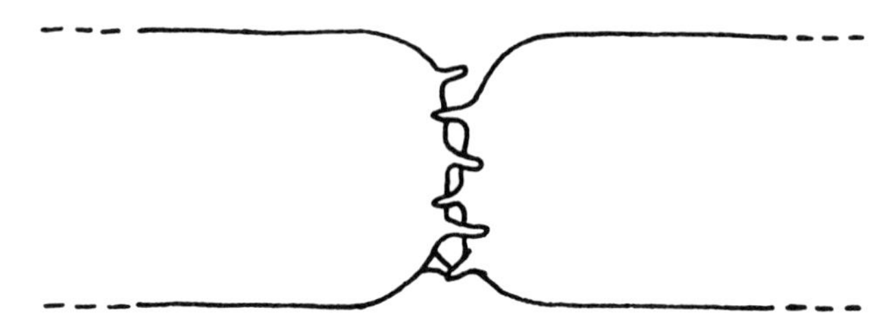

Stage 4: Formation of stable contact on smooth surfaces

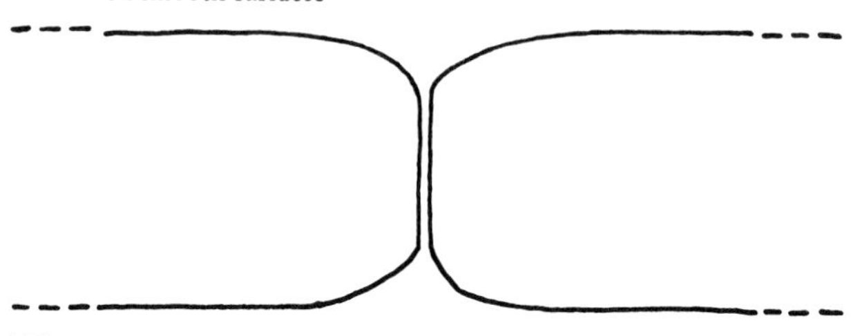

these molecules are clustered primarily at contact junctions between cells (Geltosky *et al.*, 1980). When the inhibitory effect of anti-gp150 Fab was tested against cells derived from different developmental stages, it became evident that gp150 mediates cell cohesion primarily at the post-aggregation stages (Lam *et al.*, 1981). The anti-gp150 Fab is a poor inhibitor of cell cohesion at the aggregation stage. In addition, its inhibitory effect shows cell type specificity, with prespore cells being more sensitive than prestalk cells (Lam *et al.*, 1981).

In the post-aggregation stages, cells differentiate into two major cell types, prespore cells and prestalk cells, which are organised in a unique spatial pattern in the slug, with prestalk cells in the anterior and prespore cells in the posterior. When these cells are mixed randomly, they are capable of sorting out from each other and reform the original prestalk–prespore polarity. This can result from differences in the cohesiveness of these two cell types (Steinberg, 1975; Lam *et al.*, 1981). Since prespore cells are more sensitive to anti-gp150 Fab dissociation, it is likely that gp150 is involved in the sorting out process. Indeed, when prespore cells and prestalk cells are mixed in the presence of anti-gp150 Fab, cells are able to form clumps via their EDTA-sensitive binding sites, but they remain randomly mixed and fail to sort out from each other (Siu *et al.*, 1983).

Several other cell cohesion molecules have been reported by different laboratories. A surface glycoprotein of M_r 95,000 has been implicated in mediating cell–cell binding immediately after the aggregation stage (Steinemann & Parish, 1980; Saxe & Sussman, 1982). More recently, Brodie *et al.* (1983) showed that two cell surface proteins M_r 69,000 and 73,000 are probably involved in cell cohesion during the early stages of development. These two proteins share a common epitope and may have arisen from a common precursor.

With the discovery of these cohesion molecules, and certainly more still await discovery, it becomes evident that several different and possibly independent cell cohesion systems are present on the surface of these cells. Our future research will deal with not only the mechanism of cell cohesion mediated by each of these systems, but also the interplay of these systems during morphogenesis.

Acknowledgements

The work was supported by an operating grant from the Medical Research Council of Canada.

References

Ashworth, J. M. & Quance, J. (1972). Enzyme synthesis in myxamoebae of the cellular slime mould *Dictyostelium discoideum* during growth in axenic culture. *Biochem. J.*, **126**, 601–8

Beug, H., Katz, F. & Gerisch, G. (1973). Dynamics of antigenic membrane sites relating to cell aggregation in *Dictyostelium discoideum*. *J. Cell Biol.*, **56**, 647–58

Bonner, J. T. (1959). Evidence for the sorting out of cells in the development of the cellular slime molds. *Proc. Natl. Acad. Sci. USA*, **56**, 379–83

Bonner, J. T. (1967). *The Cellular Slime Molds*. Princeton: Princeton University Press

Brodie, C., Klein, C. & Swierkosz, J. (1983). Monoclonal antibodies: Use to detect developmentally regulated antigens on *D. discoideum* amoebae. *Cell*, **32**, 1115–23

Chadwick, C. M. & Garrod, D. R. (1983). Identification of the cohesion molecule, contact site B, of *Dictyostelium discoideum*. *J. Cell Sci.*, **60**, 251–66

Choi, A. & Siu, C.-H. (1984). Preferential association of the contact site A glycoprotein with filopodia and contact areas of *Dictyostelium discoideum*. *J. Cell Biol.*, **99**, 70a

Chou, P. Y. & Fasman, G. D. (1979). Predictions of Beta turns. *Biophys. J.*, **26**, 367–84

Damsky, C. H., Richa, J., Sloter, D., Knudsen, K. & Buck, C. A. (1983). Identification and purification of a cell surface glycoprotein mediating intercellular adhesion in embryonic and adult tissue. *Cell*, **34**, 455–66

De Chastellier, C. & Ryter, A. (1980). Characteristic ultrastructural transformations upon starvation of *Dictyostelium discoideum* and their relationship with aggregation: Study of wild-type and aggregation mutants. *Biol. Cellulaire*, **38**, 121–8

Geltosky, J. E., Siu, C.-H. & Lerner, R. A. (1976). Glycoproteins of the plasma membrane of *Dictyostelium discoideum* during development. *Cell*, **8**, 391–6

Geltosky, J. E., Weseman, J., Bakke, A. & Lerner, R. A. (1979). Identification of a cell surface glycoprotein involved in cell aggregation in *Dictyostelium discoideum*. *Cell*, **18**, 391–8

Geltosky, J. E., Birdwell, C. R., Wiseman, J. & Lerner, R. A. (1980). A glycoprotein involved in aggregation of *D. discoideum* is distributed on the cell surface in a nonrandom fashion favouring cell junctions. *Cell*, **21**, 239–345

Gerisch, G. (1961). Zellfunktionen und Zellfunktionswechsel in der Entwicklung von Dictyostelium discoideum. V. Standienspezifische Zelkontaktbidung und ihre quantitative Erfassung. *Exp. Cell Res.*, **25**, 535–54

Hyafil, F., Morello, D., Babinet, C. & Jacobs, F. (1980). A cell surface glycoprotein involved in the compaction of embryonal carcinoma cells and cleavage stage embryos. *Cell*, **21**, 927–34

Lam, T. Y., Pickering, G., Geltosky, J. & Siu, C.-H. (1981). Differential cell cohesiveness expressed by prespore and prestalk cells of *Dictyostelium discoideum*. *Differentiation*, **20**, 22–8

Levitt, M. (1978). Conformational preferences of amino acids in globular proteins. *Biochemistry*, **17**, 4277–84

Loomis, W. F. (1975) *Dictyostelium discoideum: A Developmental System*. New York: Academic Press

Loomis, W. F. ed. (1982) *The Development of Dictyostelium discoideum*. New York: Academic Press

Muller, K. & Gerisch, G. (1978). A specific glycoprotein as the target site of adhesion blocking Fab in aggregating *Dictyostelium discoideum*. *Nature, Lond.*, **274**, 445–9

Murray, B. A., Yee, L. D. & Loomis, W. F. (1981). Immunological analysis of a glycoprotein (contact site A) involved in intercellular adhesion of *Dictyostelium discoideum. J. Supramolecular Structure & Cell Biochem.*, **17**, 197–211

Ochiai, H., Schwarz, H., Merkl, R. Wagle, G. & Gerisch, G. (1982). Stage-specific antigens reacting with monoclonal antibodies against contact site A, a cell surface glycoprotein of *Dictyostelium discoideum. Cell Differentiation*, **11**, 1–13

Rosen, S. D., Kafka, J. A., Simpson, D. L. & Barondes, S. H. (1973). Developmentally regulated carbohydrate-binding protein in *Dictyostelium discoideum. Proc. Nat. Acad. Sci. USA*, **70**, 2554–7

Saxe, C. L. & Sussman, M. (1982). Induction of stage-specific cell cohesion in *D. discoideum* by a plasma-membrane-associated moiety reactive with wheat germ agglutinin. *Cell*, **29**, 755–9

Shirayoshi, Y., Okado, T. S. & Takeichi, M. (1983). The calcium-dependent cell–cell adhesion system regulates inner cell mass formation and cell surface polarization in early mouse development. *Cell*, **35**, 631–8

Siu, C.-H., Des Roches, B. & Lam, T. Y. (1983). Involvement of a cell-surface glycoprotein in the cell-sorting process of *Dictyostelium discoideum. Proc. Nat. Acad. Amer. USA*, **80**, 6596–600

Siu, C.-H., Lam, T. Y. & Choi, A. H. C. (1985). Inhibition of cell–cell binding at the aggregation stage of *Dictyostelium discoideum* development by monoclonal antibodies directed against an 80,000-Dalton surface glycoprotein. *J. Biol. Chem.*, **260**, 16030–6

Springer, W. R. & Barondes, S. H. (1980). Cell adhesion molecules detection with univalent antibody. *J. Cell Biol.*, **87**, 703–7

Springer, W. R. & Barondes, S. H. (1983). Monoclonal antibodies block cell–cell adhesion in *Dictyostelium discoideum. J. Biol. Chem.*, **258**, 4698–701

Stadler, J., Bordier, C., Lottspeich, F., Henschen, A. & Gerisch, G. (1982). Improved purification and N-terminal amino acid sequence determination of the contact site A glycoprotein of *Dictyostelium discoideum.* Hoppe-Seyler's *Z. Physiol. Chem.*, **363**, 771–6

Steinberg, M. S. (1975). Adhesion-guided multicellular assembly: a commentary upon the postulates, real and imagined, of the differential adhesion hypothesis, with special attention to computer simulations of cell sorting. *J. Theoretical Biol.*, 55, 431–43

Steinemann, C. & Parish, R. W. (1980). Evidence that a developmentally regulated glycoprotein is target of adhesion-blocking Fab in reaggregating *Dictyostelium. Nature, Lond.*, **286**, 621–3

Takeuchi, I. (1969). Establishment of polar organization during slime mold development. In *Nucleic Acid Metabolism, Cell Differentiation and Cancer Growth*, eds. Cowdry, E. V. and Seno, S. pp. 297–304. Oxford: Pergamon Press.

Wong, L. M. & Siu, C.-H. (1986). Cloning of cDNA for the contact site A glycoprotein of *Dictyostelium discoideum. Proc. Natl. Acad. Amer., USA*, **83**, 4248–52

10 Multiple low affinity carbohydrates as the basis for cell recognition in the sponge *Microciona prolifera*

Gradimir N. Misevic and Max M. Burger

Biocenter of the University of Basel,
Klingelbergstrasse 70, CH-4056 Basel,
Switzerland and the Marine Biological Laboratory,
Woods Hole, Massachusetts 02543, USA

Abstract

Species-specific reaggregation of dissociated cells from the sponge *Microciona prolifera* is mediated by a proteoglycan-like aggregation factor (MAF) of M_r 2×10^7 operating via two functional domains, a cell binding and a self-association site. The use of monoclonal antibodies and the reconstitution of both functions by cross-linking isolated protein-free glycans of MAF indicated that: (i) at least two different carbohydrate structures are representing two functional domains, (ii) Ca^{2+}-dependent carbohydrate-carbohydrate interactions are responsible for the self-association of MAF and (iii) the molecular mechanism by which both functional domains are involved in the process of self-recognition is based on the novel principle of multiple low affinity interactions.

Recognition and cell interactions

The appearance of multicellular organisms during evolution was dependent on development of cell–cell recognition, adhesion, and communication phenomena. Such cellular processes ensured the integrity and enabled the synchronous functioning of tissue and organs as well as the capability of the individual to distinguish between self and non-self. In some cases it is not easy to separate recognition from adhesion processes since they could be carried out by the same molecules; or adhesion might be induced shortly after recognition has occurred. It is also unclear to what degree molecules responsible for communication between cells (like gap junctions) are contributing to adhesion. Cell–cell recognition as one of these three important processes has been shown to play a crucial role during fertilisation, morphogenesis and immune response. However, the basic principles about molecular mechanisms of cellular recognition have not yet been understood completely mainly because of their complexity and multistep nature.

Cell–cell recognition in marine sponges

The interest in marine sponges as objects for the study of cell–cell recognition is mainly due to two reasons: (i) marine sponges are descendants of primordial multicellular organisms in which cell–cell recognition became operational for the first time during evolution, and (ii) simple reaggregation assays for testing molecules which are presumably involved in the recognition process can be performed with cells dissociated without proteolytic damage of their surface components. Such *in vitro* studies have led to the discovery of a new basic concept about the mechanism of cell–cell recognition.

The assay system used to study cellular recognition in marine sponges is based on a species-specific reaggregation and/or sorting out of dissociated cells of different species see Fig. 10.1 (Wilson, 1907; Galtsoff, 1929; Humphreys, 1963). Depending on the pair of species used, cells can either reaggregate directly into separate clumps (species-specific reaggregation) or they first make mixed aggregates which eventually undergo a sorting out process (species-specific sorting). Both of these cases demonstrate that cell–cell recognition is operational in sponges, however, in the former case it is concomitant with the reaggregation and in the latter with the secondary sorting out process. It is also interesting that such clumps of reaggregated cells will, in a few days, undergo a reconstitution of the tissue organisation which results in the formation of an adult functional spongelet. This may be considered as a secondary sorting out of cell types and may be used as a simple model in the study of the contribution of cell recognition during morphogenesis.

The first step towards understanding the mechanism of cell recognition phenomenona in sponges was the identification of molecules involved in this process. Up to now at least three components have been found to be necessary for reaggregation of *Microciona prolifera* sponge cells: (i) Ca^{2+} ions, 10 mM (Wilson, 1907), (ii) a peripheral membrane proteoglycan-like molecule, called aggregation factor (Humphreys, 1963; Henkart, Humphreys & Humphreys, 1973; Cauldwell, Henkart & Humphreys, 1973), and (iii) a plasma membrane receptor for the aggregation factor (Weinbaum & Burger, 1973). Identification and purification of aggregation factors from different species has led to the finding that these large molecules ($M_r \sim 10^7$) mediate species-specific reaggregation (Humphreys, Humphreys & Sano, 1977; Müller & Zahn, 1973). Jumblatt, Schlup & Burger (1980) have further shown that the *Microciona prolifera* aggregation factor (MAF) has two functional domains, a Ca^{2+} independent cell binding site, enabling species-specific interaction with the cell surface receptor, and a Ca^{2+} dependent self-association domain. Only when both of the domains are operational, will the MAF mediate cell reaggregation.

The cell binding domain

In order to isolate the cell binding domain from the large MAF molecule which is composed of several sub-units and consists of 60% carbohydrate by weight, dissociation procedures (urea, EDTA) were combined with trypsin treatment (Misevic, Jumblatt & Burger, 1982). Two large sub-units and five fragments were obtained and separated by gel filtration (Fig. 10.2). Their cell binding activity was tested using an assay system based on species-specific

Fig. 10.1. Dissociation, species-specific reaggregation and sorting out of sponge cells. Cells of two different sponge species may be dissociated in Ca^{2+}, Mg^{2+} free sea water. Depending on the pair of species used, such cells when mixed together in the presence of 10 mM Ca^{2+} (sea water concentration) will undergo immediate species-specific reaggregation or sorting out from a mixed aggregate. Eventually, each aggregate will reconstitute an adult functional sponge. Only a few of several different cell types are schematically presented here.

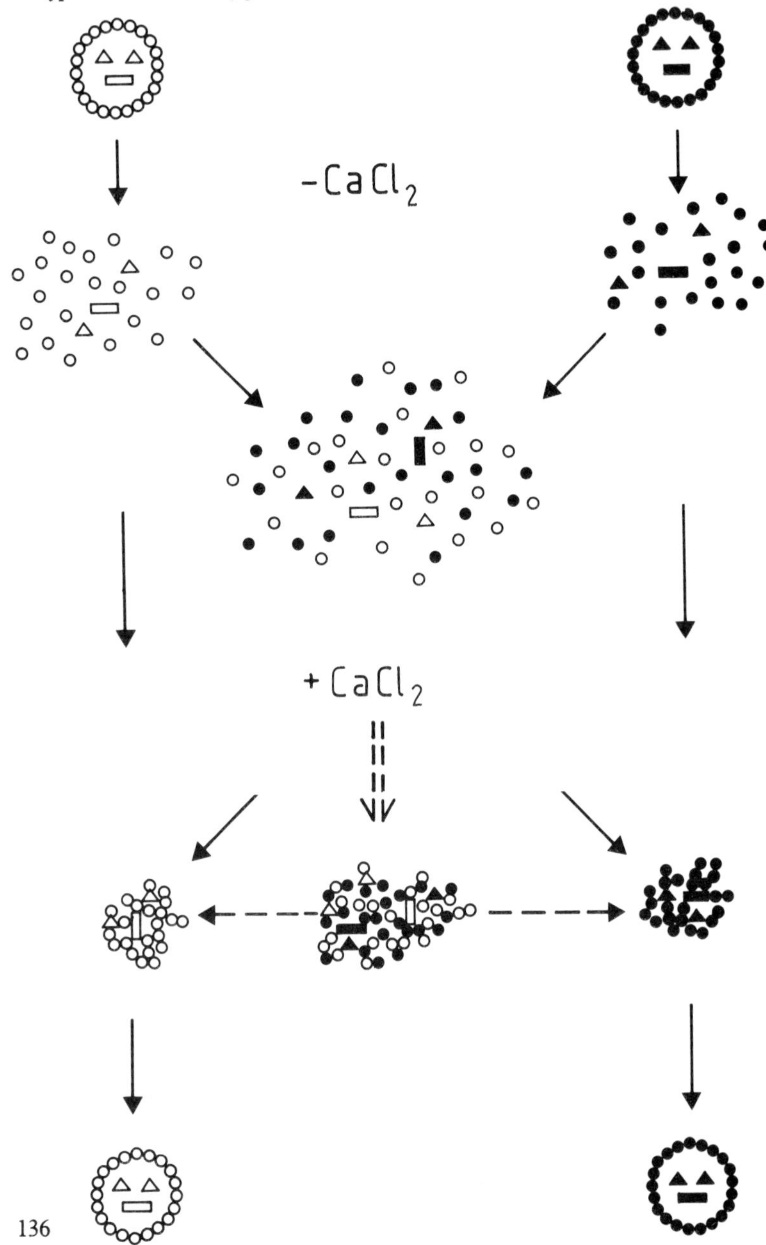

binding to homo- or heterotypic cells (Jumblatt *et al.*, 1980). As shown in Table 10.1 each of the sub-units and fragments displayed such species-specific binding indicating the presence of the intact cell binding site. The question arose whether these fragments have the same or a different binding affinity. The apparent association constants obtained from the binding assays of the labelled ligands decreased proportionally with their size, whereas the maximal number of binding sites per cell increased (Fig. 10.3). One of the explanations for this finding would be that the cell binding site is of repetitive character in MAF. The large fragments would therefore have more copies of the cell binding sites resulting in their higher binding affinity.

If the hypothesis of the highly polyvalent cell binding domain is correct it should be possible to prove this by testing the two following predictions experimentally: 1) the cell binding site carried by the smallest glycoprotein fragment of $M_r = 10{,}000$ (T–10) should be present in many copies in MAF and 2) reconstitution of the high binding affinity of MAF should be possible by cross-linking the small T-10 fragment into polymers of MAF size.

The first prediction was tested by measuring the amount of the total mass of MAF recovered in the T-10 fragment after extensive and repeated trypsin treatment. The prolonged digestion (12 h) resulted in conversion of 31% of the protein and sugar into T-10. The fractions larger than 10,000 D obtained upon gel filtration on a S-200 Sephacryl column were pooled and retrypsinised. Similarly, cleavage of this material yielded an additional quantity of T-10 fragments, corresponding to 32% of the original MAF sample. The third digestion, however, did not yield significant amounts of this smallest fragment, indicating that 63% of the MAF mass is represented by the T-10 glycoprotein. Since also the purity of T-10, tested by ion exchange chromatography, gel electrophoresis (Fig. 10.4) and analytical ultracentrifugation in the presence or absence of 6 M guanidine hydrochloride has shown very minor charge and size microheterogeneity, it may be concluded that this glycoprotein fragment is present in 1,300 copies per one MAF molecule, thus strongly supporting the hypothesis of the polyvalency of the cell binding site.

Second, an even more important criterion to satisfy was to test whether the low affinity T-10 fragment could be reconstituted into a high-binding affinity molecule by cross-linking it into polymers of MAF size. Since the T-10 glycoprotein contains 60% of carbohydrate by weight and a very low amount of basic amino acids, all attempts to cross-link it into large polymers with bifunctional reagents specific for amino or amino and carboxyl groups were unsuccessful. A procedure was therefore developed which is based on cross-linking the T-10 with a polyfunctional glutaraldehyde reagent. At pH7 glutaraldehyde undergoes an aldol condensation with itself, followed by dehydration to generate α, β-unsaturated aldehyde polymers (Richards & Knowles, 1968; Monsan, Puzo & Mazarguil, 1975). In an attempt to increase the efficiency of cross-linking even more, T-10 fragments were pretreated either with periodate resulting in generation of aldehyde groups in the carbohydrate portion of the molecule or were reacted with diepoxybutane before the addition of glutaraldehyde (Misevic & Burger, 1986). The resulting polymers of periodate treated T-10 cross-linked with glutaraldehyde (PG polymers) and of diepoxybutane glutaraldehyde cross-linked T-10 (EG polymers) were separated on a Bio-gel A 1.5 m and subsequently on an A 15 m column (Fig. 10.5).

The binding specificity of the polymers (EG1–EG5 and PG5) was tested with homotypic and heterotypic cells. As shown in Table 10.2 each of these polymers displayed species-specific binding indicating the presence of the active cell binding sites. The specificity was confirmed by competition of binding with the intact MAF for the same cell surface receptor(s), which also indicated that the activity of the cell binding site had not been altered. In order to measure whether the higher affinity of the cross-linked T-10 polymers had been reconstituted, two different approaches were made. In the first, rather indirect, technique, we found indeed that with increasing polymer size smaller numbers of polymer molecules were needed to cause 50% inhibition of ^{125}I-MAF binding to homotypic cells (Misevic & Burger, 1986). Secondly, direct measurements of the association constants of the iodinated T-10 polymers have also revealed that the increase in affinity was directly proportional with their molecular weight, whereas the maximal number of molecules bound per cell was as expected in the reverse proportion (Fig. 10.3). Since the largest polymers which were in the range of MAF

Fig. 10.2. Gel filtration on Sepharose 4B of urea, EDTA, treated [^{125}I] MAF A), and gel filtration on Sephacryl S-300 of trypsin treated [^{125}I] MAF B). A, 5 μg of [^{125}I] MAF dissociated with urea and EDTA was applied to a Sepharose 4B column (75 × 0.8 cm) and eluted with Ca^{2+}, Mg^{2+} free seawater. The 1 ml fractions were collected and counted in a Packard γ counter (Misevic *et al.*, 1982). B, 0.5 mg of [^{125}I] MAF was treated with trypsin for a period of 8 h. MAF fragments were separated by gel filtration on a Sephacryl S-300 column (90 × 1.6 cm). The 1 ml fractions were collected and counted.

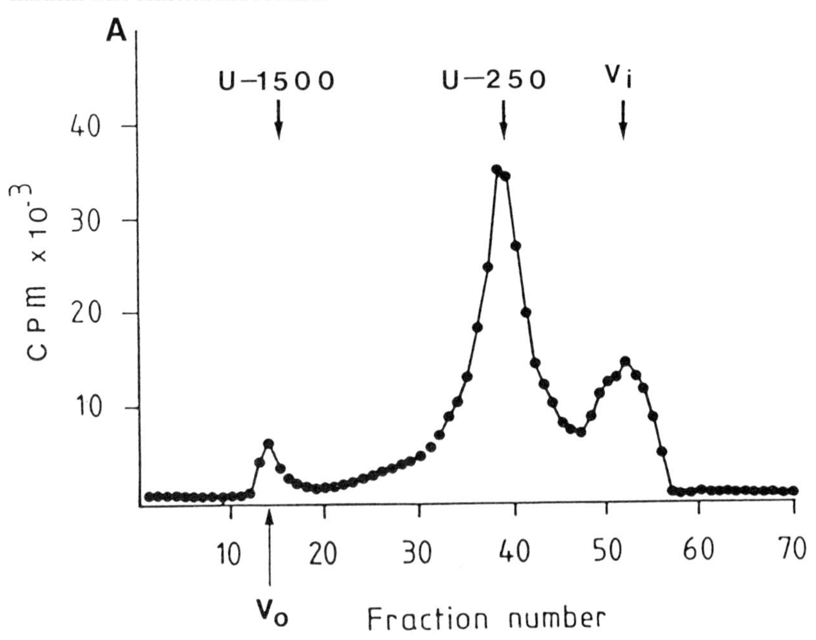

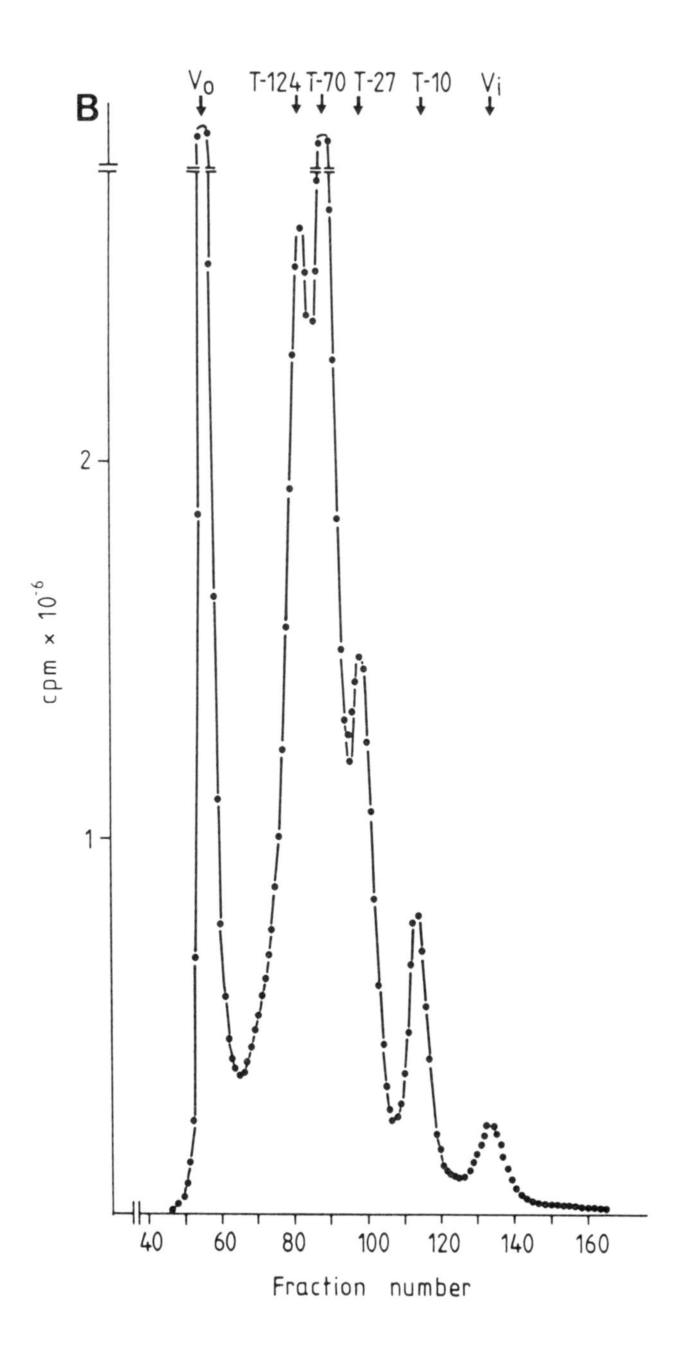
B
V_o
T-124
T-70
T-27
T-10
V_i
2
1
cpm × 10^{-6}
40
60
80
100
120
140
160
Fraction number

Table 10.1. *Binding specificity of ^{125}I-MAF and the MAF fragments.*

Binding of iodinated MAF and MAF fragments (U-urea, EDTA subunits, T-tryptic fragments) to *M. prolifera*, *C. celata* and *M. fusca* 10^7 cells/ml was carried under standard assay conditions for 20 min at room temperature (Jumblatt *et al.*, 1980; Misevic *et al.*, 1982). After incubation, cells were layered on and centrifuged through 0, 1% bovine serum albumine, 10% sucrose in Ca^{2+}, Mg^{2+} free seawater. Supernatants were aspirated and the pellet counted in a Packard γ spectrometer.

	% bound to		
MAF fragments	*M. prolifera*	*C. celata*	*M. fusca*
MAF	33	2.2	2.1
U-1500	59	2.7	2.3
U-250	54	0.6	1.6
T-124	31	8.1	7.9
T-70	29	7.3	6.8
T-27	31	7.8	6.7
T-10	45	12.0	11.1

size (EG5 and PG5 $M_r > 1.5 \times 10^7$) displayed 1 : 1 competition with MAF in binding to homotypic cells, and since their association constants (K_a) were found to be of the same order of magnitude as the native aggregation factor, cross-linking the low affinity T-10 glycoprotein fragment did indeed

Fig. 10.3. Comparison of binding data for MAF fragments and for reconstituted polymers of the T-10 fragment. Dependence of K_a, and B_{max} on molecular weights of MAF fragments and T-10 reconstituted polymers: ●——●, log K_a of MAF fragments; ○---○, log K_a, of cross-linked MAF T-10 fragments; ▲——▲, log B_{max} of MAF fragments; △---△, log B_{max} of cross-linked MAF T-10 fragments (Misevic & Burger, 1986).

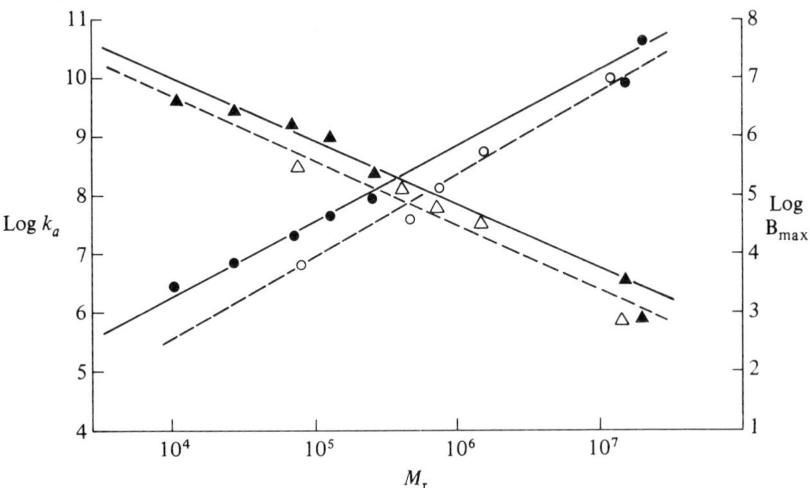

Fig. 10.4. SDS-Electrophoresis of MAF and of the T-10 fragment on a 5–20% linear polyacrylamide gel. Lane a, 20 μg of MAF and lane b, 5 μg of T-10 fragment, both stained with Stains-all; lane c, 20 μg of MAF; and lane d, 5 μg of T-10 both stained with Alcian blue. After electrophoresis, gels were washed four times in 25% isopropyl alcohol for periods of 2 h each. They were then placed in a solution of either 1. 0.0025% Stains-all (from Serva), 25% isopropyl alcohol, 7.5% formamide, 30 mM Tris, pH 8.0 in the dark for 48 h; or 2. 0.5% Alcian blue (from Serva), 25% isopropyl alcohol, 1% acetic acid for 48 h. Destaining was carried out in the same solutions without the dyes (Misevic & Burger, 1986). Molecular weight standards were obtained from Bethesda Research Laboratories.

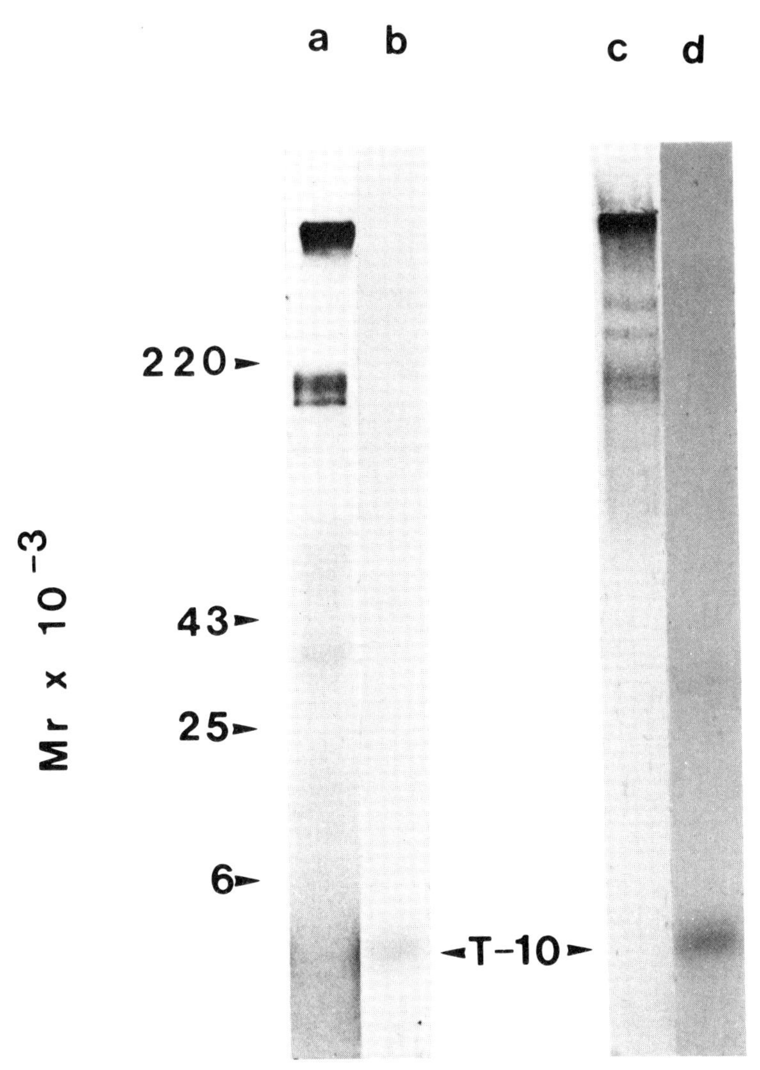

Fig. 10.5. Gel filtration of the polymerised T-10 glycopeptide fragment on a Bio-Gel A-1.5 m (column A), and a Bio-Gel A-15 m (column B). A, 200 μg of 2.5×10^4 cpm/μg of diepoxybutane and glutaraldehyde T-10 polymers (EG polymers) were applied on to a Bio-Gel A-1.5 m column (1 × 46 cm) (——). 210 μg of 2.5×10^4 cpm/μg of periodate oxidised and glutaraldehyde cross-linked T-10 (PG polymers) was applied on to the same type of column (– – –). The radioactivity of 50 μl from each fraction was measured in γ-counter. B, a Bio-Gel A-15 m column (1 × 46 cm) was used for separation of 150 μg (2.5×10^4 cpm/μg) of the void volume fractions from the Bio-Gel A-1.5 m column. Void volume fractions of diepoxybutane and glutaraldehyde cross-linked T-10 (——), void volume fractions of periodate oxidised and glutaraldehyde cross-linked T-10 (– – –). From each fraction, the radioactivity of 100 μl was measured in γ-counter (Misevic & Burger, 1986).

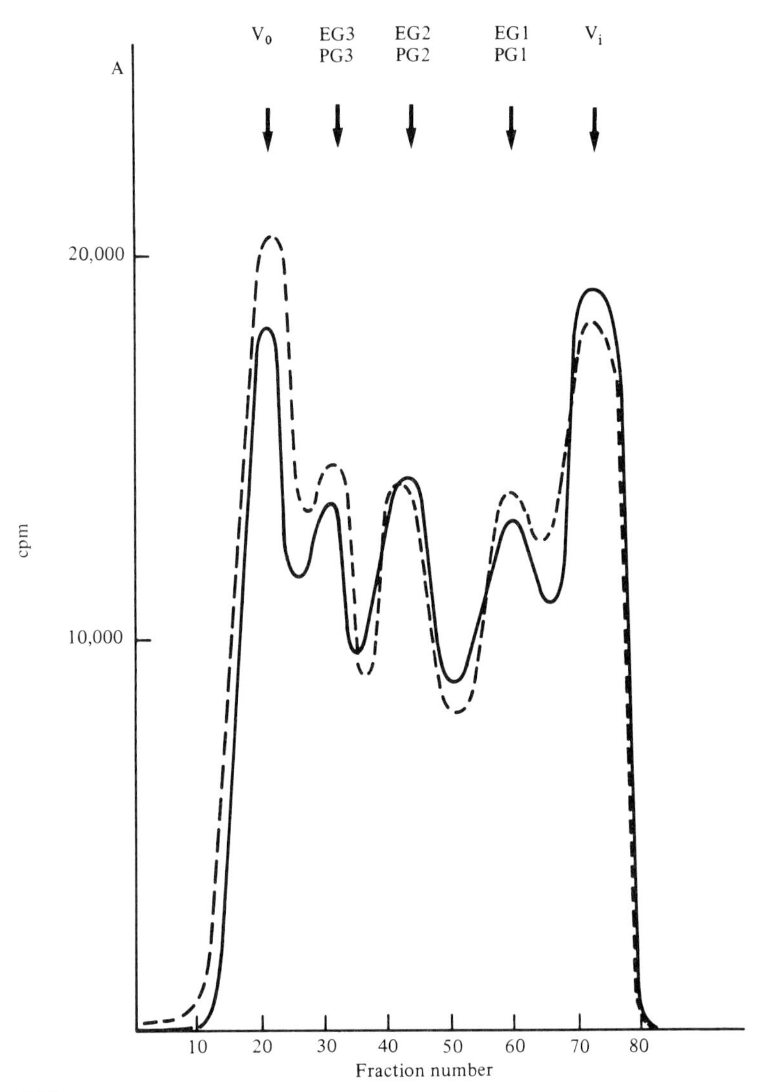

restore the cell binding function (Fig. 10.3). Thus, the prediction made for reconstitution experiments was also fulfilled, confirming the multiple low affinity nature of the cell binding site.

The cell binding site is localised in the T-10 glycoprotein fragment which is represented in 1,300 copies in the MAF molecule. Amino acid and carbohydrate analyses revealed that T-10 contains about 40% of protein and 60% of sugar by weight (Misevic *et al.*, 1982). The high, negative, charge found by ion exchange chromatography and binding of cationic dyes is due to the presence of glycuronic acid and possibly glutamic acid. Interestingly, T-10 contains a relatively high amount of fucose together with mannose, galactose and glucuronic acid, a feature not so far found in animal glyco-

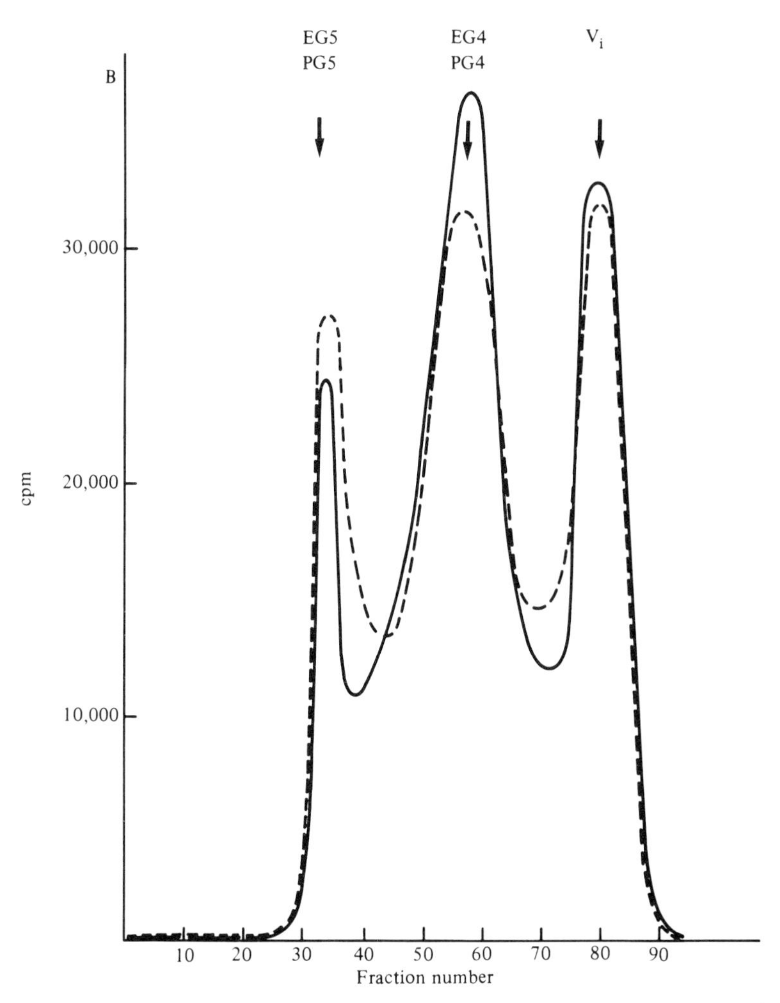

Table 10.2. *Binding specificities of the polymerised T-10 glycopeptide fragments to cells.*

Binding of the iodinated T-10 fragment, EG and PG polymers of T-10, was performed as described in the legend of Table 10.1.

	% bound to fixed cells of		
Ligand	*M. prolifera*	*H. occulata*	*C. celata*
T-10	15.9	4.7	8.9
EG1	14.1	3.9	6.0
PG1	15.9	3.5	5.9
EG2	22.5	3.2	6.1
PG2	23.1	2.8	4.1
EG3	27.2	2.6	6.0
PG3	28.7	3.1	4.3
EG4	43.6	3.2	4.0
PG4	44.4	3.4	2.9
EG5	20.0	2.9	3.2
PG5	21.2	2.1	2.9
MAF	23.1	2.6	2.1

proteins (Misevic & Burger, 1986). Whether these are present on the same or on different oligosaccharides remains an open question.

In an attempt to investigate whether the carbohydrate or the protein part represents the actual cell binding site, protein-free glycans were isolated from MAF after complete pronase digestion. Amino acid and carbohydrate analyses of these glycans have shown that essentially only one mole of asparagine is present per mole of glycan. When tested for presence of the cell binding activity in such a monomeric form, glycans did not display detectable binding to homotypic cells. However, cross-linking them with glutaraldehyde into polymers of MAF size reconstituted the cell binding activity. This indicated that the actual single cell binding site is present in the carbohydrate moiety and is of very low affinity not detectable with present methods. Since MAF contains three different classes of oligosaccharides based on their size, and since a mixture of all has been used successfully for cross-linking in reconstitution experiments, the identification and characterisation of the glycan carrying the cell binding site will have to be determined. Preliminary biochemical analyses of the T-10 glycoprotein oligosaccharide showed that it contains only two of the three MAF glycans. Both of these glycans are present in several copies in T-10, resulting in a considerably higher and measurable binding affinity of this glycoprotein than of its monomeric glycans. Reconstitution experiments will have to show whether one of these two or both oligosaccharides are carrying the cell binding site.

Our results based on biochemical evidence and reconstitution experiments suggested that the high cell binding affinity and specificity of MAF-cell interaction is mediated by highly polyvalent, low affinity cell binding sites of carbohydrate nature. Such findings indicate that small glycopeptides or oligosaccharides isolated from cell–cell interaction molecules, which did not show activity or measurable binding capacity, may indeed represent similar low affinity sites exerting their function through high polyvalency (Taylor & Orton, 1971; Yen & Ballou, 1974; Ballou, 1982; Orlean *et al.*, 1986). We suggest therefore, that it is essential to do reconstitution experiments by cross-linking the small ligands in artificial polymers with an inert backbone where high valency, and if possible, proper spacing of ligands can be achieved.

The self-interaction domain

Monoclonal antibodies (MAb) were raised against MAF in order to identify the self-interaction site. This strategy is based on the fact that the activity of self-interaction is irreversibly lost upon the dissociation of MAF. Out of 500 hybridoma clones 21 were found to produce antibodies against MAF. Five anti-MAF monoclonal IgG antibodies, which showed the strongest reactivity, were purified and tested for their inhibitory activity towards self-interaction of the native MAF. For these purposes an *in vitro* assay system, based on aggregation of MAF coated Sepharose beads, has been developed. Like the intact MAF the beads were aggregating only in the presence of 10 mM Ca^{2+}, which is necessary for the activity of the self-interaction domain. Purified Fab fragments have been used for testing their inhibitory activity towards MAF beads agglutination. Two of the IgG clones, Block 1 and Block 2, have shown a concentration dependent inhibition effect (Fig. 10.6). None of the other clones were able even with high excess (molar ratio to MAF 5,000 : 1) to block such an aggregation. One of these clones (C-16) has been chosen as a control, since it had binding characteristics very similar to those of Block 1 and Block 2 monoclonal antibodies (see below).

Two types of control experiments were carried out in order to provide the evidence for specificity of the blocking effect of the MAbs. In the first test the Block 1, Block 2 and C-16 MAbs did not show any inhibition of the activity of the cell binding domain. In the second type of experiments using MAF promoted aggregation of cells, again only Fab fragments of the Block 1 and the Block 2 antibodies caused inhibition of cell aggregation in a concentration-dependent manner as was observed with the MAF coated beads (Misevic, Finne & Burger, 1987). This indicated that these two blocking MAbs recognise the epitope(s) structures representing the self-interaction site of MAF.

The binding affinity of Block 1, Block 2 and C-16 as well as of their Fab fragments to MAF has been measured in the solid phase ELISA assay system. All three had almost identical affinity ruling out the possibility that C-16 does not bind well to MAF and therefore is incapable of showing a blocking effect (Table 10.3). The same binding experiments presented in Table 10.3 provided the evidence that the epitope structures, recognised by each of the Fabs derived from these three MAbs, are highly repetitive (over 1,000 copies per one MAF molecule). Only when all of the epitope sites have been occupied by Block 1 or Block 2, but not in the case of C-16, the complete inhibition of self-interaction could be achieved (see Fig. 10.6). This supports the notion that the self-interaction site is itself highly polyvalent. In the competition experiments none of the three MAbs or their Fabs competed with the other two heterologous MAbs or Fabs but only with themselves indicating that they recognise three different epitope structures of MAF (Table 10.3). Since C-16 recognises a distinct epitope, a structure of even higher polyvalency, and different from those of Block 1 and 2, and since its affinity is almost identical with those of the blocking MAbs, it strongly supports the interpretation that Block 1 and 2 are directed

Fig. 10.6. Inhibition of MAF-Sepharose bead aggregation by Fab fragments of monoclonal antibodies. Suspensions of 20 μl of MAF-Sepharose beads (5×10^4 beads/ml of Ca^{2+}, Mg^{2+} free seawater, 0.36 ng MAF per bead) were incubated with Fab fragments from monoclonal antibodies in CSW (Ca^{2+} Mg^{2+} free seawater supplemented with 2 mM $CaCl_2$) for 2 h at room temperature. Upon raising the Ca^{2+} concentration to 10 mM (A-I, J) beads were gently mixed in a moist chamber for 10 min at room temperature and inhibition of aggregation was examined with a Zeiss Axiomat microscope. A, 1 μg of Fab from the Block 1, B, 10 μg of Fab from the Block 1, C, 20 μg of Fab from the Block 1, D, 1 μg of Fab from the Block 2, E, 10 μg of Fab from the Block 2, F, 20 μg of Fab from the Block 2, G, 1 μg of Fab from the C-16, H, 10 μg of Fab from the C-16, I, 20 μg of Fab from the C-16, J, no Fab added, K, no Fab added, Ca^{2+} 2 mM.

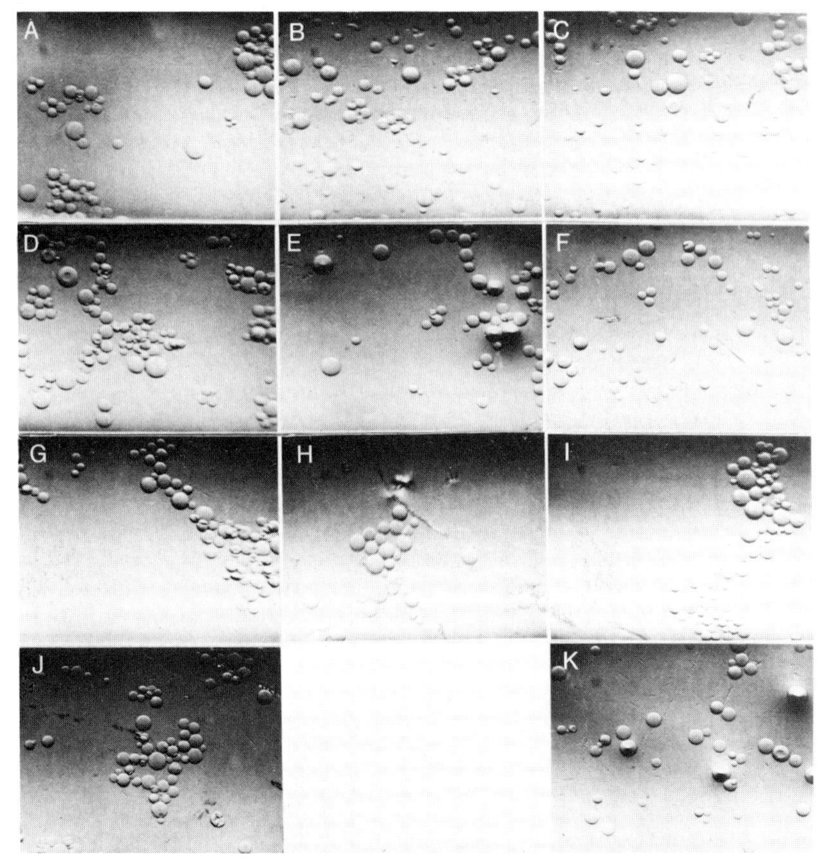

Table 10.3. *Binding characteristics of the anti-MAF monoclonal antibodies to MAF*

The number of binding sites for the MAbs and their Fab fragments as well as their association constants (K_a) were calculated from the binding data. In competition experiments, the binding of 1 μg ^{125}I-MAb (1.3–1.5 × 10^4 cpm/μg) or 10 μg of their ^{125}I-Fab fragments (2–3 × 10^4 cpm/μg) to 40 ng MAF was measured in the presence of an equal amount of homologous or heterologous antibody or their Fab fragments (Misevic *et al.*, 1987). The values are expressed as the ratio of the amount of iodinated antibody or Fab fragment bound to MAF at the subsaturating concentration to the amount of non-labelled antibody or Fab fragment causing 50% binding inhibition of the iodinated ligand.

^{125}I-antibody	No. of binding sites (moles/mole)	K_a (M)	Binding ratio		
			Block 1	Block 2	C-16
Block 1 IgG3	350	1.0×10^8	0.9	0.0	0.0[a]
Block 2 IgG2b	600	1.0×10^8	0.0	1.0	0.0
C-16 IgG2b	370	8.4×10^7	0.0	0.0	1.0
Block 1 Fab	1,100	2.0×10^6	0.9	0.0	0.0
Block 2 Fab	2,500	2.1×10^6	0.0	1.1	0.0
C-16 Fab	2,000	1.0×10^6	0.0	0.0	0.9

[a] 0.0 = no competition of binding even with 500-fold excess.

against the functionally active epitopes. The use of MAbs provided two important pieces of evidence about the characteristics of the self-interaction of MAF: 1) the self-interaction site is, as well as the cell binding site, highly polyvalent, and 2) it is represented by structures different from the cell binding site.

The high number of repeats of the antigenic epitopes for the Block 1, the Block 2 and the C-16 antibodies suggested that they are of carbohydrate nature. In order to test this prediction, MAF was subjected to extensive and complete pronase digestion. The protein-free glycans were isolated subsequently using gel filtration and ion exchange chromatography (Misevic *et al.*, 1987). To determine whether the epitopes for the blocking and the control C-16 MAb are localised in these glycans two assays were used. In the first experiment, glycans were acetylated with ^{3}H-acetic anhydride on their only remaining amino acid asparagine (Misevic *et al.*, 1987). Immunoprecipitation of these labelled ligands was performed by the polyethylene glycol method (Chard, 1980). Block 1 bound maximally 19%, Block 2 16% and C-16 15% of the total MAF glycans (Misevic *et al.*, 1987). In control experiments without antibodies or using MAb not directed against MAF no more than 2% of the counts were precipitated. A mixture of the three anti-MAF antibodies bound about 60% of the total MAF glycans at their saturating concentrations. This indicated that their epitopes are at least partially present on the different oligosaccharide chains.

The second type of experiment, used to provide the final evidence that the epitopes for the Block 1, the Block 2 and the C-16 MAbs are present in the carbohydrate portion of the MAF molecule, was based on the inhibition of binding of 0.5 μg ^{125}I-MAbs to 40 ng MAF by isolated MAF glycans. Again, in this assay system complete inhibition could be achieved only with the total MAF glycans (10–15 μg/ml) or intact MAF (1 μg/ml) but not with the monosaccharides L-fucose, D-galactose, D-mannose, N-acetyl-D-glucosamine, N-acetyl-D-galactosamine, D-glucuronic acid, D-galacturonic acid, at 0.5 M concentration. Nor were the glycosaminoglycans chondroitin sulphate A, chondroitin sulphate C, dermatan sulphate, heparin, heparan sulphate and hyaluronic acid able to inhibit binding at concentrations as high as 10 mg/ml, thus indicating very stringent specificity of MAbs towards MAF carbohydrate epitopes.

To provide support for the evidence, indirectly obtained by MAbs, that highly polyvalent carbohydrates are functional structures involved in the self-association of MAF, the total MAF glycans were directly assayed for the self-interaction activity in their monomeric and polymerised forms. As expected, neither a mixture of the total MAF glycans nor any of the three isolated types of oligosaccharides in their monomeric form have shown measurable self-binding or binding to MAF in the presence of 10 mM Ca^{2+} (Table 10.4 and Fig. 10.7). They also did not promote or inhibit Ca^{2+} induced self-aggregation of MAF. However, after cross-linking the total MAF glycans with prepolymerised glutaraldehyde into the size of the native MAF ($M_r > 1.5 \times 10^7$) the self-association activity could be reconstituted together with the cell binding activity (Table 10.4 and Fig. 10.7). This indicated that the self-interaction domain is also present in the carbohydrate portion of MAF and is based on multiple low affinity carbohydrate-carbohydrate interactions. The exact nature of the actual glycans involved in this self-association and the cell binding of MAF will have to be elucidated using immunopurification and other biochemical separation methods on a large scale together with the reconstitution experiments of the isolated fractions.

The use of biochemical and immunological approaches together with the reconstitution experiments have enabled us to elucidate the basis of the molecular mechanism of cell–cell recognition in the sponge *Microciona prolifera*. The novel principle of multiple low affinity interactions enabling both MAF-cell receptor binding and MAF self-association, was found to operate during cell–cell recognition and reaggregation in this marine sponge (Fig. 10.8). This mode of action is very different from the one of other cell–cell interaction molecules described so far, which usually have one or a few functional sites with relatively high affinity (Müller & Gerisch, 1978; Edelman, 1983; Hyafil, Babinet & Jacob, 1981; Harrison & Chesterton, 1980; Roseman, 1970; Blackburn & Schnaar, 1983; Garrod & Nicol, 1981; Mueller & Franke, 1983). The nature of these functional domains which mediate either the self-association or their binding to the cell receptors is based on protein–protein or protein–carbohydrate interactions. In contrast, it is shown here that the self-association domain of MAF acts through a carbohydrate–carbohydrate binding, not yet described in cell–cell recognition or adhesion molecules. The question arises whether such multiple low

Table 10.4. *Self-association and MAF binding activity of the cross-linked MAF glycans*

^{125}I-MAF (1 μg, 2.2 × 10^5 cpm/μg), the ^{3}H-glycans (50 μg, 5.6 × 10^4 cpm/μg) or the ^{3}H-cross-linked glycans (0.2 μg, 2.2 × 10^3 cpm/μg) were incubated in the presence of 2 mM or 10 mM Ca^{2+} in 100 μl CSW containing 1% BSA with or without 10 μg of unlabelled MAF (Misevic *et al.*, 1987). The percentage of the molecules coaggregated was determined after centrifugation at 10,000 × g.

	without MAF		with MAF	
	2 mM Ca^{2+}	10 mM Ca^{2+}	2 mM Ca^{2+}	10 mM Ca^{2+}
	% coaggregation		% coaggregation	
^{125}I-MAF	5	60	6	67
^{3}H-glycans	1	1	1	1
^{3}H-cross-linked glycans	5	24	12	50

affinity carbohydrate–carbohydrate associations also occur in other more complex systems. It seems very likely that the first sensing of the environment by a cell has to be achieved with the outermost exposed layer of the cell membrane consisting of a carbohydrate rich glycocalyx (Burger & Misevic, 1985). Thus, in the cases when a cell is approaching another cell or proteoglycans of an extracellular matrix, such carbohydrate–carbohydrate interactions of glycocalyx layers may occur. However, it was difficult until now to provide evidence for such interactions since the isolated complex carbohydrates might

Fig. 10.7. Immunodetection of self-association of MAF and the cross-linked MAF glycans. MAF (30 μg of neutral hexose), the monomeric glycans (30 μg) and the cross-linked MAF glycans of $M_r > 1.5 \times 10^7$ (0.6 μg) were incubated in 50 μl CSW containing 2 mM or 10 mM Ca^{2+}. After 30 min at room temperature the coaggregated material was centrifuged at 10,000 × g for 20 min. The supernatants and the pellets of the MAF sample and the sample of the monomeric glycans in 2 mM and 10 mM Ca^{2+} were brought up to 2.5 ml volume by CSW and 1 μl was then spotted on to a nitrocellulose filter (S and P dilution 0). 1 μl of two and four times diluted samples were also applied to the same filter (S and P dilution 2 and 4). The pellets of the cross-linked glycans in 2 mM and 10 mM Ca^{2+} were brought up to 50 μl by CSW and the supernatants were not diluted. This resulted in the same concentration of the glycans as in the case of MAF and the monomeric glycans. 1 μl was spotted on to a nitrocellulose filter (S and P dilution 0). Two- and four-fold dilutions of these samples were also applied on to the same filter (S and P dilution 2 and 4). Incubation with Block 1 MAb and peroxidase-conjugated rabbit immunoglobulins to mouse immunoglobulins was performed as described by Misevic *et al.*, 1987.

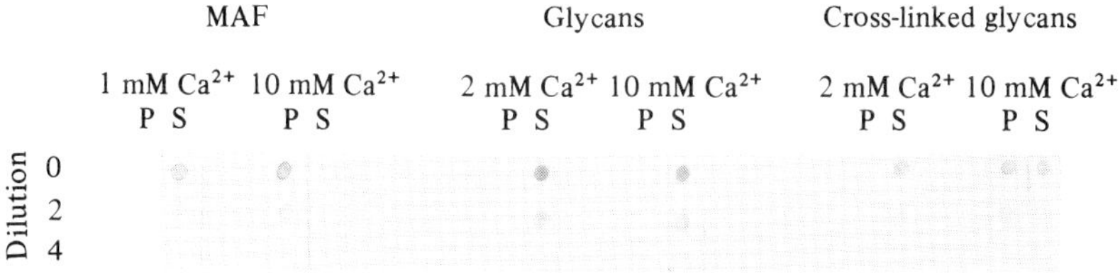

MAF Glycans Cross-linked glycans
1 mM Ca^{2+} 10 mM Ca^{2+} 2 mM Ca^{2+} 10 mM Ca^{2+} 2 mM Ca^{2+} 10 mM Ca^{2+}
P S P S P S P S P S P S
Dilution
0
2
4

not have shown any activity in their monomeric form but only in their native polymeric form. Therefore, the methods employed in this work may provide a useful tool for studying the molecules involved in cell–cell and cell–matrix interactions in more complex multicellular organisms.

Acknowledgements

This work was supported by the Swiss National Foundation for Scientific Research, Grants No. 3.269.82 and 3.169-0.85, the Ministry of the City and Canton of Basel as well as a European Molecular Biology Organisation short-term fellowship.

We are grateful to Dr Jukka Finne for critical discussion, Kjell Tullberg for his help with English grammar and Verena Schlup for technical assistance and useful comments.

References

Ballou, C. E. (1982). Yeast cell wall and cell surface. In *The Molecular Biology of the Yeast Saccharomyces Metabolism and Gene Expression,* eds J. N. Strathern, E. W. Jones & J. R. Broach, pp. 335–60. Cold Spring Harbor Laboratory: Cold Spring Harbor

Blackburn, C. C. & Schnaar, R. L. (1983). Carbohydrate-specific cell adhesion is mediated by immobilized glycolipids. J. Biol. Chem., **258**, 1180–8

Burger, M. M. & Misevic, G. N. (1985). Cell encounter: Molecular and biological aspects of initial contacts. In *Cellular and Molecular Control*

Fig. 10.8. Model of species-specific aggregation of marine sponge *M. prolifera*. Species-specific recognition and adhesion of *M. prolifera* dissociated cells are mediated by aggregation factor via its cell binding and self-association domain. Both of these functional properties of MAF are based on a highly multiple low affinity interaction of MAF carbohydrates (carbohydrate structures representing cell binding site, carbohydrate structures representing self-association site).

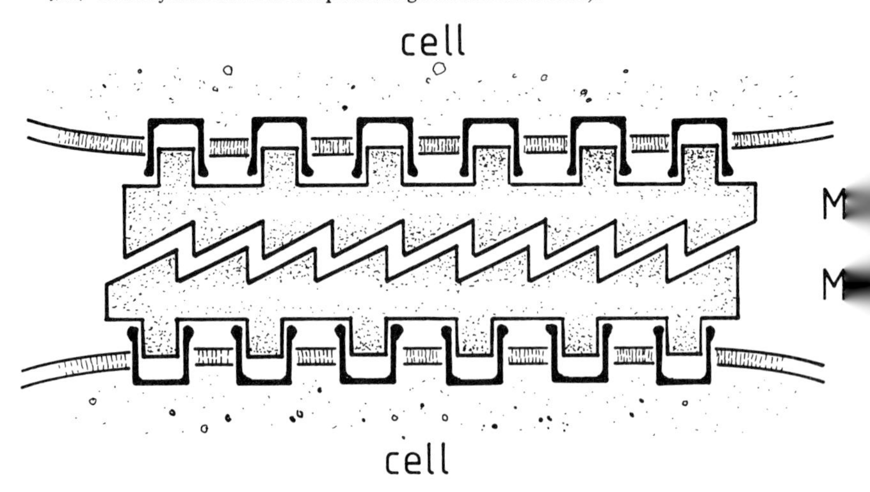

of Direct Cell Interactions, NATO ASI Series Vol. 99, ed. H. J. Marthy, pp. 3–26. New York, London: Plenum Publishing Corporation

Cauldwell, C. B., Henkart, P. & Humphreys, T. (1973). Physical properties of sponge aggregation factor. A unique proteoglycan complex. *Biochemistry*, **12**, 3051–5

Chard, T. (1980). Ammonium sulphate and polyethylene glycol as reagents to separate antigen from antigen–antibody complexes. *Methods in Enzymol.*, **70**, 280–91

Edelman, G. M. (1983). Cell adhesion molecules. *Science*, **219**, 450–7

Galtsoff, P. S. (1929). Heteroagglutination of dissociated sponge cells. *Biol. Bull.*, **57**, 250–60

Garrod, D. R. & Nicol, A. (1981). Cell behaviour and molecular mechanisms of cell–cell adhesion. *Biol. Rev. Camb. Phil. Soc.*, **56**, 199–242

Harrison, F. L. & Chesterton, C. J. (1980). Erythroid developmental agglutinin is a protein lectin mediating specific cell–cell adhesion between differentiating rabbit erythroblasts. *Nature (Lond.)*, **286**, 502–4

Henkart, P., Humphreys, S. & Humphreys, T. (1973). Characterization of sponge aggregation factor. A unique proteoglycan complex. *Biochemistry*, **12**, 3045–50

Humphreys, T. (1963). Chemical dissolution and *in vitro* reconstruction of sponge cell adhesions. *Dev. Biol.*, **8**, 27–47

Humphreys, S., Humphreys, T. & Sano, J. (1977).Organization and polysaccharides of sponge aggregation factor. *J. Supramolecular Structure*, **7**, 339–51

Hyafil, F., Babinet, C. & Jacob, F. (1981). Cell–cell interactions in early embryogenesis: A molecular approach to the role of calcium. *Cell*, **26**, 447–54

Jumblatt, J. E., Schlup, V. & Burger, M. M. (1980). Cell–cell recognition: Specific binding of *Microciona* sponge aggregation factor to homotypic cells and the role of calcium ions. *Biochemistry*, **19**, 1038–42

Misevic, G. N. & Burger, M. M. (1986). Reconstitution of high cell binding affinity of a marine sponge aggregation factor by cross-linking of small low affinity fragments into a large polyvalent polymer. *J. Biol. Chem.*, **261**, 2853–9

Misevic, G. N., Finne, J. & Burger, M. M. (1987). Involvement of carbohydrates as multiple low affinity interaction sites in the self-association of the aggregation factor from the marine sponge *Microciona prolifera*. *J. Biol. Chem.*, **262**, 5870–7

Misevic, G. N., Jumblatt, J. E. & Burger, M. M. (1982). Cell binding fragments from a sponge proteoglycan-like aggregation factor. *J. Biol. Chem.*, **257**, 6931–6

Monsan, P., Puzo, G. & Mazarguil, H. (1975). Etude du mecanisme d'établissement des liaisons glutaraldehyde-proteines. *Biochimie*, **57**, 1281–92

Mueller, H. & Franke, W. W. (1983). Biochemical and immunological characterization of desmoplakins I and II, the major polypeptides of the desmosomal plaque. *J. Mol. Biol.*, **163**, 647–71

Müller, K. & Gerisch, G. (1978). A specific glycoprotein as the target site of adhesion blocking Fab in aggregating *Dictyostelium cells*. *Nature (Lond.)*, **274**, 445–9

Müller, W. E. G. & Zahn, R. K. (1973). Purification and characterization of a species-specific aggregation factor in sponges. *Exp. Cell Res.*, **80**, 95–104

Orlean, P., Ammer, H., Watzele, M. & Tanner, W. (1986). Synthesis of an O-glycosylated cell surface protein induced in yeast by α factor. *Proc. Natl. Acad. Sci. USA*, **83**, 6263–6

Richards, F. M. & Knowles, J. R. (1968). Glutaraldehyde as a protein cross-linking reagent. *J. Mol. Biol.*, **37**, 231–3

Roseman, S. (1970). The synthesis of complex carbohydrates by multiglycosyltransferase systems and their potential function in intercellular adhesion. *Chem. & Physics of Lipids*, **5**, 270–97

Taylor, N. W. & Orton, W. L. (1971). Cooperation among the active binding sites in the sex-specific agglutinin from the yeast, *Hansenula wingei*. *Biochemistry*, **10**, 2043–9

Weinbaum, G. & Burger, M. M. (1973). Two component system for surface guided reassociation of animal cells. *Nature (Lond.)*, **244**, 510–12

Wilson, H. V. (1907). On some phenomena of coalescence and regeneration in sponges. *J. Exp. Zool.*, **5**, 245–58

Yen, P. H. & Ballou, C. E. (1974). Partial characterization of the sexual agglutination factor from *Hansenula wingei* Y-2340 type 5 cells. *Biochemistry*, **13**, 2428–37

11

Sperm surface carbohydrates in *Drosophila melanogaster*

Maria-Elisa Perotti and Antonella Riva

Dept of General Physiology and Biochemistry,
University of Milano,
Italy

Abstract

The mechanisms controlling sperm–egg encounter in insects are poorly understood and little is known of the molecular structure and function of the sperm plasma membrane. In this study, lectins have been used to analyse the distribution of carbohydrate residues in the plasma membrane of *Drosophila* sperm. In the 2 mm long spermatozoon the exposed saccharides are distributed in a very polar manner, being concentrated almost exclusively over the small acrosome (3 μm long, 0.24 μm maximum width). In this region D-mannose/glucose, L-fucose, N-acetylglucosamine, N-acetylgalactosamine and sialic acid are present. The 3 μm long tail end piece contains D-mannose/glucose moieties and is the only other region with discrete amounts of terminal carbohydrates. The data are discussed in relation to possible mechanisms of sperm–egg interactions.

Introduction

The events leading to the union of spermatozoon and egg in insects are poorly understood, despite the fact that most of the available information on the physiology of reproduction in invertebrates concerns this class. The knowledge of the process of fertilisation is restricted to gamete structure, morphology of the reproductive tracts and events which follow sperm penetration into the egg. In particular, in spite of the potential role of the sperm surface during fertilisation, nothing is known of the molecular details of sperm membrane structure or function. The paucity of information on the mechanisms which control the encounter of egg and sperm is probably related to the extensive variations existing among species in sperm structure, egg envelopes, reproductive tract morphology and to the technical difficulties which so far have prevented the development of procedures for *in vitro* fertilisation in these organisms.

We have started a cytochemical analysis of the sperm membranes of the spermatozoon of *Drosophila melanogaster* and a parallel study of the egg envelopes before and after fertilisation, with the aim of identifying possible regional specialisations of the gamete surfaces which might be involved in the gamete fusion, and therefore of obtaining indirect information on the mechanisms which control the interactions between egg and sperm. The choice of *Drosophila melanogaster* as a model is suggested by the fact that, structurally, its spermatozoon can be regarded as most representative of insect sperm (Dallai, 1979), the details of its ultrastructure and morphogenesis are well documented (Perotti, 1969; Lindsley & Tokuyasu, 1980) and ample information is available on the patterns of sperm transfer, storage and utilisation by the female (for a review see Fowler, 1973). Furthermore, several features of the genetic control of male and female gametogenesis are known, and the use of mutations with impairment of fertility may provide additional insight into the mechanisms which are involved in the interaction of sperm and egg.

During recent years the role of carbohydrates in the process of fertilisation has been established in different animal species. It seems that, in a manner similar to that in other systems where cell-to-cell recognition and adhesion occur, sperm–egg interaction is mediated by the association of sugar moieties present on the surface of one gamete with complementary molecules present at the surface of the other gamete (Peterson *et al.*, 1980; Shur & Hall, 1982; Bleil & Wasserman, 1983; Hoshi *et al.*, 1983; Lambert, 1984; Lopez *et al.*, 1985). The involvement of L-fucose in the binding of sperm to egg has been established in a variety of species: hamster, guinea-pig, man (Ahuja, 1982; Huang, Ohzu & Yanigimachi, 1982); boar (Topfer-Petersen *et al.*, 1985), rat (Shalgi *et al.*, 1986), amphibians (Denis-Donini & Campanella, 1977), ascidians (Rosati *et al.*, 1980) and sea-urchin (SeGall & Lennarz, 1979). In hamster and rat, other carbohydrates such as α-methyl-mannoside and D-mannose co-operate with L-fucose in the sperm–egg interactions, whereas a more specific interaction seems to be at play in mouse, where in three different strains three different sugars, i.e. N-acetylglucosamine, α-methyl-mannoside and sialic acid appear to be important for gamete recognition (Shur & Hall, 1982; Lambert, 1984). In a mouse strain (Shur & Hall, 1982) and in an ascidian (Hoshi *et al.*, 1983), it has been suggested that the specific carbohydrate responsible for sperm–egg binding is a component of a glycoprotein sperm-receptor present on the egg surface and which interacts with a specific enzyme on the sperm surface. On the other hand, in hamster (Ahuja, 1982) and in a different strain of mouse (Lambert, 1984; Lambert & Van Le, 1984) the presence of glycoproteins on the sperm plasma membrane have been reported and these may interact with lectin-like molecules on the zona pellucida.

In this study the presence and topographical distribution of saccharide residues in the plasma membrane of *Drosophila melanogaster* spermatozoon were examined by the use of a variety of lectins conjugated with fluorescein isothiocyanate and with specificity for different terminal sugar residues.

Materials and methods

Spermatozoa used in this research were collected from the seminal vesicles of young adult males and were therefore completely differentiated and motile (Perotti, 1969; Fowler, 1973; Lindsley & Tokuyasu, 1980). The lectins used were: Concanavalin A (Con A), specific for α-D-Mannose and α-D-glucose residues; wheat germ agglutinin (WGA), specific for terminal N-acetyl-D-glucosamine and sialic acid; succinylated WGA (S-WGA), specific for terminal N-acetyl-D-glucosamine; soybean agglutinin (SBA), specific for terminal N-acetyl-D-galactosamine; castor bean agglutinin I (RCA_I), specific for β-D-galactose residues; fucose binding protein from *Lotus tetragonolobus* (FBP), specific for terminal L-fucose; and *Limulus polyphemus* agglutinin (LPA), specific for sialic acid residues. Stock solutions of FITC-SBA, -WGA, -S-WGA, -FPB were diluted to 100 μg/ml in PBS, whereas those of FITC-Con A were brought to the same dilutions with PBS containing 2 mM $CaCl_2$ and $MgCl_2$ and those of LPA with Tris buffer containing 0.15 M NaCl and 0.01 M $CaCl_2$, pH 8.

Sperm were fixed with 2% paraformaldehyde in PBS, pH 7.2, rinsed extensively with PBS and subsequently with 0.1 M NH_4Cl in PBS to quench free aldehyde groups, and finally incubated in diluted FITC-lectins for 15 min at room temperature. After rinsing in the appropriate buffer they were mounted in a polyvinyl alcohol medium, pH 8.4 (Mowiol 4-88) (Osborn & Weber, 1982) and observed with a Zeiss III Photomicroscope equipped for epifluorescence.

Controls were performed by preincubating the lectins before use with the appropriate hapten sugars at 0.2 M. Sperm were also incubated in FITC-lectins without prior fixation: incubations were then performed at 4 °C and at room temperature.

To obtain preliminary information on the sperm–egg interactions, mature eggs obtained from the ovary were treated with horseradish peroxidase conjugated with FITC (FITC-HRP) and prepared for fluorescence microscopy as indicated above.

Results and discussion

As described previously by several authors (Perotti, 1969; Stanley *et al.*, 1972; Tokuyasu, 1974) the *Drosophila melanogaster* sperm is approximately 1.8 mm long. The acrosome is 3 μm long, roughly cylindrical in shape and its posterior end is inserted on one side of the nucleus. In a more anterior region its profile shows a keel which accounts for the acrosome widest diameter (0.24 μm) whereas at the apex the keel disappears and the acrosome shape becomes that of a slightly concave lamina. The nucleus is 10 μm long, cylindrical in shape and with a major cross diameter of 0.3 μm. The remainder of the sperm is represented by the tail, which is 0.4 μm in cross diameter, and presents the same structure along its length except for the very short connecting piece and the end piece. Because of their size and shape, the acrosomal and tail ends cannot be recognised from each other in conventional light microscopy.

The predominant patterns of FITC binding were different for each of the lectins used. The FITC-lectin binding patterns of unfixed sperm were the same as those of sperm fixed in 2% paraformaldehyde.

With FITC-Con A the sperm showed a distinct fluorescence at two different extremities: one, 2.5 μm long, had an irregular elongated profile with some bulges along its extent (Fig. 11.1), whereas the other was tapering and 5 μm long (Fig. 11.2). On the basis of their morphology and length the two extremities were identified as the acrosome and the tail end piece respectively. The acrosome was followed by a tract approximately 10 μm long and devoid of Con A binding which could be identified as the nuclear region, whereas the remainder of the tail was weakly positive (Fig. 11.1). The binding pattern indicates the presence of a large amount of D-mannose/D-glucose residues on the plasma membrane overlying almost the whole length of the acrosome and along the tail end piece and the complete absence of these terminal carbohydrates from the plasma membrane covering the nucleus.

After FITC-WGA treatment the acrosome was intensely fluorescent along its entire length, i.e. 3 μm, (Figs. 11.3, 11.4), its overall shape was even better defined than after Con A binding and comparable to the one usually revealed only in electron microscope images. When observed at higher magnification, the acrosome's most caudal part, probably the one laterally inserted on the nucleus, showed a weaker labelling (Fig. 11.4). The nuclear region was negative, the tail main region showed a very negligible fluorescence (Fig. 11.3) and the tail end piece was not labelled. A large amount of the terminal carbohydrates recognised by this lectin, sialic acid and/or N-acetylglucosamine, seems therefore to be concentrated in the plasma membrane region overlying the whole acrosome.

Fig. 11.1, 11.2. Labelling with FITC-Con A. The surface over the acrosome (A) is intensely labelled, the nuclear region (N) is negative, the tail main piece (MP) is weakly positive and the end piece of the tail is significantly fluorescent (arrow).
Fig. 11.3, 11.4. Labelling with FITC-WGA. The whole acrosome region is labelled, with a decreasing intensity over the posterior region (arrow). The nuclear region is negative, the tail shows an extremely weak fluorescence.

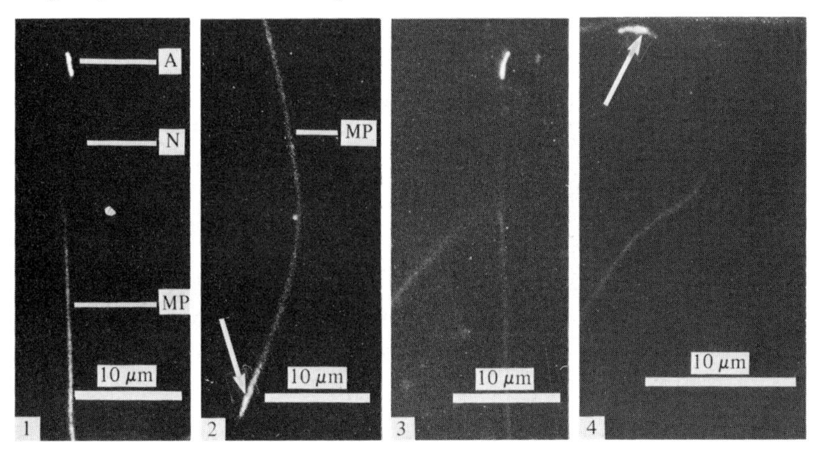

After incubation in FITC-S-WGA the only sperm component labelled was the acrosome, which exhibited a fluorescence weaker than the ones evident after Con A and WGA treatment. The labelling was not homogeneous along the acrosome extent, being more intense at its anterior part and weaker in the caudal region (Fig. 11.5), whereas the middle part seemed to bear no binding sites for this lectin. It can be therefore assumed that the N-acetylglucosamine residues are more concentrated in the anterior than in the posterior segment of the acrosome and that they are absent from the acrosome middle part.

Labelling with FITC-LPA complemented the information provided by WGA and S-WGA. The spermatozoon showed a bright fluorescent spot in an area roughly corresponding to the acrosome middle part (Fig. 11.6), the nuclear region was again negative and the tail exhibited an extremely weak, ill-defined fluorescence. Sialic acid residues appear therefore to be restricted to a small part of the acrosomal region and present in extremely low amount, if any, along the tail.

Treatment with FITC-FBP labelled an area which, according to its shape and extent, can be tentatively identified as the middle and caudal part of the acrosome (Fig. 11.7). However, at this level of resolution it is difficult to rule out the possibility that L-fucose residues might extend shortly from the acrosome caudal region down over the initial part of the nuclear region. The tail showed an extremely weak binding.

Fig. 11.5. Labelling with FITC-S-WGA. Only the acrosome is positive, with a more intense fluorescence over the anterior part and a weak labelling over the posterior segment. The acrosome middle part is negative.
Fig. 11.6. Labelling with FITC-LPA. Only the acrosome middle part appears stained.
Fig. 11.7. Labelling with FITC-FBP. The acrosome region shows the middle part more intensely stained whereas the posterior segment is weakly fluorescent.
Fig. 11.8. Labelling with FITC-SBA. A weak fluorescence is present over the acrosome tip and over the tail.

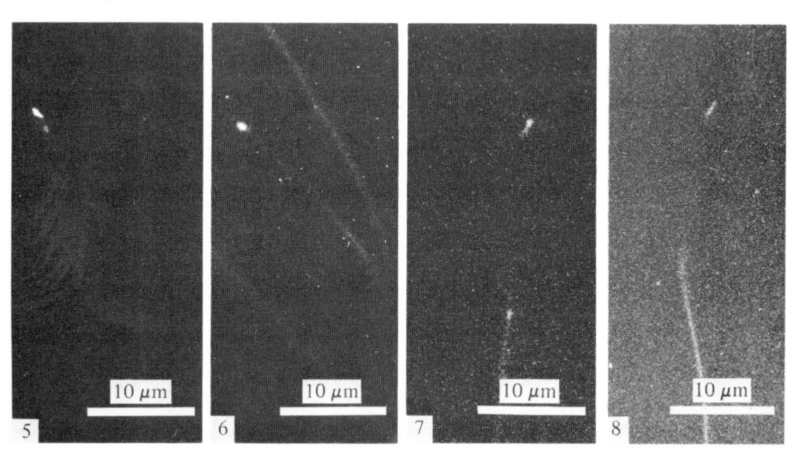

Incubation with FITC-SBA resulted in a weak fluorescence of the most anterior part of the acrosome, whose overall size appeared thinner than after labelling with Con A and WGA (Fig. 11.8). N-acetylgalactosamine residues seem therefore confined to this acrosome region. The intensity and the extent of the labelling pattern suggest that they are less abundant than the terminal carbohydrates recognised by Con A and WGA. An extremely weak fluorescence was also observed along the tail main piece.

Treatment with FITC-RCA_I gave consistently negative results, indicating that β-D-galactose residues are entirely absent from the sperm surface.

The results demonstrates that the distribution of carbohydrates residues in the plasma membrane is not uniform and these findings parallel those reported in other animal species. When observed at electron microscope level the plasma membrane of *Drosophila* sperm shows a smooth thickening of the outer layer, typical of the 'fruit-fly' type spermatozoon and present not only in many insects but also in molluscs and vertebrates (Baccetti & Afzelius, 1976). Whereas the plasma membrane ultrastructure is uniform along the whole sperm length (Perotti, 1969), its chemical composition has regional differentiations, as shown in this study. The carbohydrates here investigated, which seem to be the most relevant in modulating the sperm activity (for reviews: Millette, 1977; Nicolson & Yanagimachi, 1979; Koehler, 1981; Peterson & Russell, 1985) are all concentrated over the acrosome and some of them are exclusive of this region. In *Drosophila* sperm the distribution of terminal carbohydrates in the plasma membrane seems therefore extremely polarised when compared to the other spermatozoa so far studied, either from invertebrates or vertebrates. Furthermore, despite the small size of the acrosome the surface saccharides are not homogeneously distributed within it, but they seem to be organised in subdomains. Differences among the acrosome carbohydrates seem to exist, not only in topographical distribution, but also in relative amounts. From the comparative point of view it is interesting to notice the presence in the acrosome plasma membrane of two carbohydrates which have been seldom observed in the sperm of other species, i.e. L-fucose and sialic acid. The former has been detected only in ascidians, where however it is distributed over the whole sperm surface (Rosati, De Santis & Monroy, 1978). Sialic acid has been observed only in a few mammalian spermatozoa, i.e., mouse (Lambert & Van Le, 1984), ram (Malik & Bartoov, 1985) and man (Levinski *et al.*, 1983) and in all these cases it was restricted to the acrosomal and nuclear regions. Also, the total lack of β-D-galactose seems to be a rather unique finding as the presence of this saccharide is widespread in vertebrate and invertebrate spermatozoa.

The absence of surface carbohydrates from the nuclear region suggests that the plasma membrane of this area might not be involved in the events leading to gamete fusion. If this lack of biochemical specialisation is confirmed by further analysis, then the fertilisation mechanisms of *Drosophila* could be regarded as more like those reported for species with external fertilisation than those reported in species with internal fertilisation. In fact, in the former the most relevant compartments for gamete fusion are the plasma membrane overlying the acrosome and the acrosome membranes (for reviews: Epel & Vacquier, 1978; Rosati, 1985), whereas in the latter the final events leading to sperm penetration into the ooplasm involve fusion between egg plasma membrane and the sperm plasma membrane over the acrosome equatorial segment and/or the plasma membrane over the post-equatorial region (for a review, Bedford, 1982).

An intriguing result is represented by the distinct binding of Con A to the tail end piece, similar in intensity to the one observed in the acrosome and indicating the presence of a high concentration of D-mannose/D-glucose residues. The significance of the differentiation in chemical composition of this short tract of the tail cannot conceivably be related to the sperm movement, as in this region the typical axonemal structures disappear (Perotti, 1969). On the other hand, ultrastructural studies on fertilisation in *Drosophila* showed that, as in most species, the spermatozoon enters the egg with its head pointing forward (Perotti, 1974). When mature eggs were incubated in FITC-HRP, binding sites for HRP were found only on the tip of the egg micropyle (Fig. 11.9). Binding of HRP was blocked when eggs were pre-incubated with mannose. As HRP contains both mannose and N-acetylglucosamine groups (Clarke & Shannon, 1976) it can be assumed that on the surface of the micropyle lectin-like molecules exist which can interact with the D-mannose residues on the sperm surface, so that the sperm would be able to bind to the outer egg envelope both with the acrosome and the tail end piece. How this double possibility of interaction might control the encounter between egg and sperm is presently only a matter of speculation. The spermatozoon penetrates into the egg when the latter, descending from the common oviduct, places its micropyle in front of the narrow opening of the sperm storage organ, the so-called seminal receptacle. (Fig. 11.10) (for a review, Fowler, 1973). Utilisation of the stored sperm for fertilisation is in *Drosophila* very efficient, with very little or no waste of sperm (for a review, Fowler, 1973) and almost all the eggs laid by an inseminated female are fertilised (Boulétreau-Merle, 1977). The sperm kept in the seminal receptacle are in a continuous state of circulation (Fowler, 1973) and it has been demonstrated that in the proximity of the opening of the receptacle some sperm are moving toward the opening, whereas others are pointing in the opposite direction (Tokuyasu, 1974). The presence at the surface of both acrosome and tail end piece of carbohydrates which might interact with complementary molecules present on the micropyle surface would favour the preliminary adhesion to the egg of any sperm whose anterior or caudal extremities are located at the opening of the receptacle while an egg is moving down the uterus and would therefore ensure egg fertilisation.

Acknowledgements

This research was supported by the CNR project 'Biology of Reproduction and Differentiation' and by a MPI grant.

References

Ahuja, K. K. (1982). Fertilization studies in the hamster: the role of cell surface carbohydrates. *Exp. Cell Res.*, **140**, 353–62

Baccetti, B. & Afzelius, A. (1976). *The Biology of the Sperm Cell.* Basel: Karger

Bedford, J. M. (1982). Fertilization. In *Reproduction in Mammals – Germ Cells and Fertilization*, 2nd edn, eds C. R. Austin & R. V. Short, pp. 128–63. Cambridge, London: Cambridge University Press

Bleil, J. D. & Wassarman, P. L. (1983). Sperm–egg interactions in the mouse: sequence of events and induction of the acrosome reaction by a zona pellucida glycoprotein. *Dev. Biol.*, **95**, 317–24

Boulétreau-Merle, J. (1977). Role des spermathèques dans l'utilization du spérme et la stimulation de l'ovogénèse chez *Drosophila melanogaster*. *J. Insect Physiol.*, **23**, 1009–104

Fig. 11.9. Diagram of the structure of the egg micropyle. The asterisks indicate the distribution of the mannose-binding sites over the chorion of the micropylar region, as evidenced by FITC-HRP binding. ch = chorion, mc = micropylar channel, oo = ooplasm, vm = vitelline membrane. The oolemma, closely adherent to the vitelline membrane, is not shown.

Fig. 11.10. Diagram of the site of the sperm–egg encounter. The fertilising sperm are stored in the seminal receptacle, the egg is in the uterus (u) with the micropyle closely apposite to the opening of the seminal receptacle. The accessory glands and the additional sperm storing organs (spermathecae) are not shown. a = respiratory appendages of the chorion, co = common oviduct, m = micropyle, sr = seminal receptacle, u = uterus. Diagram based on Nonidez (1920) and Lefèvre & Jonsson (1962).

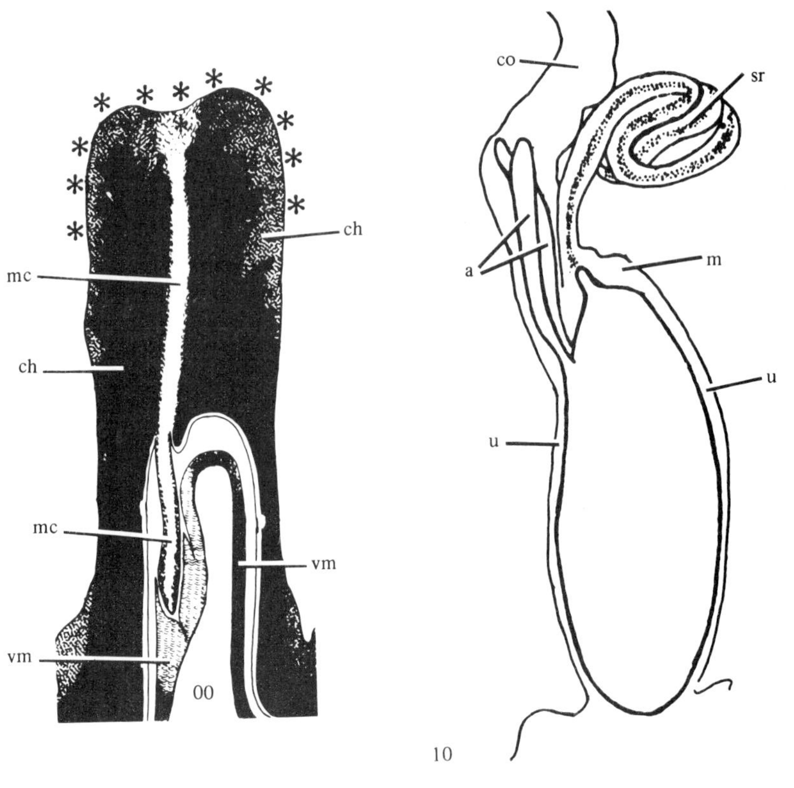

Clarke, J. & Shannon, L. M. (1976). The isolation and characterization of the glycopeptides from horseradish peroxidase isoenzyme C. *Biochim. & Biophys. Acta*, **427**, 428–42

Dallai, R. (1979). An overview of atypical spermatozoa in insects. In *The Spermatozoon*, eds D. W. Fawcett & J. M. Bedford, pp. 253–65. Baltimore, Munich: Urban & Schwarzenberg

Denis-Donini, S. & Campanella, C. (1977). Ultrastructural and lectin binding changes during the formation of the animal dimple in oocytes of *Discoglossus pictus* (Anura). *Dev. Biol.*, **61**, 140–52

Epel, D. & Vacquier, V. (1978). Membrane fusion events during invertebrate fertilization. In *Cell Surface Reviews*, vol. 5, eds G. Poste & G. L. Nicolson, pp. 1–65. Amsterdam: North-Holland Publishers

Fowler, G. L. (1973). Some aspects of the reproductive biology of *Drosophila*: sperm transfer, sperm storage and sperm utilization. *Adv. Genetics*, **17**, 293–360

Hoshi, M., De Santis, R., Pinto, M. R., Cotelli, F. & Rosati, F. (1983). Is sperm L-fucosidase responsible for sperm-egg binding in *Ciona intestinalis*? In *The Sperm Cell*, ed J. André, pp. 107–10. The Hague: Martinus Nijhoff Publishers

Huang, T. T. F., Ohzu, E. & Yanagimachi, R. (1982). Evidence suggesting that L-fucose is part of a recognition signal for sperm-zona pellucida attachment in mammals. *Gamete Res.*, 5, 355–61

Koehler, J. K. (1981). Lectins as probes of the spermatozoon surface. *Arch. Androl.*, **6**, 197–217

Lambert, H. (1984). Role of sperm-surface glycoprotein in gamete recognition in two mouse species. *J. Reproduction & Fertility*, **70**, 281–4

Lambert, H. & Van Le, A. (1984) Possible involvement of a sialylated component of the sperm plasma membrane in sperm-zona interaction in the mouse. *Gamete Res.*, 153–63

Lefèvre, G. & Jonsson, U. B. (1962). Sperm transfer, storage, displacement and utilization in *Drosophila melanogaster*. *Genetics*, **47**, 1719–36

Levinski, H., Singer, R., Malik, Z., Sagiv, M., Cohen, A. M., Servadio, C. & Allalouf, F. (1983). Distribution of sialic acid in human sperm membrane. *Arch. Androl.*, **10**, 209–12

Lindsley, D. L. & Tokuyasu, K. T. (1980). Spermatogenesis. In *The Genetics and Biology of Drosophila*, vol. 2d, eds M. Ashburner & T. R. F. Wright, pp. 226–94. London, New York: Academic Press

Lopez, L. C., Bayne, E. M., Litoff, D., Shaper, N. L., Shaper, J. H. & Shur, B. D. (1985). Receptor function of mouse sperm surface galactosyltransferase during fertilization. *J. Cell Biol.*, **101**, 1501–10

Malik, Z. & Bartoov, B. (1985). Sialyl glycoprotein distribution on the plasma membrane of ejaculated ram spermatozoa. *Biol. of Cell*, **54**, 93–100

Millette, C. F. (1977). Distribution and mobility of lectin binding sites on mammalian spermatozoa. In *Immunobiology of Gametes*, eds M. Edidin & M. H. Johnson, pp. 51–72. Cambridge, London: Cambridge University Press

Nicolson, G. L. & Yanagimachi, R. (1979). Cell surface changes associated with the epididymal maturation of mammalian spermatozoa. In *The spermatozoon*, eds D. W. Fawcett & J. M. Bedford, pp. 171–84. Baltimore,Munich: Urban & Schwarzenberg

Nonidez, J. F. (1920). The internal phenomena of reproduction in *Drosophila*. *Biol. Bull.*, **39**, 207–30

Osborn, M. & Weber, K. (1982) Immunofluorescence and immunocytochemical procedures with affinities purified antibodies: tubulin-containing structures. *Methods in Cell Biol.*, **24**, 97–132

Perotti, M. E. (1969). Ultrastructure of the mature sperm of *Drosophila melanogaster* Meig. *J. Submicroscopic Cytol.*, **1**, 171–96

Perotti, M. E. (1974). Ultrastructural aspects of Fertilization in *Drosophila*. In *The Functional Anatomy of the Spermatozoon*, ed. B. Afzelius, pp. 57–68. Oxford: Pergamon Press

Peterson, R. N. & Russell, L. D. (1985). The mammalian spermatozoon: a model for the study of regional specificity in plasma membrane organization and function. *Tissue & Cell*, **17**, 69–91

Peterson, L. D., Russell, L. D., Bundman, D. & Freud, M. (1980). Sperm–egg interaction: direct evidence for boar plasma membrane receptors for porcine zona pellucida. *Science*, **207**, 73–4

Rosati, F. (1985). Sperm–egg interaction in Ascidians. In *Biology of fertilization*, vol. 2, eds C. B. Metz & A. Monroy, pp. 361–88. New York, London: Academic Press

Rosati, F. & De Santis, R. (1980). Role of surface carbohydrates in sperm–egg interaction in *Ciona intestinalis*. *Nature (Lond.)*, **283**, 762–4

Rosati, F., De Santis, R. & Monroy, A. (1978). Studies on fertilization in the Ascidians. II. Lectin binding to the gametes of *Ciona intestinalis*. *Exp. Cell Res.*, **116**, 419–27

Sander, K. (1985). Fertilization and egg cell activation in Insects. In *Biology of Fertilization*, vol. 2, B. B. Metz & A. Monroy eds, pp. 409–30. New York, London: Academic Press

SeGall, G. K. & Lennarz, W. J. (1979). Chemical characterization of the components of the jelly coat from sea urchin eggs responsible for induction of the acrosome reaction. *Dev. Biol.*, **71**, 33–8

Shalgi, R., Matityahu, A. & Nebel, L. (1986). The role of carbohydrates in sperm-egg interaction in rats. *Biol. Reproduction*, **34**, 446–52

Shur, B. D. & Hall, N. G. (1982). Sperm surface galactosyltransferase activities during *in vitro* capacitation. *J. Cell Biol.*, **95**, 567–73

Stanley, H. P., Bowman, J. T., Romrell, L. J., Reed, S. C. & Wilkinson, R. F. (1972). Fine structure of normal spermatid differentiation in *Drosophila melanogaster*. *J. Ultrastructure Res.*, **41**, 433–66

Tokuyasu, K. T. (1974). Dynamics of spermiogenesis in *Drosophila melanogaster*. III. Relation between axoneme and mitochondrial derivatives. *Exp. Cell Res.*, **84**, 239–50

Topfer-Petersen, E., Friess, A. E., Nguyen, H. & Schill, W.-B. (1985). Evidence for a fucose-binding protein in boar spermatozoa. *Histochemistry*, **83**, 139–45

12

The kinetics and biochemistry of mammalian sperm–egg interactions

H. D. M. Moore, C. A. Smith and T. D. Hartman

Gamete Biology Unit,
MRC/AFRC Comparative Physiology Group,
Institute of Zoology,
Zoological Society of London,
Regent's Park, London NW1 4RY, UK

Abstract

Mammalian fertilisation involves species specific binding of the spermatozoon to the surface of the zona pellucida and to the oolemma. These events are essential prerequisites for sperm passage through the zona matrix and fusion with the vitellus respectively. The sperm surface components involved in these processes have not been determined fully but are thought to be glycoproteins first expressed during sperminogenesis which undergo subtle modification during epididymal sperm maturation. Using a library of monoclonal antibodies in combination with *in vitro* fertilisation techniques it has been possible to identify and characterise sperm surface moieties involved in fertilisation processes. A complete elucidation of these components should now be possible using appropriate gene cloning methods and the analysis of sugar residues.

Introduction

As a consequence of the evolutionary transition from external to internal fertilisation, the need to synchronise the fertilising capacity of sperm populations with the release of the oocyte, and the requirement for sperm storage, gamete recognition mechanisms in mammals have become inextricably linked to the physiology and biochemistry of the whole animal. (For reviews see Yanagimachi, 1981; Fraser, 1984; Moore & Bedford, 1983.) Even so, the species-specific mechanisms which in primitive marine metazoa ensure incompatibility between the surface of gametes of heterologous species remain, and, although behavioural and anatomical differences in mammals could largely prevent interspecies fertilisation, their gametes still exhibit highly restrictive interactions (Bedford, 1981). With the advent of *in vitro* fertilisation techniques, the nature of many fertilisation processes, culminating in sperm/egg fusion and genetic exchange have been elucidated (see Hartmann, 1983). More recently, the biochemical foundation of these events have begun to be probed with monoclonal antibody technology. (For reference see Moore, 1985.)

To understand how recognition processes between the spermatozoon and egg may operate, it is important to consider how and when relevant gamete surface determinants are expressed, and the sequential changes they undergo in the male and female genital tracts. Although the oocyte contains essential receptors (complementary to those on spermatozoa) at the surface and within its investments, these entities are, in general, expressed early in development and thereafter remain biologically active and relatively stable (Dunbar, 1983). By contrast, the mammalian spermatozoon undergoes continual development as components which recognise the oocyte are incorporated into the sperm plasmalemma. Thus immature oocytes can be penetrated and fertilised by mature spermatozoa, even though they may not have attained the potential for future embryonic development (Overstreet & Hembree, 1976; Moore & Bedford, 1978). But the converse is not possible because immature spermatozoa leaving the testis have not at this stage acquired components recognising the egg surface (for reviews see Bedford, 1975; Orgebin-Crist, Danzo & Davies, 1975).

This chapter will cover the mechanisms by which the mammalian spermatozoon develops its recognition for the egg and the nature of the components involved. Particular reference will be made to the use of monoclonal antibodies which have been generated in our laboratory and elsewhere to probe fertilisation processes *in vitro*.

Spermatogenesis

In the mammalian testes, the primary diploid germ cell (spermatogonium) proliferates by mitosis and then by meiotic division to form a haploid spermatocyte (Steinberger & Steinberger, 1975; Bellve & O'Brien, 1983). This cell type undergoes extensive morphological reorganisation, developing the essential organelles for future fertilisation events. These include the acrosome, the elongated condensed nucleus and the flagellum. Perhaps the most salient feature related to recognition processes is the formation of a multidomain plasma-membrane (Holt, 1982). This provides the spermatozoon with discrete regions, particularly over the sperm head, with which sperm–egg interactions will be initiated. At this stage it should be appreciated that on leaving the testes, the spermatozoa are totally infertile and fail to bind to the oocytes. This does not mean however that development changes to the sperm surface during spermatogenesis do not have a bearing on fertilisation events.

During haploid development (spermatogenesis), a variety of new membrane components are synthesised and expressed. This can be demonstrated by immunolocalisation studies using monoclonal antibodies. For instance, in our laboratory, monoclonal antibody 97.5 (Moore *et al.*, 1985) recognises an antigen over the acrosomic granule and spreads over the rim of the acrosome, concomitant with the morphogenic changes to the spermatid

(Fig. 12.1). Other workers have also detected specific antigens on pachytene or later stage germ cells (O'Rand & Romrell, 1977; Bechtol, Brown & Kennett, 1979). Some of these determinants may result from stored mRNA although recent evidence is accumulating that there is post-meiotic gene transcription to account for these new proteins. (Kleene, Distel & Hecht, 1983; Dudley *et al.*, 1984). Glycosylation processes may also be taking place and in this regard Fenderson and co-workers, (1984) have shown with elegant sugar hapten assays that monoclonal antibodies to haploid germ cells may be recognising antigens associated with polylactosamine glycoconjugates. Neuraminidase action on these components suggests that cell surface sialyation accompanies meiosis. Such gene products may control the temporal and spatial co-ordination of the seminiferous epithelium, but they may also be involved in fertilisation recognition processes as monoclonal antibody 97.5 will specifically inhibit the binding of fully mature spermatozoa to the zona pellucida of the oocyte. (Moore, *et al.*, 1985.)

Acquisition by the spermatozoa of recognition receptors for the oocyte

As seen from antibody inhibition studies, stage-specific surface determinants expressed during spermatogenesis may clearly play an important role during

Fig. 12.1. Immunofluorescent localisation on elongated hamster spermatids of the antigen recognised by monoclonal antibody 97.5. Antibody is bound to membrane over developing acrosomes. Mag = × 800 (from Moore *et al.*, 1985).

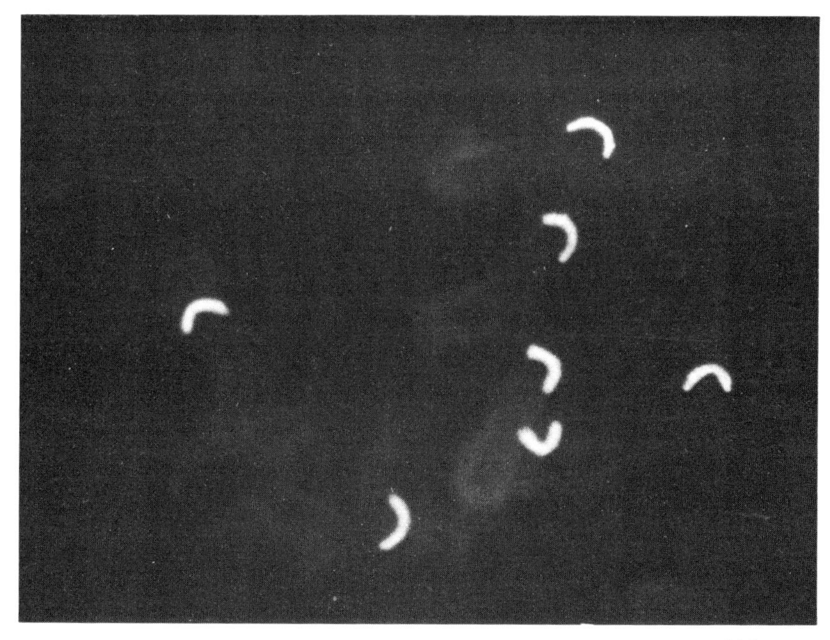

fertilisation; however, not until mammalian spermatozoa have passed through at least the proximal excurrent duct leading from the testis (efferent ducts and head of the epididymis) do they acquire fertilising capacity. This functional change to spermatozoa is accompanied by a number of intrinsic alterations to the sperm surface which result in the ability to bind to the zona pellucida and oolemma of oocytes. Such crucial maturation has been observed in every eutherian mammal examined. A good example is the more recent study by Saling (1982) in the mouse. She demonstrated with experiments to control the effect of the sperm motility (which also develops during maturation) that mouse spermatozoa must pass into the distal corpus epididymidis before they will attach to the zona pellucida. Similarly, in primates, using a heterologous assay with zona-free hamster oocytes, the acquisition of sperm receptors for the oolemma have been assessed in both the human and marmoset monkey (Hinrichsen & Blaquier, 1980; Moore, Hartman & Pryor, 1983; Moore, 1981*a*). In both species, spermatozoa apparently acquire receptors when passing through the corpus epididymidis. This change is associated with a modification in the binding of lectin probes (Fig. 12.2) such as ferritin-labelled Concanavalin A to the spermatozoa (Moore, Hartman & Holt, 1984). Given the condensed nature of epididymal sperm chromatin and the lack of organelles for protein synthesis, the key questions are how do new surface determinants become expressed during epididymal maturation and what is the nature of these epitopes?

Over the last few years, several lines of evidence point to the epididymal epithelium, under the control of androgens providing factors for sperm maturation. A complete review of these reports is beyond the scope of this chapter and the reader is directed to other reviews (Moore, 1983; Eddy *et al.*, 1985). Briefly, both antisera and monoclonal antibodies have been generated against epididymal epithelial secretory products which bind irreversibly to spermatozoa during epididymal transit (Lea, Petrusz & French, 1978; Moore & Hartman, 1984). These antigenic moieties have various molecular weights and it is unclear at present whether this exchange of components is the result of covalent binding or is mediated through, for instance, sugar

Fig. 12.2. Electron micrograph of a longitudinal section of a marmoset spermatozoon from the cauda epididymidis after incubation with ferritin-labelled Concanavalin A. The plasmalemma has an even distribution of ferritin particles over the head and tail. Mag = × 20,000 (from Moore, Hartman & Holt, 1984).

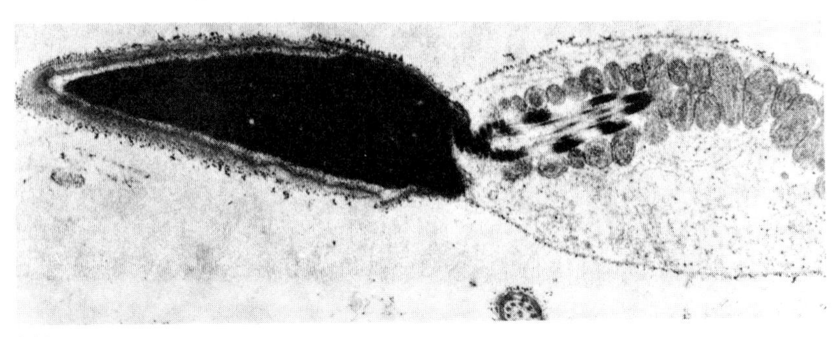

transferases which have been shown to be present in epididymal fluid and synthesised by the epithelial cells (Hamilton, 1980). Recent experiments in our laboratory involving the incubation of hamster immature spermatozoa in co-culture with epididymal epithelium indicate a highly specific transfer of antigen (34 kD) from the apical surface of the epithelial principal cells (Fig. 12.3) to sperm membrane (Fig. 12.4) of the post-acrosomal region and annulus (Smith, Hartman & Moore, 1986). Other studies involving *in vitro* fertilisation assays implicate some of these secretions as recognition factors since antibodies will specifically block sperm binding to the zona pellucida

Fig. 12.3. An electron micrograph of hamster epididymal epithelial cells after *in vitro* culture for five days. A 34 K determinant is immunolocalised at the apical cell surface by monoclonal antibody (C5). Mag = × 12,000 (from Smith, Hartman & Moore, 1986).

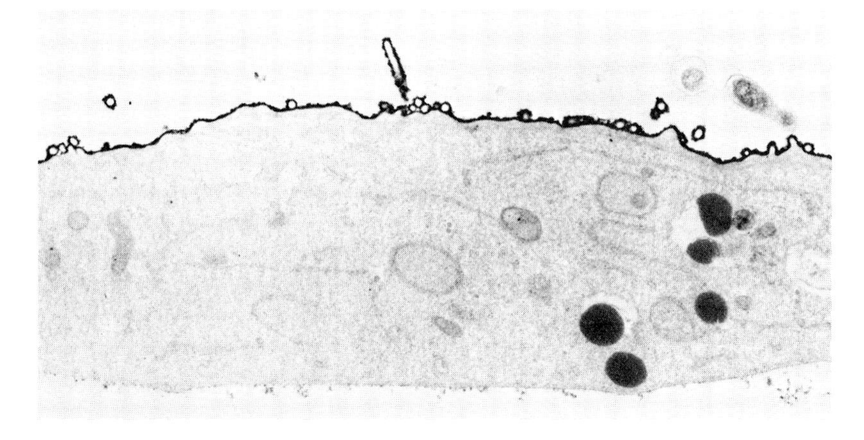

Fig. 12.4. An electron micrograph of an immature hamster spermatozoon following co-culture with hamster epididymal epithelial cells *in vitro*. The same determinant localised in Fig. 12.3 on the epithelial cells is bound to the post-acrosomal region of the sperm head as visualised by immunoperoxidase staining. Mag = × 13,000 (from Smith, Hartman & Moore, 1986).

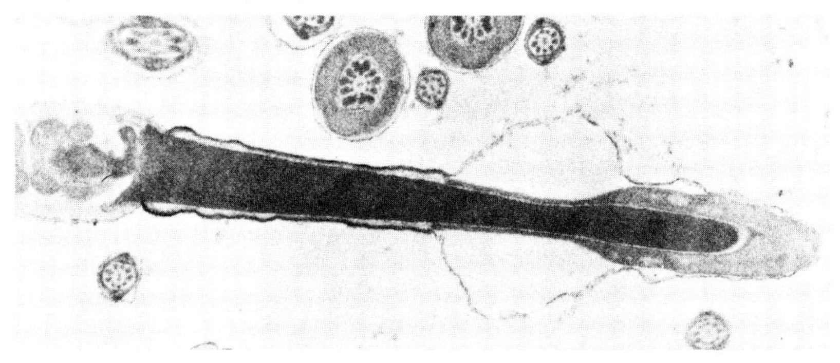

or oolemma (Moore, 1981; Moore & Hartman, 1984; Saling, Irons & Waibel, 1985).

Capacitation and the acrosome reaction

Mammalian spermatozoa have the potential for fertilisation on reaching the distal excurrent duct (i.e. cauda epididymis and vas deferens) but they must reside in the female tract or in appropriate culture medium *in vitro* before fertilisation can be realised. This period of conditioning is termed capacitation and immediately precedes the acrosome reaction; a fusion of the sperm plasmalemma and the outer acrosome membrane which results in breakdown of the anterior acrosome and release of its content. Both processes are prequisites for fertilisation and most likely confer on spermatozoa additional recognition sites for the oocyte (Yanagimachi, 1981; Moore & Bedford, 1983; Talbot, 1985).

Non-capacitated spermatozoa in some species can bind to the zona-pellucida *in vitro* (Bedford, 1967; Hartmann & Hutchinson, 1976) but this unnatural attachment will not allow sperm penetration of the zona matrix. At the site of fertilisation *in vivo* it is unlikely that any non-capacitated spermatozoa would reach the zona surfaces since the female tract and the cumulus mass surrounding the oocyte contain potent capacitating factors (see Moore & Bedford, 1983). Although little is known about capacitation, it is clear from studies with monoclonal antibodies that it may evoke a redistribution of antigen over the sperm surface, for instance, the flagellum and head (Myles & Primakoff, 1984; Ellis, Hartman & Moore, 1985). These changes are correlated with the development of increased sperm motility (hyperactivation) but it is unknown whether recognition components are also altered (Yanagimachi, 1970).

A more profound effect on sperm–egg recognition is the acrosome reaction such that in some species (i.e. guinea pig) the acrosome must be absent before spermatozoa can attach to the zona-pellucida (Huang, Flemming & Yanagimachi, 1981). The exact site at which the acrosome reaction occurs may be species specific. In the mouse, there is considerable evidence that it is initiated at the zona surface possibly in response to factors emanating from the oocyte (Saling, Sowinski & Storey, 1979). On the other hand, the hamster acrosome reaction may first occur as the spermatozoon traverses the cumulus mass (Cummins & Yanagimachi, 1982). Perhaps of more importance is the fact that the acrosome is, or soon becomes, fenestrated on presentation at the zona surface. This permits the spermatozoa to become attached and aligned so that they can subsequently penetrate the zona matrix.

Persuasive evidence of this type of recognition process comes from scanning electron microscopes observation of hamster-sperm attachment, and the similar angle of zona penetration slits (Yanagimachi & Phillips, 1984).

Sperm–oolemma interaction

The acrosome-reacted spermatozoon passes through the zona pellucida and enters the perivitelline space to then undergo fusion with the oolemma. The latter process occurs in a highly specific manner and involves an initial contact and fusion interaction of the sperm plasmalemma over the equatorial region of the head (Fig. 12.5) with microvilli of the oocyte (Moore & Bedford, 1978). To protect the fusogenic membrane of the equatorial segment during zona penetration, spermatozoa appear to have adopted a number of morphological strategies (e.g. a shoulder or rim around the anterior equatorial segment, or a depression of the distal equatorial region) to minimise sperm membrane contact with hard zona matrix (Bedford, Moore & Franklin, 1979). The exact nature of sperm–egg fusion is not readily understood. Most probably, surface recognition receptors are involved since once again specific antibodies will inhibit fusion (Moore & Hartman, 1984; Saling, Irons & Waibel, 1985) and even when the zona pellucida is removed oocytes will normally reject spermatozoa of a foreign species (Yanagimachi, 1984). The hamster oocyte is exceptional in this respect. When its zona pellucida is removed it will permit entry of spermatozoa of many mammalian species (Yanagimachi, 1984) and even avian spermatozoa (Samour, Moore & Smith, 1986).

Initial contact between the spermatozoa and the vitellus must be followed by a clearing of membrane protein in order to provide lipid interfaces at which fusion can occur. Why eutherian mammals have evolved such a highly

Fig. 12.5. An electron micrograph of the initial phase of the interaction of the fertilising hamster sperm head with the surface of an ovulated ovum in an *in vitro* fertilisation system. Z = Zona Pellucida, O = Ooplasma, P = Perinuclear Material. Mag = × 22,000 (from Bedford, Moore & Franklin, 1979).

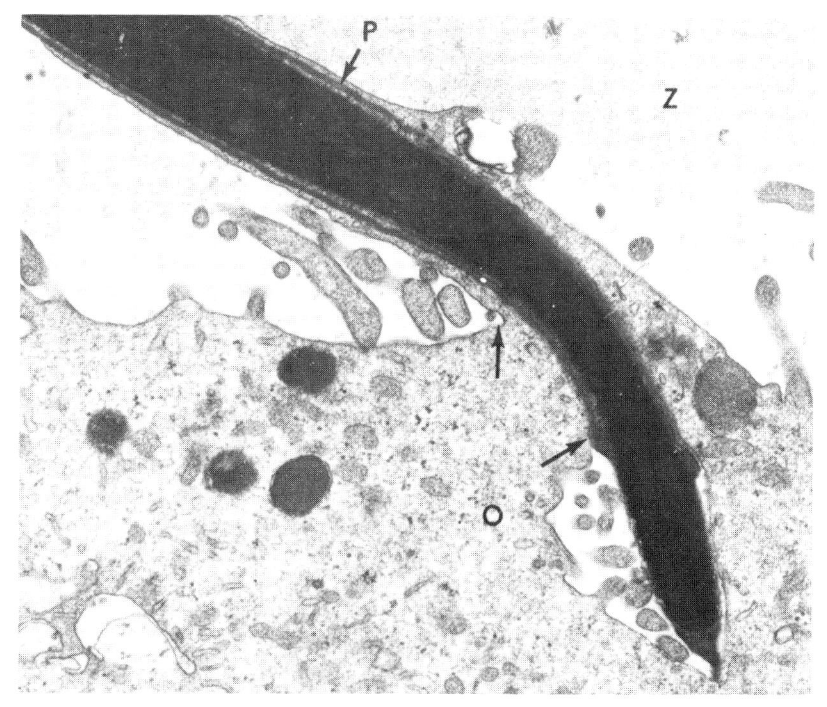

specific gamete mechanisms is not known. Spermatozoa from metatherians (marsupials, echidna) do not have a region equivalent to the equatorial segment and undergo a more direct fusion procedure (Rodger & Bedford, 1982).

Molecular biology of gamete recognition

Despite considerable effort over the last few years to characterise (particularly with antibodies) the components mediating gamete interaction, the molecular basis for these events is still uncertain. At the present time, the general interpretation of the results suggest that surface glycoproteins of the plasmalemma and the zona pellucida are of primary importance in regulating gamete adhesion. Since several recent reviews discuss in detail these processes, (Hartmann, 1983; Ahuja, 1985) the most important results will be just briefly mentioned here.

Three lines of evidence suggests carbohydrate involvement. Firstly, specific saccharides, especially those ending in a fucose or N-acetylgalactosamine and N-acetylglucosamine (i.e. fucoidan, ovomucoid) will inhibit fertilisation in rodents (Ahuja, 1982). Secondly, exoglycosidase (galactosidase, fucosidase) treatment of oocytes or neuraminidase treatment of spermatozoa can also reduce sperm–egg binding (Lambert & Le, 1984). Finally, an active carbohydrate component on mouse zona pellucida which recognises sperm (ZP3) has been isolated and purified (Wasserman & Bleil, 1982). Studies in the mouse also implicate the role of galactosyl transferase in sperm–egg binding although it is unlikely that this enzyme is the sole factor involved (Shur & Hall, 1982).

Summary

Gamete recognition in mammals involves the development of species specific receptors on both the spermatozoon and oocyte. The expression of such receptors on spermatozoa is determined by membrane changes initiated during spermatogenesis which are subsequently modified by factors secreted by the epididymal epithelium. Further alterations to these receptors may also occur during capacitation and the acrosome reaction. These sequential events have probably evolved to ensure that a population of fertile spermatozoa will be present in the ampulla of the oviduct soon after ovulation. Studies with antibodies and oligosaccharides suggest that glycoproteins are involved in sperm–egg interactions but the exact nature of these determinants have not yet been elucidated.

Acknowledgements

Studies by the authors have been supported by grants from the Ford Foundation and Elf-Aquitaine PLC and an MRC/AFRC programme grant.

References

Ahuja, K. K. (1982). Fertilization studies in the hamster. The role of cell surface carbohydrates. *Ex. Cell Res.*, **140**, 353–62

Ahuja, K. K. (1985). Carbohydrate determinants involved in mammalian fertilization. *Am. J. Anat.*, **174**, 207–23

Bechtol, K. B., Brown, S. C. & Kennett, R. H. (1979). Recognition of differentiation antigens of spermatozoa in the mouse by using antibodies from spleen-cell myeloma hybrids after syngenic immunization. *Proc. Natl Acad. Sci. USA*, **76**, 363–7

Bedford, J. M. (1967). Experimental requirement for capacitation and observations on ultrastructural changes in the rabbit spermatozoa during fertilization. *J. Reproduction & Fertility*, **2**, 35–48

Bedford, J. M. (1975). Maturation, transport and fate of spermatozoa in the epididymis. In *Handbook of Physiology*, Section 7, Endocrinology V, pp. 303–18, eds R. O. Greep & D. W. Hamilton. American Physiological Society, Washington

Bedford, J. M. (1981). Why mammalian gametes don't mix. *Nature (Lond)*, **291**, 286–7

Bedford, J. M. & Cooper, G. W. (1978). Membrane fusion events in the fertilization of vertebrate eggs. In *Cell Surface Reviews: Membrane Fusion*, pp. 65–125, eds G. Poste & G. L. Nicolson. Elsevier/North Holland, Amsterdam

Bedford, J. M., Moore, H. D. M. & Franklin, L. E. (1979). Significance of the equatorial segment of the acrosome of the spermatozoa in eutherian mammals. *Exp. Cell Res.*, **119**, 119–26

Bellve, A. R. & O'Brien, D. A. (1983). The mammalian spermatozoon: Structure and temporal assembly. In *Control of Mammalian Fertilization*, pp. 55–137, ed. J. F. Hartmann. Academic Press, New York.

Cummins, J. M. & Yanagimachi, R. (1982). Sperm–egg ratios and the site of the acrosome reaction during *in vivo* fertilization in the hamster. *Gamete Res.*, **5**, 239–56

Dudley, K., Potter, J., Lyons, M. F. & Willison, K. R. (1984). Analysis of male sterile mutations in the mouse using haploid stage expressed DNA probes. *Nucleic Acid Res.*, **12**, 4281–93

Dunbar, B. S. (1983). Morphological, biochemical and immunochemical characterisation of the mammalian zona pellucida. In Mechanism and control of animal fertilization, pp. 139–175, ed. J. F. Hartmann. Academic Press, New York

Eddy, E. M., Vernon, R. B., Muller, C. H., Hahnel, A. C. & Fenderson, B. A. (1985). Immunodissection of sperm surface modifications during epididymal maturation. *Am. J. Anat.*, **174**, 225–37

Ellis, D. H., Hartman, T. D. & Moore, H. D. M. (1985). Maturation and function of the hamster spermatozoon probed with monoclonal antibodies. *Journal of Reproductive Immunol.*, **7**, 299–314

Fenderson, B. A., O'Brien, D. A., Millette, C. F. & Eddy, E. M. (1984). Stage specific expression of three cell surface carbohydrates during spermatogenesis detected with monoclonal antibodies. *Dev. Biol.*, **103**, 117–28

Fraser, L. R. (1984) Mechanisms controlling mammalian fertilisation. In *Oxford Reviews of Reproductive Biology*, pp. 174–225, ed C. A. Finn. Oxford University Press, Oxford

Hamilton, D. W. (1980). UDP-galactose:N-acetylglucosamine galacosyltransferase in fluids from rat rete testis and epididymis. *Biol. of Reproduction*, **23**, 377–85

Hartmann, J. F. (1983). Gamete surface interactions *in vitro*. In *Mechanism and Control of Animal Fertilization*, pp. 325–63, ed. J. F. Hartmann. Academic Press, New York

Hartmann, J. F. & Hutchinson, C. F. (1976). Surface interactions between mammalian sperm and egg. Variation of spermatozoa concentration as a probe for the study of binding *in vitro*. *Cell Physiol*, **88**, 219–26

Hinrichsen, M. J. & Blaquier, J. A. (1980). Evidence supporting the existence of sperm maturation in the human epididymis. *J. Reproduction & Fertility*, **60**, 291–4

Holt, W. V. (1982). Functional development of the mammalian sperm plasma membrane. In *Oxford Reviews of Reproductive Biology*, pp 195–240, ed. C. A. Finn. Clarendon Press, Oxford

Holt, W. V. (1983). Membrane heterogenicity in the mammalian spermatozoon. In *International Review of Cytology*, pp. 159–94, eds G. H. Bourne & J. F. Danielli. Academic Press, Orlando

Huang, T. T. F., Flemming, A. D. & Yamagimachi, R. (1981). Only acrosome reacted spermatozoa can bind to and penetrate the zona pellucida: A study using the guinea-pig. *J. Exp. Zool.*, **217**, 287–90

Kleene, K. C., Distel, R. J. & Hecht, N. B. (1983). DNA clones encoding cytoplasmic poly(a) RNAs which first appear at detectable levels in haploid phases of spermatogenesis in the mouse. *Dev. Biol.*, **98**, 455–64

Lambert, H. & Le, A. V. (1984). Possible involvement of a sialyated component of the sperm plasma membrane in sperm–zona interactions in the mouse. *Gamete Res.*, **10**, 153–63

Lea, O. A., Petrusz, P. & French, F. S. (1978). Purification and isolation of acidic epididymal gylcoprotein (AEG); A sperm coating protein secreted by the rat epididymis. *Int. J. Androl.*, **2**, 592–607

Moore, H. D. M. (1981*a*). An assessment of the fertilising ability of spermatozoa in the epididymis of the marmoset monkey (*Callithrix jacchus*). *Int. J. Androl.*, **4**, 321–30

Moore, H. D. M. (1981*b*). Glycoprotein secretions of the epididymis in the rabbit and hamster: Localisation on epididymal spermatozoa of specific antibodies on fertilisation *in vivo*. *J. Exptl. Zool.*, **215**, 77–85

Moore, H. D. M. (1983). Physiological and *in vitro* models of sperm maturation. In *In vitro Fertilization and Embryo Transfer*, pp. 9–38, eds. P. G. Crosigiani & B. L. Rubin. Academic Press, New York

Moore, H. D. M. (1985). The mammalian sperm surface – a bibliography. *Bibliog. Reproduction*, **46(1)**, A1–A12

Moore, H. D. M. & Bedford, J. M. (1978). Ultrastructure of the equatorial segment of hamster spermatozoa during penetration of oocytes. *Ultrastructural Res.*, **63**, 110–17

Moore, H. D. M. & Bedford, J. M. (1983). The interaction of mammalian gametes in the female. In *Mechanism and Control of Animal Fertilisation*, pp. 435–97, ed. J. F. Hartman. Academic Press, New York

Moore, H. D. M. & Hartman, T. D. (1984). Localization by monoclonal antibodies of various surface antigens of hamster spermatozoa and the effect of antibody on fertilization *in vitro*. *J. Reproduction & Fertility*, **70**, 175–83

Moore, H. D. M., Hartman, T. D., Brown, A. C., Smith, C. A. & Ellis, D. H. (1985). Expression of sperm antigens during spermatogenesis and maturation detected with monoclonal antibodies. *Exptl. & Clin. Immunogenetics*, **2**, 84–96

Moore, H. D. M., Hartman, T. D. & Holt, W. V. (1984). The structure and epididymal maturation of the spermatozoon of the common marmoset (*Callithrix jacchus*). *J. Anat.*, **138**, 227–35

Moore, H. D. M., Hartman, T. D. & Pryor, J. P. (1983). Development of the oocyte-penetrating capacity of spermatozoa in the human epididymis. *Int. J. Androl.*, **6**, 310–18

Myles, D. G. & Primakoff, P. (1984). Localised surface antigens of guinea-pig migrate to new regions prior to fertilisation. *Cell Biol.*, **99**, 1634–42

O'Rand, M. G. & Romrell, L. J. (1977). Appearance of cell surface auto and isoantigens from rabbit testes. *Immunology*, **122**, 1248–54

Orgebin-Crist, M. C., Danzo, B. J. & Davis, J. (1975). Endocrine control of the development and maintenance of sperm fertilising ability in the epididymis. In *Handbook of Physiology*, Section 7, *Endocrinology*, pp. 319–38, eds D. W. Hamilton & R. O. Greep. American Physiological Society, Washington

Overstreet, J. W. & Hembree, W. C. (1976). Penetration of the zona pellucida of non-living oocytes by human spermatozoa *in vitro*. *Fertility & Sterility*, **27**, 815–31

Rodger, J. C. & Bedford, J. M. (1982). Separation of sperm pairs, and sperm–egg interaction in the opossum *Didelphis virginiana*. *J. Reproduction & Fertility*, **64**, 171–9

Saling, P., Sowinski, J. & Storey, B. (1979). An ultrastructural study of epididymal mouse sperm binding to the zona pellucida *in vitro*: sequential relationship to the acrosome reaction. *J. Exp. Zool.*, **209**, 229–38

Saling, P. M. (1982). Development of the ability to bind to zonae pellucidae during epididymal maturation: reversible immobilization of mouse spermatozoa by lanthanum. *Biol. Reproduction*, **26**, 429–36

Saling, P. M., Irons, G. & Weibel, K. (1985). Mouse sperm antigens that participate in fertilization. 1. Inhibition of sperm fusion with the egg plasm membrane using monoclonal antibodies. *Biol. Reproduction*, **33**, 525–36

Samour, J., Moore, H. D. M. & Smith, C. A. (1986). Avian spermatozoa penetrate zona-free hamster oocytes *in vitro*. *J. Exp. Zool.*, **239**, 295–8

Shur, B. D. & Hall, N. G. (1982). A role for mouse sperm surface galactosyl transferase in sperm binding to the zona pellucida. *Cell Biol.*, **95**, 574–9

Smith, C. A., Hartman, T. D. & Moore, H. D. M. (1986). A determinant of Mr, 34,000 expressed by hamster epithelium binds specifically to spermatozoa in co-culture. *J. Reproduction & Fertility*, 78 (in press)

Steinberger, E. & Steinberger, A. (1975). Spermatogenic function of the testes. In *Handbook of Physiology*, Section 7, *Endocrinology*, Vol. V., pp. 1–20, eds D. W. Hamilton & R. O. Greep. American Physiological Society, Washington

Talbot, P. (1985). Sperm penetration through oocyte investments in mammals. *Amer. J. Anat.*, **174**, 331–46

Wasserman, P. M. & Bleil, J. D. (1982). The role of zona pellucida glycoprotein as regulator of sperm–egg interactions in the mouse. In *Cellular Recognition*, pp. 845–56, eds. L. Glaser, D. Gotteib & W. Frazier. AR liss, New York

Yanagimachi, R. (1970). The movement of golden hamsters spermatozoa before and after capacitation. *J. Reproduction & Fertility*, **23**, 193–6

Yanagimachi, R. (1981). Mechanisms of fertilisation in mammals. In *Fertilisation and Embryonic Development in vitro*, pp. 81–182, eds L. Mastroianni Jr & J. D. Biggers. Plenum, New York

Yanagimachi, R. (1984). Zona free hamster eggs. Their use in assessing fertilising capacity and examining chromosomes of human spermatozoa. *Gamete Res.*, **10**, 187–232

Yanagimachi, R. & Phillips, D. M. (1984). The status of acrosomal caps of hamster spermatozoa immediately before fertilization *in vivo*. *Gamete Res.*, **9**, 1–20

13

Sperm cell determinants and control of fertilisation in plants

R. Bruce Knox[1], Darlene Southworth[1,2] and Mohan B. Singh[1]

[1]Plant Cell Biology Research Centre,
School of Botany, University of Melbourne,
Parkville, Victoria 3052, Australia;
[2]Southern Oregon State College,
Ashland, OR 97520, USA

Abstract

The sperm cells of flowering plants are natural protoplasts held within the vegetative cell of mature viable pollen or pollen tubes. Three-dimensional reconstruction of serial thin sections in four genera shows that the pair of sperm cells are linked, and one is associated with the vegetative nucleus, forming a male germ unit. These cells move together through the pollen tube, apparently pre-programmed for cellular fusion with the egg or the central cell initiating double fertilisation. This paper addresses two questions. What is the nature of the sperm cells and the delivery systems and how do sperm cells recognise the egg and central cell? In *Brassica* and *Gerbera*, cell fractionation, following physical disruption of pollen, provides preparations of osmotically active elongate sperm cells. Monoclonal antibodies are being used to characterise their cellular determinants.

Introduction

In flowering plants, many species fail to achieve full reproductive capacity, manifested by a reduction in seed set. A framework in which we can determine the component processes involved is reproductive success. This is defined by population biologists as the rate of success of an individual plant in fathering as many viable embryos as possible (see Willson, 1983). In terms of cell biology, it is possible to quantitatively estimate certain components of reproductive success, namely:

- pollen quantity and viability;
- cytological monitoring of pollen germination and rate of tube growth through the pistil;
- pollen gene expression;
- success of fertilisation.

The first three components have recently been reviewed (Knox *et al.*, 1986; Knox, 1987). The last parameter will be considered in this chapter in terms of the effectiveness of the male gametes, the sperm cells.

Little is known of these male gametes. The haploid pollen grains of most angiosperms carry two cells. These are a large vegetative cell that fills the grain, and a smaller cell – the progenitor of the gametes – the generative cell, which lies wholly within the vegetative cell. Division of the generative cell to form two sperm cells usually occurs several hours after pollen germination, during tube growth through the style (see Knox, *et al.*, 1986). In about 30% of angiosperm families, the grains are tricellular, containing the sperm cells at maturity. These occur in such families as the grasses and cereals (*Gramineae*), cabbages (*Cruciferae*) and sunflower (*Compositae*). In both cases, the sperm cells are considered to be carried passively via the pollen tube through the tissues of the pistil to the ovules, where fertilisation occurs.

In angiosperms, fertilisation is shown to be a double event by several cytologists in the last years of the nineteenth century. Apparently at random, one sperm cell fertilises the egg cell to form the diploid embryo, while the other fuses with the central cell to give the usually triploid endosperm. The cell biology of these events has been explored in the pioneering ultrastructural studies of Jensen (1974) in several plant systems. Briefly, he showed that when the pollen tube is in the style, one of the synergids undergoes autolysis. It is this cell that the tube enters (Fig. 13.1A) and the sperm cells are discharged through a terminal pore in the pollen tube. The sperms pass up through a window in the cell wall of the synergid, and enter the periplasmic space between the plasma membranes of the two female gametes (Fig. 13.1B). Fertilisation occurs initially by cellular fusion between the plasma membranes of the pairs of gametes, and later after migration of the male nuclei, is followed by nuclear fusion between both male and female gametes (Fig. 13.1C).

The evidence is fragmentary, and available for only a few species. The role of the sperm cells in double fertilisation remains an enigma. On the one hand, there is the possibility that cellular fusion is a random process. On the other hand, the potential specificity of determinants in cellular membranes suggests that some form of specialisation of reproductive function may occur, and the sperms may be pre-programmed for double fertilisation. This article reviews some of the possible options, and presents new findings that show how this problem is being approached with the range of technologies currently available in plant cell biology.

What is the nature of the sperm cells and the delivery system?

In appearance, the sperm cells of higher plants are elongate or spindle-shaped (Table 13.1). This is their characteristic form as determined by ultrastructural observations of mature pollen grains or tubes. In size, the sperm cells are small and reduced compared to the vegetative cell (Table 13.1), for example, *Brassica campestris*, 2×7 μm within a grain of 35 μm diameter and *Zea mays* 5×35 μm within a grain of 120 μm diameter.

In structure, most sperm cells comprise two regions that appear to be functionally independent; an enlarged nuclear region which also contains some heritable organelles, and one or more usually narrow cytoplasmic extensions. Sperm cell nuclei contain densely staining heterochromatic granular chromatin, suggesting an inactive state in terms of gene transcription. This is in keeping with their reduced state.

The heritable organelles have a characteristic morphology or organisation that may be associated with the specialised function of sperm cells. They show a close association with the nucleus. In *Zea mays*, three-dimensional reconstruction indicates the existence of one of more filamentous mitochondrial complexes in the sperms of mature pollen (McConchie *et al.*, 1987). In *Brassica* sperms the mitochondria are spherical and are smaller than their counterparts in the vegetative cell (Dumas *et al.*, 1985).

The sperm cells also possess one or more extensions, that are much narrower, for example, in *B. campestris* the extension may be 10 μm long, and comprise paired plasma membranes and an array of microtubules (McConchie *et al.*, 1985*a*). The occurrence of arrays of microtubules in these sperm cells may account for their shape. An extension up to 30 μm long is reported in *Plumbago* sperm cells (Russell, 1984).

The most obvious difference from animal sperms is in motility, which is a characteristic of animal sperm and those of lower plants. In order to achieve this, mammalian sperm cells have two functionally different regions; a head containing the nucleus and acrosome (an apparatus for egg penetration), and a tail that generates the flagellar wave and assists in egg penetration (see Holt, 1984). The tail of mammalian sperm comprises two separate components, a mitochondrial spiral in the middle piece, and axial filaments and fibres of the posterior region of the flagellum.

The entire sperm cell of mammals is surrounded by its plasma membrane, although some of the individual regions are also separately enveloped: the nucleus, mitochondria and acrosome (see Holt, 1984). Prior to fertilisation, during epididymal transit, the acrosome becomes smaller, and the outer acrosomal and plasma membranes fuse, resulting in the release of the acrosomal enzymes and exposure of a new surface membrane, the inner acrosomal membrane.

The sperm cells of higher plants are bounded by paired plasma membranes: their own and that of the inner vegetative cell, for example, *Brassica campestris* (Fig. 13.2). They are natural protoplasts, except for the few cases where a wall has been reported (Table 13.1). The plasma membranes may be closely appressed or loosely associated with regions of close binding and pockets where a periplasm exists between the cells (Table 13.1). Plasmodesmata-like connections or trabeculae, membranous connections, may cross between the two plasma membranes, and uranophylic vesicles and osmiophilic droplets may also occur in the periplasmic space (Dumas *et al.*, 1985). These observations suggest that some kind of communication is maintained between the vegetative cell and its descendants, the sperm cells and vice versa.

In several systems, the sperm cells are paired together, within a common periplasm by the inner vegetative cell plasma membrane. In *Brassica campestris*, this is achieved by a network of finger-like evaginations (McConchie, *et al.*, 1985*a*; 1986). In *Plumbago zeylanica* it is by means of a common lateral wall, traversed by plasmodesmata (Russell & Cass, 1981).

The dimorphism of flowering plant sperms is largely caused by the

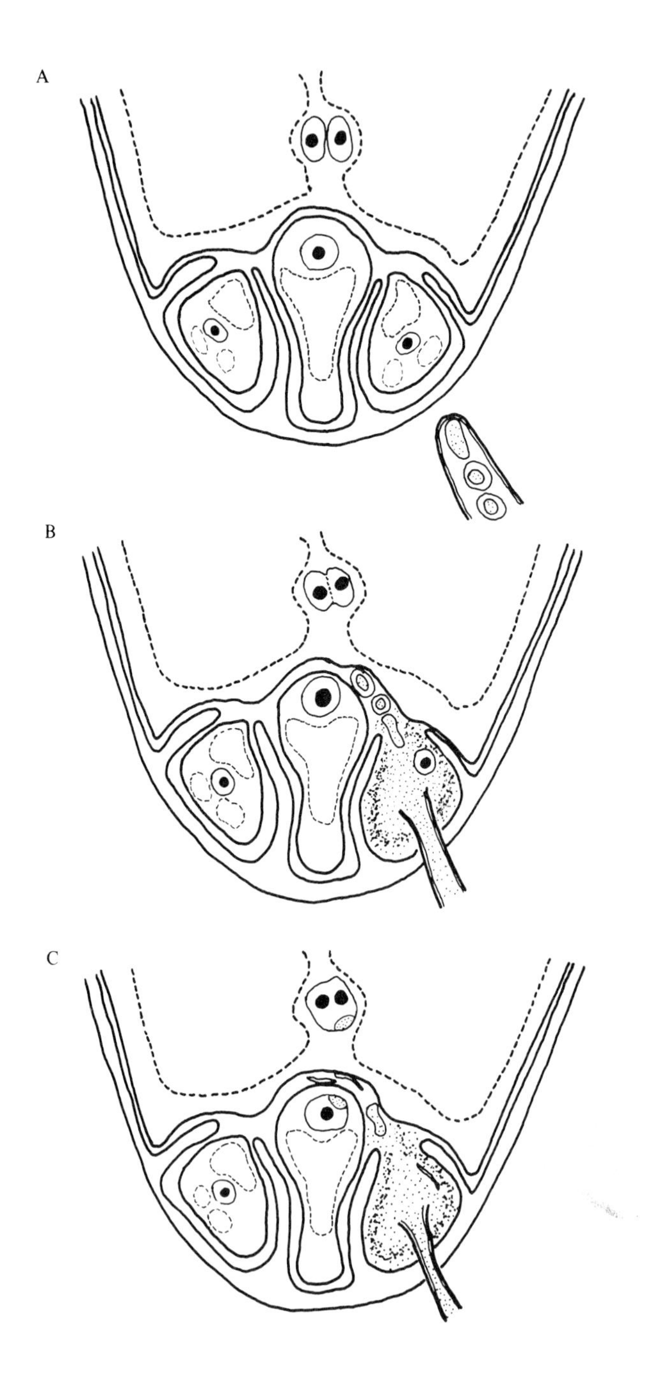
A
B
C

Fig. 13.1. Schematic diagrams showing the key ultrastructural features of double fertilisation based on the work of Jensen (1974). A, Pollen tube (*lower RHS*) approaches the embryo sac. Ovular envelopes are not shown. B, Pollen tube has penetrated the embryo sac wall and entered the RHS synergid, which had previously undergone autolysis. The sperm cells are discharged into the periplasmic space between the egg and central cell. Cellular fusion follows. C, The sperm nuclei have migrated to the female gamete nuclei and nuclear fusion has occurred. In this system, the cytoplasm of the sperm cells aborts in the periplasm, and does not contribute to inheritance. (from Knox *et al.*, 1986). Explanation of symbols: e = egg cell; cc = central cell; cw = wall of embryo sac; sy = synergid; pt = pollen tube; sc = sperm cell; vn = vegetative nucleus; sn = sperm nucleus.

extension that links one of the sperms to the vegetative nucleus, for example in *Brassica* and *Plumbago*. One of the sperm cells is associated by means of the extension with the vegetative nucleus (Svn), while the other is not associated (Sua). In *Brassica campestris* this has been demonstrated by use of the technique of 3-dimensional reconstruction of serial thin sections (Fig. 13.3). The extension appears to be appressed to the surface of the envelope of the vegetative nucleus. In thin sections, there is a gap of several nm between them. Essentially similar observations have been made in *Plumbago zeylanica* and *Brassica oleracea* (Dumas *et al.*, 1985; McConchie *et al.*, 1985*a*,*b*).

In *B. campestris*, the extension of Svn, strengthened by a microtubular array, passes into enclaves in the vegetative nucleus, often emerging at the other side, forming a distinct protuberance (Fig. 13.4). The extension is accompanied by cytoplasmic components of the vegetative cell, especially RER. This physical association of the sperm cells and vegetative nucleus constitutes the male germ unit. This refers to the linkage of all the DNA of male heredity, both cytoplasmic and nuclear, into a single unit for transmission to the embryo sac (Dumas *et al.*, 1984*a* and *b*).

Male germ units probably occur in at least 12 genera of angiosperms (Table 13.2). The nature of the evidence supporting these findings is most convincing in three of the four cases where three-dimensional reconstruction of serial thin sections has been carried out (Fig. 13.5). In *Brassica*, *Plumbago* and *Spinacia*, the unit is maintained by physical contact. In the fourth genus, *Zea*, there is no apparent attachment of the male gametes, although one sperm cell is sited within a few nm of the vegetative nuclear envelope (McConchie *et al.*, 1987).

Sperm dimorphism also involves an unequal distribution of heritable cytoplasmic organelles (see Russell, 1986). Relative numbers of mitochondria and plastids in Svn and Sua of several systems are given in Table 13.3. Mitochondrial numbers per sperm cell vary from 6 to 336 (but only to a maximum of 23 in *Brassica*), and plastid numbers from 0 to 24. Plastids are apparently absent in *Brassica*, *Hordeum* and *Spinacia* sperm cells where mitochondria represent the only heritable organelles.

The question arises whether these mitochondria contain any DNA in view of their restricted quantity? The mitochondria of *Brassica* are small and spherical, as shown by three-dimensional reconstruction (McConchie *et al.*, 1985*a*). It is known that less than half the spherical mitochondria of cucurbit leaves contain a complete genome (Bendich & Gauriloff, 1984). Since the

Table 13.1. *Selected list of characteristic features of angiosperm sperm cells, based on ultrastructural observations.*

Species	Size and shape (μm)	Dimorphic	Cell wall (W) or Periplasm (P)	Heritable organelles	Reference
Beta vulgaris	Spindle NR2	ND	P, paired membranes very close	M	Hoefert, 1969
Brassica campestris	2 × 7, Ext 10 × 0.5, NR 1.5–2	+	P, 0.1–0.2 variable, incl connections	M	McConchie *et al.*, 1985*a*, 1986
Brassica oleracea	2 × 3 × 10	+	P, 0.1–0.5, incl vesicles, trabeculae	M	Dumas *et al.*, 1985; McConchie *et al.*, 1986
Gossypium hirsutum	8 × 14 in pollen tube	ND	P	M	Jensen & Fisher, 1969
Hippeastrum vitatum	Ext present pollen tube	–	P	M	Mogensen, 1986
Hordeum vulgare	Ext present in one sperm cell	–	P, 0.6	M	Mogensen & Rushe, 1985
Plumbago zeylanica	Spindle 4 × 8 ext 30 × 1	+	W, rudimentary 0.05–0.2 incl plasmodesmata and vesicles	M, Pl	Russell, 1983 1984, 1985
Secale cereale	Elongate	ND		M	Karas & Cass, 1976
Spinacia oleracea	NR 2, 10–12 long	ND	P, 0.05 with cross-connections ('plasmodesmata channels'). W common cross-wall	M	Wilms & van Aelst, 1983; Wilms, 1985, 1986
Triticum with tail *aestivum*	Elongate	ND	P ('wall')	M, Pl	Chu *et al.*, 1980; Chu & Hu, 1981
Triticale sp.	Elongate	ND	P membranes paired	M, Pl	Schroder, 1983
Zea mays	1 × 5 × 35 3.5 × 4.5 × 10	+	P	M	McConchie *et al.*, 1987

Abbreviations: NR nuclear region diameter; Ext extension; ND not determined; M mitochondria; Pl plastids.

mitochondria of *Brassica* sperm are similar in shape and volume it follows that these mitochondria may contain little or no DNA. This may be a means by which naturally occurring cytoplasmic hybridity in this species may be prevented (McConchie *et al.*, 1986).

A highly branched filamentous chondriome has been found in *Zea mays* sperm cells in mature pollen (McConchie *et al.*, 1987). This is analogous to the situation in several animal systems, for example, the *nebenkern*, the highly branched interlocking pair of mitochondrial complexes found in spermatid metamorphosis in insects (Favard & Andre, 1970; Tokuyasu, 1975). Similar structures are to be found in the sperm of lower plants, for example, the green alga *Bryopsis* (Burr & West, 1970), developing sperm cells of bryophytes (Duckett, 1974) and the multiflagellate sperms of *Equisetum* (Duckett, 1975) and ferns (Myles & Bell, 1975).

During transit through the pollen tube, the sperm cells may show two characteristic features:

- changes in size, for example in *Brassica campestris* (Williams *et al.*, submitted for publication);
- elongation of the extension when the sperm cells are associated as a germ unit, as shown in *Plumbago* (Russell & Cass, 1981).

This period, ends at the moment of entry into the embryo sac. In *Plumbago*, Russell (1983, 1984) has recorded that after arrival within the synergid, the pollen tube deposits the sperm cells, via a terminal aperture, into the periplasm between the egg and central cell in preparation for fertilisation. However, during this process, the outer membrane surrounding the male germ unit (i.e. the plasma membrane of the inner vegetative cell) is shed, so that the sperm cells are surrounded only by their own plasma membrane. This process certainly has parallels with the acrosome reaction in mammalian sperm. Russell (1984) has commented on its similarity with the process of capacitation. The properties of the mammalian sperm plasma membrane when exposed to the environment of the female reproductive tract are changed in terms of membrane permeability, membrane fluidity and receptor specificity (see Holt, 1984). These changes prepare the sperms for initiation of the acrosome reaction and hence for fertilisation.

It is evident that we currently know very little about sperm cell biology in the lead up to fertilisation in higher plants. However, if Russell's (1984) observations prove a general phenomenon in other systems, it seems that the loss of the outer membrane of the pair represents an important step in establishing sperm membrane specificity, that is essentially analogous to the acrosome reaction in its effect.

How do the sperm cells recognise the egg and central cell?

Double fertilisation requires that one sperm fuse with the egg while the other fuses with the central cell. These two fusions occur at adjacent cells close together in time and without competition from other sperm since in most cases, only one pollen tube discharges into a synergid. Without both fertilisations, the embryo will not survive: without egg–sperm fusion, the endosperm is not nourished and will die. Since fertilisation in animals (echinoderms) has been shown to involve cell surface recognition substances that bind egg and sperm together and lead to membrane fusion, it is reasonable that plant gametes might also have cell surface determinants that function in fertilisation. With the process of double fertilisation, two alternative

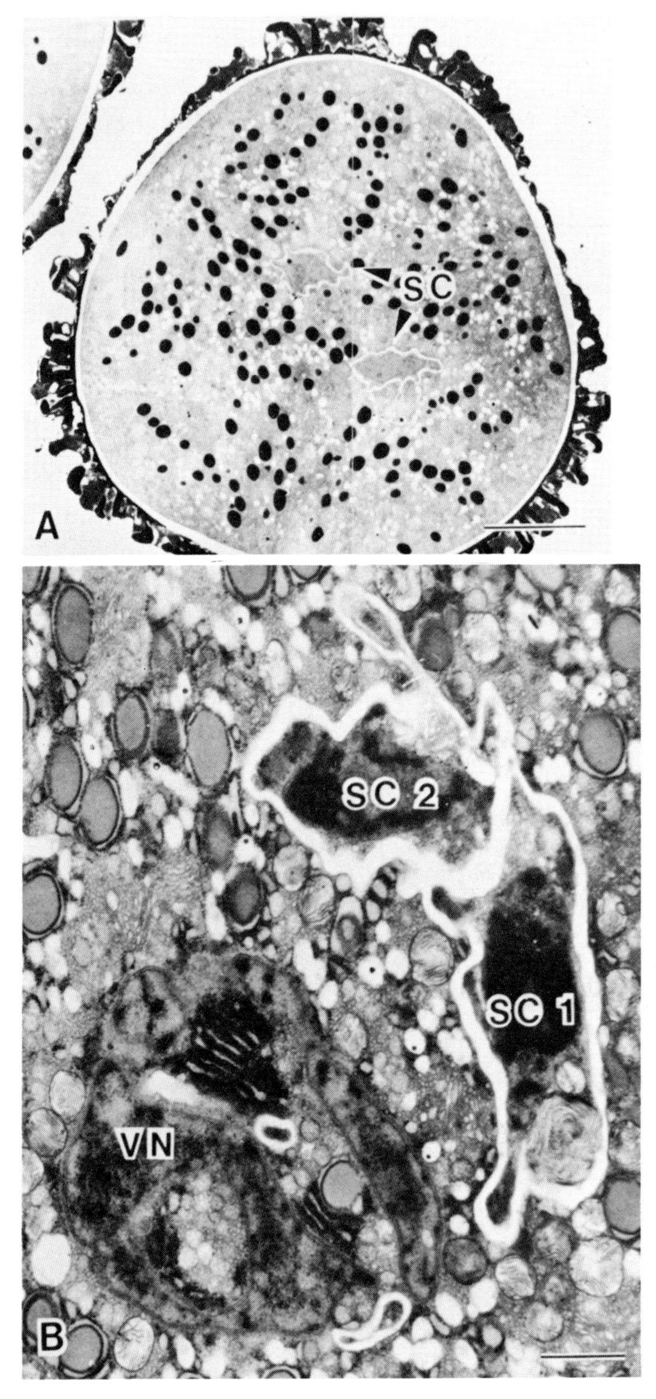
SC
A
SC 2
SC 1
VN
B

Fig. 13.2. Thin sections of the pollen grain of *Brassica campestris* showing: A, the pair of sperm cells; B, pair of sperm cells (SC1 and SC2) linked together, with an extension of one of them (SC1) within enclaves of cytoplasm within the vegetative nucleus (VN) (from McConchie *et al.*, 1985*a*).

hypotheses may be proposed: (i) the specific receptor hypothesis and (ii) the chance hypothesis.

In the specific receptor hypothesis the fusion of one sperm cell with the egg or central cell might be predetermined by specific cell surface molecules. The surfaces of the two sperm cells would possess different determinants as would the surfaces of the egg and central cell (Fig. 13.6). A combination of Brownian motion, slight sperm movement or other forces would bring the sperms into contact with the egg and central cell and agglutination or cell fusion would follow an initial recognition event.

In the chance hypothesis, the sperm cell surfaces would be identical and could bind equally well to either egg or central cell. The sperm that fertilised the egg would be determined by chance or by factors other than surface substances. Once one sperm cell bound, a reaction blocking polyspermy would occur. This could be partly a physical blockage by the attached sperm

Fig. 13.3. Computer-assisted three dimensional reconstructions of the male germ unit of *Brassica campestris*. A, shows the pair of sperm cells (SC1 and SC2), with branched extensions of SC1 traversing the vegetative nucleus (VN) and entering enclaves unit. B, same reconstruction, but data for vegetative nucleus omitted, showing the structure of the extensions of SC1 (from McConchie *et al.*, 1985*a*).

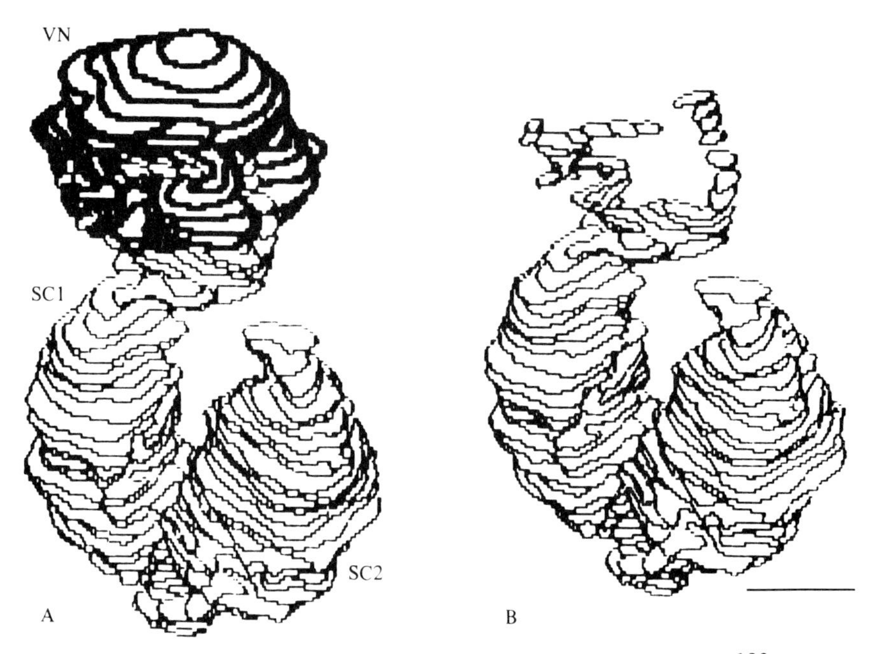

or an electrochemical block involving a transient ion exchange and voltage drop as in sea urchin eggs. Evidence of sperm cell surface dimorphism would favour the specific determinants hypothesis while absence of surface dimorphism would favour the chance and polyspermy hypothesis. Evidence needs to be obtained from experimental studies to distinguish between these two hypotheses.

Examination of free-living sperm cells outside the vegetative cell would facilitate the study of cell surfaces by a number of methods including freeze fracture electron microscopy and specific cytochemical probes, including production of monoclonal antibodies to surface molecules. Although sperm cells are closely associated with the vegetative cell plasma membrane, sperm cells have been observed in disrupted pollen grains and pollen tubes, using Nomarski DIC optics (Karas & Cass, 1976) and fluorescence microscopy with DNA-specific fluorochromes (Hough *et al.*, 1985).

Pollen grains of *Plumbago* (Russell, 1986) have been disrupted by osmotic shocks and spherical structures interpreted as osmotically swollen sperm cells are released. These protoplast-like structures excluded Evan's blue, but gave a negative FCR test. Their functional capacity is unknown. Similar spherical

Fig. 13.4. Schematic diagram of a reconstructed male germ unit of *Brassica campestris*, showing the paired sperm cells (SC1 and SC2) and the extension of SC1 that enters enclaves in the vegetative nucleus. Note the longitudinally oriented arrays of microtubules and the small spherical mitochondria adjacent to the sperm nuclei (from McConchie *et al.*, 1985*a*).

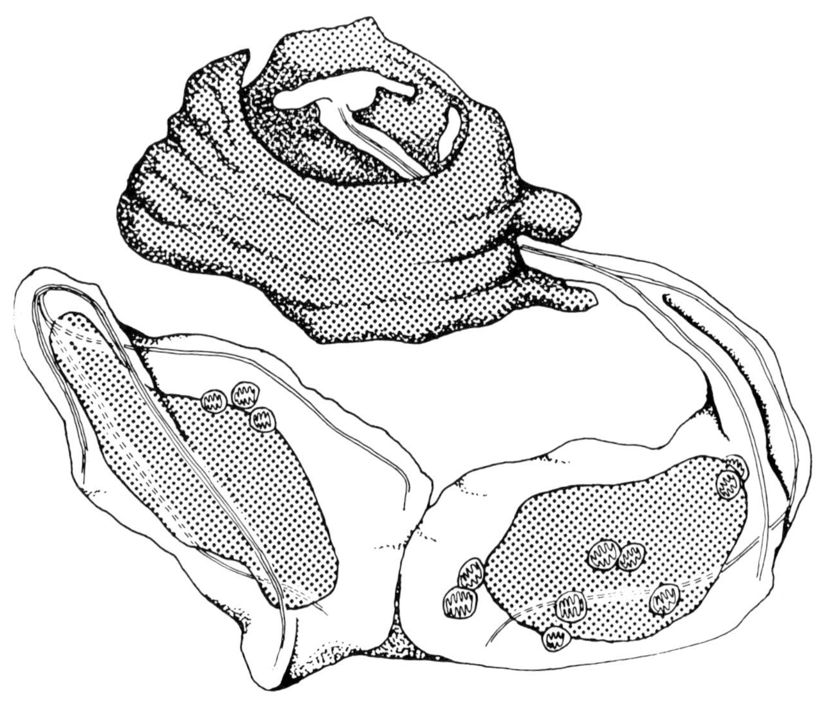

Table 13.2. *Selected list of pollen systems in which the presence (+) or absence (−) of male germ units is reported or inferred from micrographs (extended from McConchie* et al., *1986).*

Family	Species	Evidence*	Reference
A. DICOTYLEDONS			
Bicellular:			
Ericaceae	*Rhododendron laetum*	1 − pg	Theunis *et al.*, 1985
		1 + pt	Kaul *et al.*, 1987
Leguminosae, Mimosoideae	*Acacia retinodes*	2 + pg	McCoy & Knox, unpublished data
Malvaceae	*Gossypium hirsutum*	3 + pt	Jensen & Fisher, 1968
Ranunculaceae	*Ranunculus macranthus*	3 + pg	Larson, 1965
	Helleborus foetidus	4 − pg	Helop-Harrison *et al.*, 1986
Rosaceae	*Prunus avium*	3 + pt	Cresti *et al.*, 1985
Solanaceae	*Nicotiana alata*	3 + pt	Cresti *et al.*, 1985
	Petunia hybrida	2 + pt	Wagner & Mogensen, pers. comm.
Tricellular:			
Brassicaceae	*Brassica campestris*	1 + pg	McConchie *et al.*, 1985
		1 + pg	McConchie & Knox, 1986
	Brassica oleracea	1 + pg	McConchie & Knox, 1986
		2 + pg	Dumas *et al.*, 1984, 1985
Chenopodiaceae	*Beta vulgaris*	2 − pg	Hoefert, 1969
	Spinacia oleracea	2 + pt	Wilms & van Aelst, 1983
		2 + pt	Wilms, 1985
		1 + pg	Wilms *et al.*, 1986
Plumbaginaceae	*Plumbago zeylanica*	2 + pg, t	Russell & Cass, 1981
		2 + pg	Russell & Cass, 1983
		1 + pg	Russell, 1984
B. MONOCOTYLEDONS			
Bicellular:			
Amaryllidaceae	*Hippeastrum vitatum*	2 + pt	Mogensen, 1986
Tricellular:			
Poaceae	*Alopecurus pratensis*	4 − pt	Heslop-Harrison and Heslop-Harrison, 1984
	Hordeum vulgare	1 − pg	Mogensen & Rusche, 1985
	Zea mays	1 − pg	McConchie *et al.*, 1987

* Evidence classifications: 1 = Three-dimensional reconstruction of serial transmission electron micrographs; 2 = Two-dimensional projection derived from non-serial transmission electron micrographs; 3 = Inferred from published transmission electron micrographs; 4 = Fluorescence microscopy; + = germ unit present; − = germ unit absent; pg = data on pollen grain; pg, t = data on both pollen grain and tube; pt = data on pollen tube.

sperm cells have been isolated from pollen of *Zea mays* by Matthys-Rochon *et al.* (1986). These authors used osmotic shock to release the sperms, which are initially elongate, but become spherical during the isolation procedure. Cell fractionation technology has been employed successfully to separate the sperm cells from vegetative cell organelles such as starch grains. Matthys-Rochon *et al.* (1986) report a yield of 18%, and sperm cell concentrations of 3×10^6 per ml. The cellular nature of these sperms has been established by observations by phase contrast microscopy, transmission electron microscopy and freeze-fracture. The isolated cells gave a positive FCR test indicating cell viability, at least in terms of effectiveness of the plasma membranes.

In Melbourne, we have developed methods for sperm cell isolation using gentle physical grinding. This results in release of sperm cells that are osmotically active, and that retain their native elongate shape. This has been achieved for sperm cells of *Brassica campestris* (Hough *et al.*, 1986), *Zea*

Fig. 13.5. Schematic diagrams of male germ units of four angiosperms: a, *Zea mays*; b, *Spinacia oleracea*; c, *Brassica campestris* and d, *Plumbago zeylanica* based on three-dimensional reconstructions described in text (from McConchie & Knox, 1986).

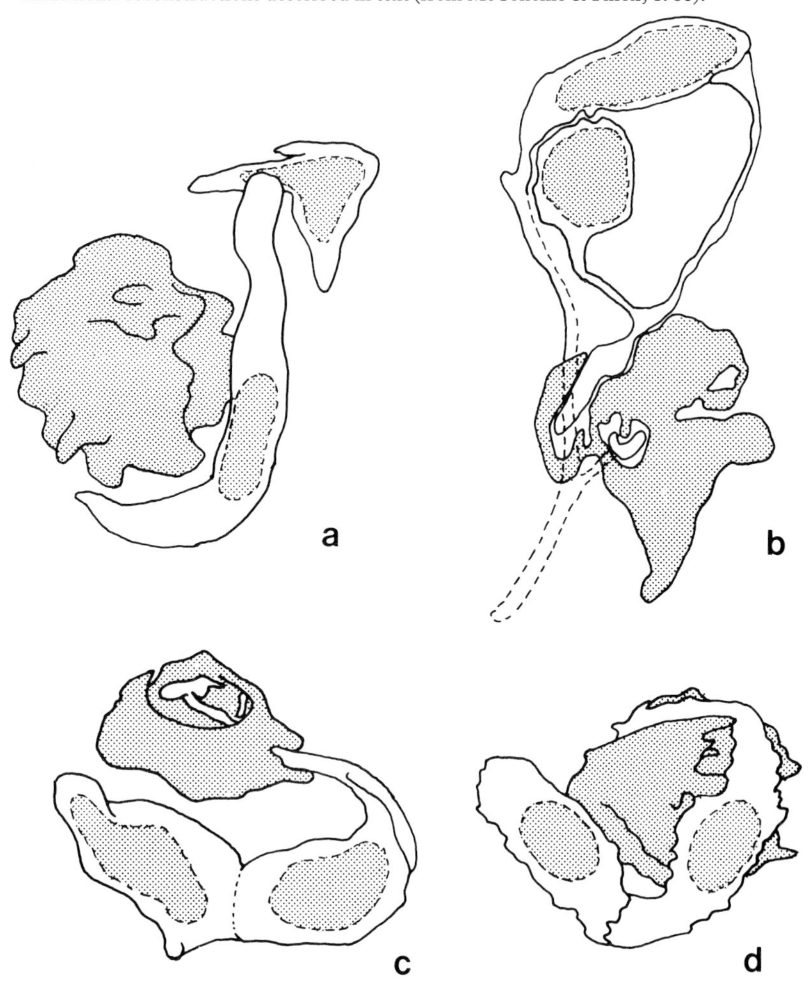

Table 13.3. *Quantitative differences in number of heritable organelles in sperm cells of* Brassica *and* Plumbago; S_{vn}, *sperm cell attached to vegetative nucleus;* S_{ua}, *sperm cell not associated with vegetative nucleus.*

Species	No. of mitochondria S_{vn}	No. of mitochondria S_{ua}	No. of plastids S_{vn}	No. of plastids S_{ua}	Reference
Brassica campestris	23.4 ±3.3	6.4 ±1.1	0	0	McConchie *et al.*, 1986
B. oleracea	13.2	9.8	0	0	McConchie *et al.*, 1986
Hordeum vulgare	34.0	27.5	0	0	Mogensen & Rusche, 1985
Plumbago zeylanica	256.2	39.8	0.5	24.3	Russell & Cass, 1983; Russell, 1984
Spinacia oleracea	20.3 ±4.5	24.7 ±6.1	0	0	Wilms, 1986

mays (Cass *et al.*, 1986) and *Gerbera jamesonii* (Southworth *et al.*, submitted for publication).

Fig. 13.6. Scheme showing the male germ unit (model: *Brassica campestris*) and its sperm cell specificities at double fertilisation. a, unit in pollen grain, showing linkage is maintained by the inner plasma membrane of the vegetative cell (vpm). b, hypothetical diagram showing sperm cells released from pollen tube, loss of the vegetative cell plasma membrane (vpm) and exposure of surface determinants of the sperm cell plasma membranes (spm). In this model, each sperm cell (Svn and Sua) carries different determinants. These are shown as receptors for either the central cell (cc) or egg cell (e).

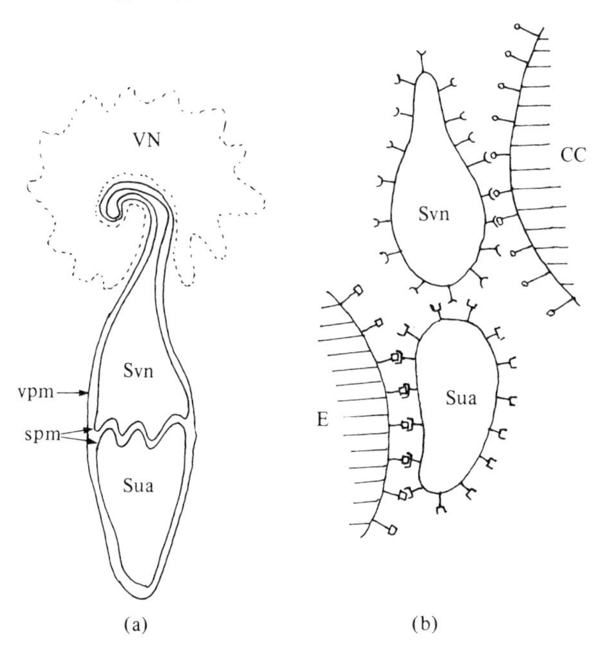

The isolation procedure employed is carried out as follows:

- Extract surface pollen coat with acetone;
- Hydrate pollen in atmosphere of 95% RH;
- Grind pollen in medium containing buffer, osmoticum, protectants and ions (see Hoekstra, 1979);
- Filter through nylon to remove large cell debris;
- Collect sperm cells on polycarbonate filter to which they bind (Hough *et al.*, 1986) or after centrifugation.

Sperm cells are detected most readily by fluorescence microscopy. The DNA fluorochrome, DAPI, can be added to the isolation medium, and the state of the cells determined by combined fluorescence/Nomarski DIC optics.

In *Gerbera*, elongate structures resembling the sperm cells in intact and recently burst pollen cytoplasm are obtained with grinding. These cells are osmotically active and have a small amount of cytoplasm that extends beyond the nucleus. They appear to contain organelles and they have an extension. With SEM these appear intact, but in TEM of the sperm pellet, the plasma membranes can be incomplete. Isolated sperm cells generally are not attached to each other or to the vegetative nucleus although occasionally two isolated sperm cells are held together by a cytoplasmic extension. Some vegetative cell cytoplasm often remains attached to the sperm cells. Further development of this method should allow for a larger, scaled up preparation for further analysis.

Monoclonal antibody technology offers the potential to determine the nature of the cell surface determinants of the isolated sperm cells. The polycarbonate filter method has been used to screen sperm-specific antibodies hybridoma lines (Hough *et al.*, 1986). Isolated sperm cells remain attached to the filters, and the specificity of the hybridoma supernatants can be readily assessed as positive or negative by indirect immunofluorescence microscopy, for example in sperms of *Brassica campestris* (Fig. 13.7)

SDS-polyacrylamide gel electrophoresis of homogenised sperm preparations has shown that isolated sperm cell preparations of both *Brassica* and *Gerbera* contain a suite of proteins and glycoproteins not present in the corresponding vegetative cell. Gels have been stained for proteins, and western blots with the peroxidase-labelled lectin Concanavalin A to detect glycoproteins (Fig. 13.8). *Brassica* sperm cells contain a major protein of M_r 66 kD that is apparently cell-specific. *Gerbera* sperm cells contain a range of proteins varying in M_r from 30 to 70 kD. After staining for glycoproteins, the *Brassica* sperm preparation shows the presence of at least five glycoproteins, while there is one major glycoprotein in the *Gerbera* preparation.

Conclusions

Isolated sperm cells may be examined using a range of physiological techniques, and their cell biology established. Questions can be asked concerning sperm movement, the nature of their translational machinery, and their unique condensed chromatin, but at present there is no evidence of their functional capacity in fertilisation. The availability of isolated sperm may lead to the development of new *in vitro* fertilisation techniques.

For example, isolated viable sperm cells may be stimulated to divide during *in vitro* culture to create haploid plants, or hybrids by *in vitro* fertilisation (see Chapman, 1986). If such methods can be developed, it might become possible to transform sperm cells with exogenous DNA and to use such transformed sperm for *in vitro* fertilisation.

Fig. 13.7. Binding of a hybridoma cell line directed against the sperm cells of *Brassica campestris*, revealed by immunofluorescence screening using the indirect method with FITC-labelled secondary antibody. The sperm cells (brightly fluorescent) are attached to a polycarbonate filter (photograph by Terryn Hough).

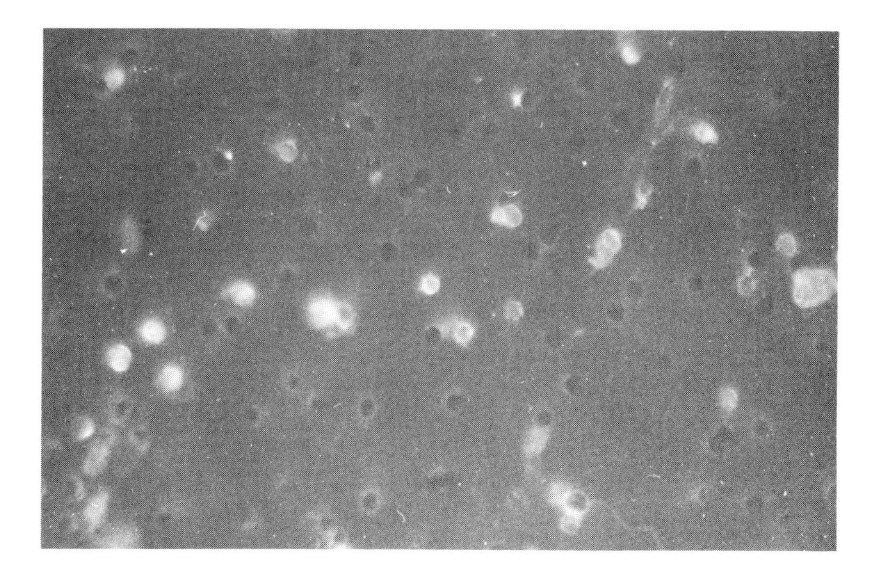

References

Bendich, A. J. & Gauriloff, L. P. (1984). Morphometric analysis of cucurbit mitochondria: The relationship between chondriome volume and DNA content. *Protoplasma*, **119**, 1–7

Burr, F. A. & West, J. A. (1970). Light and electron microscope observations on the vegetative and reproductive structures of *Bryopsis hypnoides*. *Phycologia*, **9**, 17–37

Chu, C. & Hu, S. (1981). The development and ultrastructure of wheat sperm cell. *Acta Bot. Sin.*, **23**, 85–91

Chu, C., Hu, S. & Xu, L. (1980). Ultrastructure of sperm cell in mature pollen grain of wheat. *Sci. Sin.*, **23**, 371–9

Cass, D. D. (1973). An ultrastructural and Nomarski-interference study of the sperms of barley. *Can. J. Bot.*, **51**, 601–5

Cass, D. D. & Karas, I. (1975). Development of sperm cells in barley. *Can. J. Bot.*, **53**, 1051–62

Fig. 13.8. Characterisation of the proteins and glycoproteins of sperm cells of *Brassica campestris* and *Gerbera jamesonii* compared with their respective pollen preparations, as shown by SDS-polyacrylamide gel electrophoresis of homogenates. Standards are shown at LHS, ranging from M_r 14 to 94 kD. Lanes 1,5: sperm cells of *B. campestris*; Lanes 2,6: pollen of *B. campestris*; Lanes 3,7: sperm cells of *G. jamesonii*; Lanes 4,8: pollen of *G. jamesonii*. Proteins are stained in the gel by Coomassie Blue G; glycoproteins are visualised by binding of peroxidase labelled ConA to western blot of same gel.

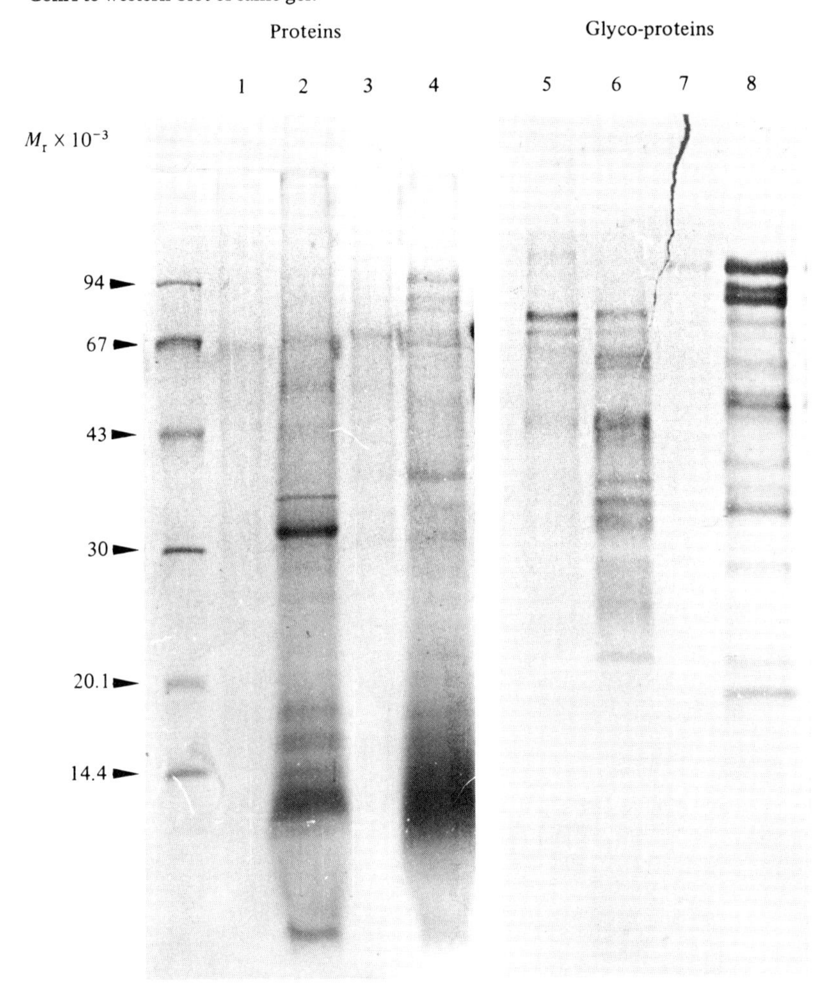

Cass, D., Hough, T., Knox, B., McConchie, C. & Singh, M. (1986). Isolation of sperm from pollen of corn and oilseed rape. *Plant Physiol.* April Supp., **80**, Abstract 694

Chapman, G. P. (1986). Modified fertilisation in plants: a strategy for directed genetic change. *Outlook on Agriculture*, **15**, 27–32

Cresti, M., Ciampolini, F., Mulcahy, D. L. M. & Mulcahy, G. (1985). Ultrastructure of *Nicotiana alata* pollen, its germination and early tube formation. *Am. J. Bot.*, **72**, 719–27

Duckett, J. G. (1974). An ultrastructural study of spermatogenesis in *Anthoceros laevis* L. with particular reference to the multilayered structure. *Bull. Soc. Fr. Coll. Bryologie*, **121**, 81–91

Duckett, J. G. (1975). Spermatogenesis in pteridophytes. In: *Biology of the male gamete*. eds. J. G. Duckett & P. A. Racey, pp. 97–128. Academic Press, London

Dumas, C., Knox, R. B. & Gaude, T. (1984*a*). Pollen–pistil recognition: new concepts from electron microscopy and cytochemistry. *Int. Rev. Cytol.*, **90**, 239–72

Dumas, C., Knox, R. B., McConchie, C. A. & Russell, S. D. (1984*b*). Emerging physiological concepts in fertilization. In: *What's new in plant physiology*, **15**, 17–20

Dumas, C., Knox, R. B. & Gaude, T. (1985). The spatial association of the sperm cells and vegetative nucleus in the pollen grain of *Brassica*. *Protoplasma*, **124**, 168–71

Favard, P. & Andre, J. (1970). The mitochondria of spermatazoa. In: *Comparative Spermatology*. ed. B. Baccetti, pp. 415–29. Academic Press, London

Heslop-Harrison, J. & Heslop-Harrison, Y. (1984). The disposition of gamete and vegetative cell nuclei in the extending pollen tubes of a grass species, *Alopecurus pratensis* L. *Acta. Bot. Neerl.*, **33**, 131–4

Heslop-Harrison, J., Heslop-Harrision, Y., Cresti, M., Tiezzi, A. & Ciampolini, F. (1986). Actin during pollen germination. *J. Cell Sci.*, **86**, 1–8

Hoefert, L. L. (1969). Fine structure of sperm cells in pollen grains *Beta*. *Protoplasma*, **68**, 237–40

Hoekstra, F. A. (1979). Mitochondrial development and activity of binucleate and trinucleate pollen during germination *in vitro*. *Planta*, **145**, 25–36

Holt, W. V. (1984). Membrane heterogeneity in the mammalian spermatozoan. *Int. Rev. Cytol.*, **87**, 159–93

Hough, T., Bernhardt, P., Knox, R. B. & Williams, E. G. (1985). Applications of fluorochromes to pollen biology. II. The DNA probes ethidium bromide and Hoechst 33258 in conjunction with the callose-specific aniline blue fluorochrome. *Stain Tech.*, **60**, 155–62

Hough, T., Singh, M. B., Smart, I. J. & Knox, R. B. (1986). Immunofluorescent screening of monoclonal antibodies to surface antigens of animal and plant cells bound to polycarbonate membranes. *J. Immun. Methods*, **92**, 103–7

Jensen, W. A. (1974). Reproduction in flowering plants. In: *Dynamic Aspects of Plant Ultrastructure*. ed. A. W. Robards, pp. 481–503. McGraw-Hill, New York

Jensen, W. A. & Fisher, D. B. (1968). Cotton embryogenesis: the entrance and discharge of the pollen tube in the embryo sac. *Planta*, **78**, 158–83

Jensen, W. A. & Fisher, D. B. (1969). Cotton embryogenesis: the sperm. *Protoplasma*, **65**, 277–86

Karas, I. & Cass, D. D. (1976). Ultrastructural aspects of sperm cells forma-

tion in rye: evidence for cell plate involvement in generative cell division. *Phytomorphology*, **26**, 36–45
Kaul, V., Theunis, C., Palser, B. , Knox, R. B. & Williams, E. G. (1987). Association of the generative cell and vegetative nucleus in pollen tubes of *Rhododendron*. *Ann. Bot.*, **59**, 227–35
Knox, R. B. (1987). Pollen differentiation patterns and male function. In: *Differentiation Patterns in Higher Plants*. ed. K. Urbanska. Academic Press, London
Knox, R. B. & McConchie, C. A. (1986). Structure and function of compound pollen. In: *Pollen Structure and Function*. ed. S. Blackmore, pp. 265–82. Academic Press, London
LaFountain, K. L. & Mascarenhas, J. P. (1972). Isolation of vegetative and generative nuclei from pollen tubes. *Exp. Cell Res.*, **73**, 233–6
Larson, D. A. (1965). Fine structural changes in the cytoplasm of germinating pollen. *Am. J. Bot.*, **52**, 139–54
McConchie, C. A., Hough, T. & Knox, R. B. (1987). Three-dimensional reconstruction of the sperm cells of mature pollen of maize, *Zea mays*. *Protoplasma*, **139**, 9–19
McConchie, C. A., Jobson, S. & Knox, R. B. (1985*a*). Computer-assisted reconstruction of the male germ unit in pollen of *Brassica campestris*. *Protoplasma*, **127**, 57–63
McConchie, C. A., Jobson, S. & Knox, R. B. (1985*b*). The structure of sperm cells in *Brassica*. In: *Sexual Reproduction of Seed Plants, Ferns and Mosses*. eds. M. T. M. Willemse & J. L. Van Went. Pudoc, Wageningen, pp. 147–8
McConchie, C. A. & Knox, R. B. (1986). The male germ unit and prospects for biotechnology. In: *Biotechnology and Ecology of Pollen*. eds. D. L. Mulcahy, G. B. Mulcahy & E. Ottaviano, pp. 289–96. Springer, New York
McConchie, C. A., Russell, S. D., Dumas C. & Knox, R. B. (1986). Quantitative cytology of the sperm cells of *Brassica campestris* and *Brassica oleracea*. *Planta*, **170**, 446–52
Matthys-Rachon, E., Roeckel, P. & Dupuis I. (1986). Pollens et gametes males: faits et perspectives. *Proc. 9th Int. Congress Cytobiology of sexual reproduction*. Reims, France (in press)
Mogensen, H. L. (1982). Double fertilization in barley and the cytological explanation for haploid embryo formation, embryoless caryopses and ovule abortion. *Carlsberg Res. Commun.*, **47**, 313–54
Mogensen, H. L. (1986). On the male germ unit in an angiosperm with bicellular pollen, *Hippeastrum vitatum*. In: *Pollen Ecology and Biotechnology*. eds. D. Mulcahy & E. Ottaviano. Springer, Berlin, Heidelberg, New York
Mogensen, H. L. & Rusche, M. L. (1985). Quantitative ultrastructural analysis of barley sperm I. Occurrence and mechanism of cytoplasm and organelle reduction and the question of sperm dimorphism. *Protoplasma*, **128**, 1–13
Myles, D. G. & Bell, P. R. (1975). An ultrastructural study of the spermatozoid of the fern *Marsilea vestita*. *J. Cell Sci.*, **17**, 633–45
Russell, S. D. (1983). Fertilization in *Plumbago zeylanica*: gametic fusion and fate of the male cytoplasm. *Am. J. Bot.*, **70**, 416–34

Russell, S. D. (1984). Ultrastructure of the sperm of *Plumbago zeylanica*. II. Quantitative cytology and three-dimensional organization. *Planta*, **162**, 385–91

Russell, S. D. (1985*a*). Preferential fertilization in *Plumbago*: Ultra-structural evidence for gamete-level recognition in an angiosperm. *Proc. Natl. Acad. Sci. USA*, **82**, 6129–32

Russell, S. D. (1985*b*). Sperm specificity in *Plumbago zeylanica*. In: *Sexual reproduction of seed plants, ferns and mosses*. eds. M. T. M. Willemse & J. L. Van Went. Pudoc, Wageningen. pp. 145–6

Russell, S. D. (1986). Isolation of sperm cells from the pollen of *Plumbago zeylanica*. *Plant Physiol.*, **81**, 317–19

Russell, S. D. & Cass, D. D. (1981). Ultrastructure of the sperms of *Plumbago zeylanica*. I. Cytology and association with the vegetative nucleus. *Protoplasma*, **107**, 85–107

Russell, S. D. & Cass, D. D. (1983). Unequal distribution of plastids and mitochondria during sperm cell formation in *Plumbago zeylanica*. In: *Pollen: Biology and implications for plant breeding*. eds. D. L. Mulcahy & E. Ottaviano. Elsevier, Amsterdam. pp. 135–40

Schroder, M. B. (1983). The ultrastructure of sperm cells in *Triticale*. In: *Fertilization and Embryogenesis in Ovulated Plants*. ed. O. Erdelska. Slovak Academy of Sciences, Czechoslovakia. pp. 101–4

Theunis, C. H., McConchie, C. A. & Knox, R. B. (1985). Three-dimensional reconstruction of the generative cell and its wall connection in mature bicellular pollen of *Rhododendron*. *Micron & Microscop. Acta*, **16**, 225–31

Tokuyasu, K. T. (1975). Dynamics of spermiogenesis in *Drosophila melanogaster*. VI. Significance of 'onion' nebenkem formation. *J. Ultrastruct. Res.*, **53**, 93–112

Willson, M. F. (1983). *Plant Reproductive Ecology*. Wiley-Interscience, John Wiley & Sons, Inc., New York

Wilms, H. J. (1981). Pollen tube penetration and fertilization in spinach. *Acta Bot. Neerl.*, **30**, 101–22

Wilms, H. J. (1985). Behaviour of spinach sperm cells in pollen tubes prior to fertilization. In: *Sexual Reproduction of Speed Plants, Ferns and Mosses*. eds. M. T. M. Willemse & J. L. Van Went. Pudoc, Wageningen. pp. 143–4

Wilms, H. J. (1986). Dimorphic sperm cells in the pollen grain of *Spinacia*. In: *Biology of Reproduction and Motility in Plants and Animals*. eds. M. Cresti & R. Dallai. pp. 193–8. University of Siena, Italy

Wilms, H. J. & van Aelst, A. C. (1983). Ultrastructure of spinach sperm cells in mature pollen. In: *Fertilization and Embryogenesis in Ovulated Plants*. ed. O. Erdelska. Slovak Academy of Sciences, Czechoslovakia. pp. 105–239

14 The role of fucosylated glycoconjugates in cell–cell interactions of the mammalian pre-implantation embryo

Susan J. Kimber

Experimental Embryology and Teratology Unit,
MRC Laboratories, Woodmansterne Road,
Carshalton, Surrey SM5 4EF, UK

Abstract

Close cell–cell interaction in the 8- to 16-cell pre-implantation mouse embryo (compaction) is crucial to differentiation of the inner cell mass from which the fetus proper is derived. The molecular mechanisms involved in this process are reviewed including current knowledge about the uvomorulin/cadherin/L-Cam system. The existence of a second cell–cell interaction mechanism involving fucosylated cell surface glycoconjugates in the stabilisation of membrane-membrane contact is proposed. The evidence for this includes the increase in incorporation of ^{3}H-fucose into glycoproteins and glycolipids between the 4-cell stage and the blastocyst, enrichment of fucosylated glycoconjugates at the cell surface, particularly cell surfaces adjacent to other cells, and the reversal of intimate cell–cell contact by an oligosaccharide containing α 1–3 linked fucose. The SSEA-1 determinant, carrying this saccharide, is known to be present from the 8-cell stage in the mouse embryo. Receptors for the fucose binding lectin from *Lotus tetragonolobus* increase over the period of increased cell interaction and fucose also promotes the interaction of embryonic cells with substrate-bound wheat germ agglutinin.

Introduction

Changes in the relationships of cells with one another occur frequently during embryonic development. For instance, cells become more closely associated with one another in the formation of epithelia, while the mesoderm cells emerging from the primitive streak of the epiblast at gastrulation, and the neural crest cells derived from the neural epithelium are more loosely associated than the cells of their progenitor tissues. The first change in cell–cell interactions occurs very early in development of the mammalian embryo (at the 8-cell stage in the mouse) when the cells which were previously rounded and loosely associated spread on one another. Cell–cell contact is maximised and cell outlines can no longer be clearly distinguished under the dissecting microscope (Figs 14.1 and 14.2). This process, known as compaction is one of the first morphogenetic events in the embryo and has been intensively studied. Other changes occurring around this stage include the development of asymmetry of surface and cytoplasmic components by the cells (Handyside, 1980; Ziomek & Johnson, 1980; Johnson & Ziomek, 1981; Johnson, 1985; Maro *et al.*, 1985) and the beginning of gap junction formation (Ducibella *et al.*, 1975; Magnuson, Demsey & Stackpole, 1977; Lo & Gilula 1979; Goodall & Johnson, 1982, 1984; McLachlin, Caveney & Kidder, 1983; McLachlin & Kidder, 1986).

At the EM level, various degrees of cell-interaction can be observed over the period of compaction from little more than repeated 'point contacts' between processes of neighbouring cells, through close membrane–membrane approximation of still rounded cells, to cells arranged in a quasi close-packed structure at the 16-cell stage (Fig. 14.2) with subsequent formation of tight junctional contact. Compaction, is a crucial event in development since it is essential for orderly morphogenesis to form the blastocyst (Ducibella & Anderson, 1975; Surani, Kimber & Handyside, 1981; Pratt *et al.*,

Fig. 14.1. Diagrammatic representation of the sequence of stages in development of the mouse embryo before implantation. Fertilisation of the ovulated oocyte (at metaphase II of the second meiotic devision) occurs in the oviduct and the embryo subsequently undergoes a series of asynchronous cleavage divisions as it passes down the oviduct. At the distal end of the oviduct, at the 8-cell stage, the embryo undergoes compaction and enters the uterus where it forms the blastocyst containing an inner clump of cells, the inner cell mass (ICM) and the outer layer of trophectoderm cells. The foetus and most of the extra-embryonic membranes are derived from the ICM. For the work described in this chapter female mice (MFI) were 'super-ovulated' by injecting pregnant mare's serum gonadotrophin, to induce follicle growth, followed 48 h later by human choronic gonadotrophin (hCG) to stimulate ovulation. This increases the yield of oocytes. The female mice were mated with C57BL/6J/Lac × CBA/Ca/Lac F_1 males and checked the following morning for a vaginal plug to confirm mating.

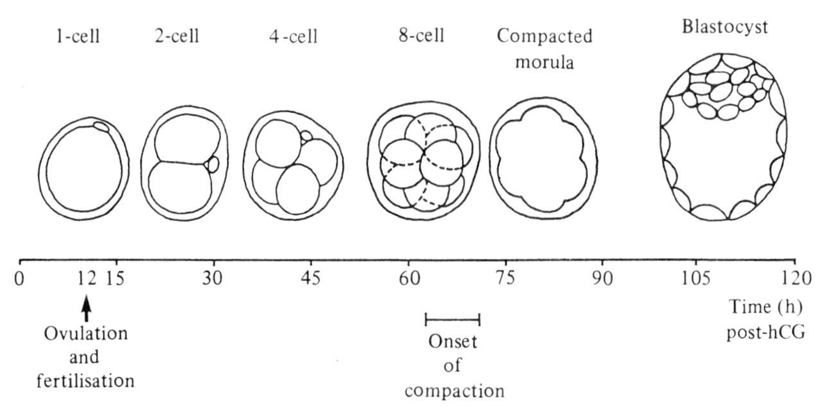

1982) with an enclosed clump of inner cell mass cells from which the embryo proper is derived, and the outer trophectoderm (Fig. 14.1). The generation of ICM and trophectoderm lineages may depend on cell contact patterns influencing the cleavage plane of asymmetrical (polarised) cells in the 8- and 16-cell embryo (Johnson & Ziomek, 1981; Johnson, 1985).

Compaction is a complex multistep cell interaction process continuing through the 8-cell and 16-cell stages which can be perturbed by various agents. Some of these are effective over a specific portion of this developmental period and presumably act on the cell adhesion mechanisms which are dominant at that time. The instigation of compaction is known to be calcium dependent and prevented when calcium is omitted from the culture medium (Ducibella & Anderson, 1979; Ogou, Okada & Takeichi, 1982; Kimber & Surani, 1982). It is also prevented or reversed by inhibitors of microfilament polymerisation such as Cytochalasin D (Ducibella & Anderson, 1975; Surani, Barton & Burling, 1980; Kimber & Surani, 1981; Pratt, Chakraborty & Surani, 1981), thus microfilaments appear to be involved in both the initiation and maintenance of compaction. In contrast, the effect of agents which influence microtubule polymerisation suggests that these cytoskeletal elements function in maintaining the precompact state (Maro & Pickering, 1984). Compaction is also known to be inhibited, or reversed, by Tunicamycin (Surani, 1979; Atienza-Samols, Pine & Sherman, 1980; Surani, Kimber & Handyside, 1981; Sutherland & Calarco-Gillam, 1983) which interferes with dolichol-mediated N-linked glycosylation of protein (Takatsuki *et al.*, 1971). In the presence of Tunicamycin embryos compact briefly but then decompact. Cell surface glycoproteins, for instance as assessed by receptors for lectins, are depeleted in the presence of Tunicamycin (Surani *et al.*,

Fig. 14.2. Electron micrographs of an uncompacted (A, and compacted (B, mouse embryo. In the uncompacted 8-cell embryo processes contact neighbouring cells but there is little close membrane–membrane contact. In the compacted 16-cell morula the cells are arranged in a quasi-close-packed arrangement with maximum cell–cell contact. Scale bars = 2 μm.

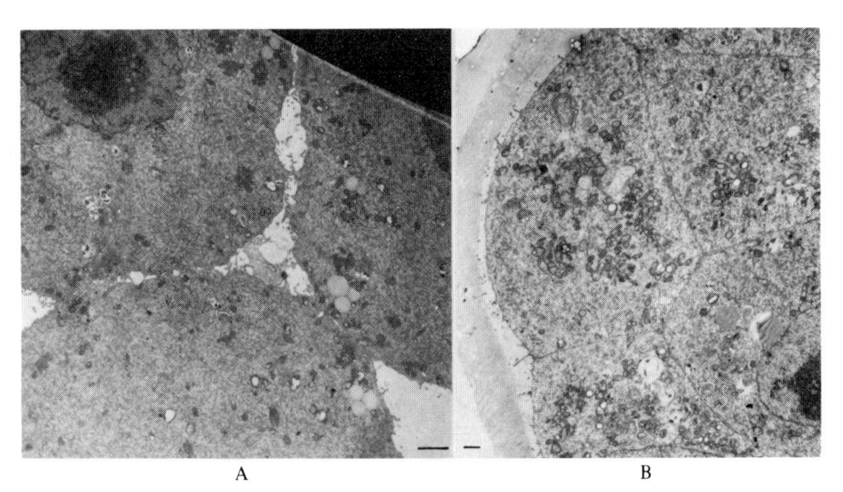

1981). In support of the importance of approximately concurrent N-linked glycosylation for compaction, is the finding that inhibition of the synthesis of non-sterol isoprenes, like dolichol, also prevents the process (Surani, Kimber & Osborn, 1983). In contrast, concurrent protein synthesis does not seem to be necessary for compaction to occur (Kidder & McLachlin, 1985; Levy *et al.*, 1986).

A fruitful approach to the study of compaction has been the use of poly- and mono-clonal antibodies (or their Fab fragments) against embryonal carcinoma cells (EC cells), the stem cells of teratocarcinomas. Different EC cell lines share properties with pre-implantation embryonic cells to different extents, but appear generally to correspond to a slightly later stage of development than the 8- to 16-cell embryo (see Papaioannou & Rossant, 1983; Rossant & Papaioannou, 1984). Antibodies have been produced to these cells which will decompact embryos (Kemler *et al.*, 1977; Johnson *et al.*, 1979; Ducibella, 1980; Shirayoshi, Okada & Takeichi, 1983). A major target molecule for these decompacting antisera is a 120 kD glycoprotein present on EC cells and embryos, which is trypsin-sensitive in the absence of calcium and known variously as uvomorulin or cadherin (Hyafil *et al.*, 1980; Hyafil, Babinet & Jacob, 1981; Yoshida & Takeichi, 1982; Peyriéras *et al.*, 1983; Shirayoshi *et al.*, 1983; Yoshida-Noro, Suzuki & Takeichi, 1984; Ekblom, Vestweber & Kemler, 1986). It is a fucoprotein carrying N-linked oligosaccharides of the complex type (Peyriéras, Louvard & Jacob, 1985), and there is also evidence for it being a hydrophilic and extrinsic cell surface component (Peyriéras, *et al.*, 1983; Vestweber & Kemler, 1984*b*). However, since it is phosphorylated and trypsinisation leaves a 35 kD fragment associated with EC cell surfaces it has been suggested that it may be integrated in the plasma membrane (Vestweber & Kemler, 1984*a*). Certain other proteins recognised by the antisera may be different polypeptides with shared epitopes (Peyriéras *et al.*, 1985) or post-translational modifications of a single precursor, as indicated by the lack of heterogeneity of uvomorulin mRNA (Vestweber & Kemler, 1984*b*; Schuh *et al.*, 1986). The functional site of uvomorulin/cadherin in cell–cell interaction has been reported to reside in a 26 kD proteolytic fragment that does not appear to include sugar residues (Vestweber & Kemler, 1984*a*, 1985). Thus, although uvomorulin/cadherin is a fucosylated glycoprotein, the saccharide moieties do not appear to be directly involved in cell–cell adhesion.

Antibodies to uvomorulin/cadherin interfere with the calcium-dependent cell adhesion system (Hyafil *et al.*, 1981; Ogou *et al.*, 1982). The presence of both calcium-dependent and calcium-independent mechanisms of cell–cell adhesion in pre-implantation embryos mirrors similar systems of other interacting cells (Takeichi *et al.*, 1979, 1982; Grunwald, Geller & Lilien, 1980; Brackenbury, Rutishauser & Edelman, 1981; Magnani, Thomas & Steinberg, 1981, Edelman *et al.*, 1985; see chapter by Sui and Misevic and colleagues, this volume). Indeed, the calcium-dependent cell adhesion component of the early mouse embryo appears to be similar or identical to other cell adhesion molecules such as cell CAM 120/80 of mammary carcinoma cells (Damsky *et al.*, 1983; Richa *et al.*, 1985), the embryonic chick liver cell adhesion molecule L-CAM (Gallin *et al.*, 1983) and canine Arc-1 or rr1 (Behrens *et al.*, 1985; Gumbiner & Simons, 1985) which share many of its properties. Antibodies to cell-CAM 120/80 are able to inhibit compaction and their effect can be blocked by the 80 kD portion of cell-CAM 120/80 which is released into the medium by human mammary tumour cells (Damsky *et al.*, 1983). Nor is uvomorulin/cadherin confined to the early embryo; related molecules are present in epithelial tissues of the developing peri-implantation and post-implantation rodent embryo (Vestweber & Kemler, 1984*b*; Damjanov *et al.*, 1986) and the adult (Ogou *et al.*, 1983; Vestweber & Kemler 1984*b*; Boller *et al.*, 1985; Vestweber, Kemler & Ekblom, 1985) where there is evidence it is present in junctional complexes, probably the intermediate type junctions (Boller *et al.*, 1985; Gumbiner & Simons, 1985). Other cell adhesion glycoproteins such as cell-CAM 105 (Ocklind & Öbrink, 1982; Vestweber *et al.*, 1985), N-CAM and Ng-CAM (Edelman *et al.*, 1985; Edelman, 1986) and the LFA-1/Mac-1 family of glycoproteins (Springer, Sastre & Anderson, 1985; Mentzer, Burakoff & Faller, 1986) are clearly distinct from uvomorulin.

Uvomorulin/cadherin does not appear to be the only component involved in maintaining cell–cell adhesion in the pre-implantation mouse embryo. For instance, antibodies to this glycoprotein lose their ability to prevent compaction between the 16-cell and 32-cell stage (Shirayoshi *et al.*, 1983; Damsky *et al.*, 1983; Vestweber & Kemler, 1985). Around this time embryos also become refractory to decompaction in calcium-free medium and tight junctions start to be established (Ducibella & Anderson, 1975, 1979; Ducibella *et al.*, 1975; Magnuson, Demsey & Stackpole, 1977). Uvomorulin is present throughout the pre-implantation period (Fig. 14.3), yet compaction does not occur until the 8-cell stage: other permissive changes seem to be required for its initiation (Levy *et al.*, 1986). In any case, 4-cell stage, or earlier, embryos do not provide a substrate which is capable of supporting the spreading of cells from 8-cell embryos (Kimber, Surani & Barton, 1982; Ducibella, 1982), so other cell surface components, or a reorganisation of those already present, must be involved even in the early phase of compaction.

Other molecules which may function in this process might be identified by considering known cell surface components. Figure 14.3 shows a number of antigens which have been demonstrated on pre-implantation embryos. All the determinents shown here with the exception of receptors for the lectins *Dolichos biflorus* agglutinin (DBA) and the fucose binding protein from *Lotus tetragonolobus* (LTA) are recognised by monoclonal antibodies and all are carried either by glycoprotein or glycolipid. The carbohydrate portion of these antigens (Fig. 14.3) generally falls into one of two categories.

(i) based on type 2 polylactosamine chains (open circles, Fig. 14.3)
(ii) based on globoside type chains (closed circles, Fig. 14.3)

A number of these antigens, such as SSEA-1 (Gooi *et al.*, 1981; Hakomori *et al.*, 1981) cadherin and the 100 kD glucose regulated protein (McCormick *et al.*, 1979; McCormick & Babiarz, 1984) are also known to be fucosylated. However, carbohydrate moieties other than those based on type 2 oligosaccharide chains or globoside are also known to be present. Examples are the complex type oligosaccharide chains of uvomorulin/cadherin (Peyriéras, op. cit.) and sialylated short chain gangliosides (Pennington *et al.*, 1985).

Most of the surface carbohydrate containing antigens which have been demonstrated are present throughout the pre-implantation period. Others appear or disappear, particularly around the time of compaction, for instance, SSEA-1, which starts to appear at the 8-cell stage (Solter & Knowles, 1978), while the Forssman glycolipid appears first in the compacted morula (Willison & Stern, 1978; Willison *et al.*, 1982). Both these antigens have been suggested to arise by post-translational modification, possibly by the action of surface glycosyl transferases (see Conclusions) on other components present from an earlier stage; the I antigen and globoside respectively (Gooi *et al.*, 1981; Knowles *et al.*, 1982; Willison *et al.*, 1982). Antibodies which recognise SSEA-1 have been shown to have high affinity for the sequence Gal β(1–4) [Fuc α(1–3)] GlcNAc- (Gooi *et al.*, 1981; Hakomori *et al.*, 1981) and the appearance of this antigen just prior to compaction lead to the idea that it might have a role in this process (Gooi *et al.*, 1981). As development continues in the mouse embryo, SSEA-1 becomes restricted to the inner cell mass of the blastocyst, and the egg cylinder. By day 8 it is present on the yolk sac, gut, neurectoderm, primordial germ cells and embryonic endoderm (Solter & Knowles, 1978; Fox *et al.*, 1981; Pennington *et al.*,

Fig. 14.3. Diagrammatic representation of various carbohydrate cell-surface antigens present on mouse embryos in relation to stage of development. Dotted line indicates presence at a low level, solid line indicates presence is easily detectable. Open circles, determinants related to lactosamine; solid circles, determinants related to globoside; squares, fucosylated antigens. Data from: Solter & Knowles, 1978; Shevinsky *et al.*, 1982; Kannagi *et al.*, 1983; Knowles *et al.*, 1982; Pennington *et al.*, 1985; Kemler *et al.*, 1979; Marticorena *et al.*, 1983; Randle, 1982; Hyafil *et al.*, 1981; Shirayoshi, *et al.*, 1983; Kimber & Bird, 1985; Fenderson *et al.*, 1983; Sato *et al.*, 1984; Willison *et al.*, 1982; McCormick & Babiarz, 1984.

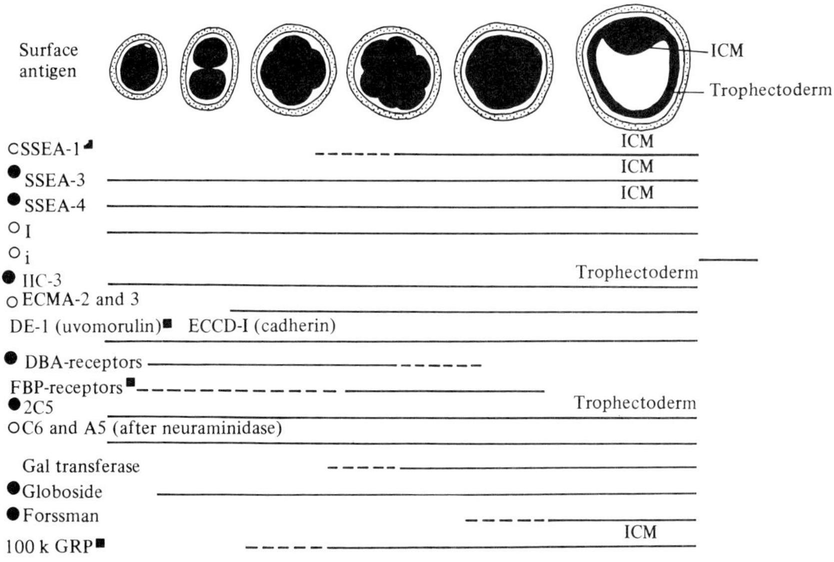

Surface antigen
ICM
Trophectoderm
○SSEA-1■ — ICM
●SSEA-3 — ICM
●SSEA-4 — ICM
○I
○i
●IIC-3 — Trophectoderm
○ECMA-2 and 3
DE-1 (uvomorulin)■ ECCD-I (cadherin)
● DBA-receptors
FBP-receptors■
●2C5 — Trophectoderm
○C6 and A5 (after neuraminidase)
Gal transferase
●Globoside
●Forssman — ICM
100 k GRP■

1985) and later on other specific cell types (Fox *et al.*, 1981, 1982; Yamamoto, Boyer & Schwarting, 1985). SSEA-1 may be carried on the oligosaccharide chains of either glycoproteins or glycolipids. In EC cells it is predominantly protein bound (Andrews *et al.*, 1982; Child *et al.*, 1983) the major carriers being the exceptionally large polyactosamine chains of 'embryoglycans' (Ozawa, Muramatsu & Solter, 1985), which are characteristic of embryonic cells both in early mouse embryos and EC cells (Muramatsu *et al.*, 1978, 1980, 1983). On various other cell types, including erythrocytes, SSEA-1 is present on glycolipids (Hakomori *et al.*, 1981; Watkins, 1980; Feizi, 1982).

Further data which stimulated our interest in fucosylated molecules in respect of embryonic cell interactions, was a report that a cell surface carbohydrate-binding component from EC cells was able to prevent reaggregation of the dispersed cells and its effect was neutralised by polysaccharides containing fucose and mannose (Grabel, Rosen & Martin, 1979: Grabel *et al.*, 1981). More recently an endogenous fucan/mannan binding lectin has been preliminarly reported, which may mediate divalent cation independent cell–cell adhesion in teratocarcinoma stem cells (Grabel, 1984). Several carbohydrate binding receptors appear to recognise the similar stereostructure of L-fucose and D-mannose (Kameyama, Oishi & Aida, 1979; Townsend & Stahl, 1981; Stahl & Gordon, 1982). Indeed there is also now good evidence for a phosphomannosyl/fucoidin binding receptor on the surface of mouse lymphocytes which is involved in their calcium dependent adhesion to peripheral lymph node high endothelial venules (Yednock, Stoolman & Rosen, 1987).

The question of whether fucosylated components are functioning in the close cell–cell association of the compacted mouse embryos is therefore a pertinent one.

Use of saccharides in hapten-inhibition of cell interactions

In initial experiments we decided to investigate whether a binding reaction involving saccharides is important in the interaction between the cells of the embryo. Extrapolating from work on cell-association in many systems including sponges, slime moulds, the erythroid system, chick neural retina and hepatocytes (see Harrison & Chesterton, 1980; Barondes, 1981; Bozzaro, 1985, for reviews), we undertook hapten-inhibition experiments using glycosylated probes in the experimental regime shown in Fig. 14.4. The rationale behind these experiments was that, if a cell surface oligosaccharide–receptor interaction is normally involved in compaction, free saccharides could compete for the receptor and thereby block compaction. Thus, we hoped to be able to identify the sugars involved.

Embryos were selected at the early 8-cell stage when some cells had started to show signs of compaction. By using this strategy we knew that the embryos were compaction competent. Zona-free embryos were decompacted completely in medium lacking added calcium and magnesium to allow maximum exposure of the membrane to test saccharides. In initial experiments, we disaggregated embryos completely, pooled the cells and then reassembled them in groups of eight. Because results were identical if embryos were merely decompacted, we did not disaggregate embryos in later experiments.

We looked at the effects of 19 mono- and di-saccharides, at concentrations

Fig. 14.4. Diagrammatic representation of the method used for the hapten inhibition assays. Embryos were flushed from the oviduct at the 2-cell stage and cultured overnight to the 8-cell stage. As the first cells started to show signs of compaction embryos were removed for a 30 s incubation in acid tyrode to solubilise the zona pellucida. They were decompacted in medium lacking added calcium and magnesium and either dissociated by pipetting, the cells pooled, and then regrouped in clumps of eight in medium containing the test sugar, or cultured directly in the test sugar after decompaction.

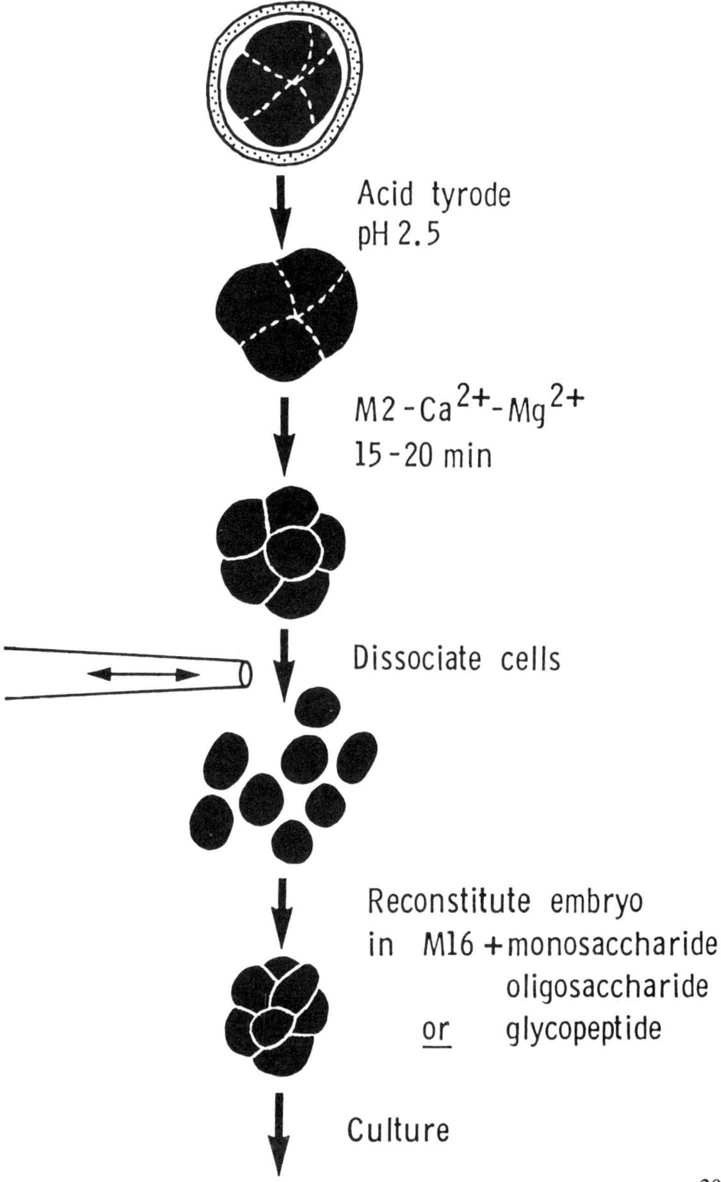

from 1 mM up to a very non-physiological 50 mM, as well as combinations of GlcNAc, Man, Gal, Fuc and lactose. None of these simple saccharides inhibited compaction (Kimber & Surani, 1982). We carried out similar experiments using various polysaccharides and glycopeptides which were available commercially. These substances were also, in general, ineffective in perturbing compaction, but fucoidin, a sulphated polymer of fucose, did cause partial decompaction of embryos. Although the effects of fucoidin might have been indirect or through its sulphate moieties, this result suggested that it was worthwhile pursuing the role of cell surface fucose.

Most receptor–hapten interactions involve recognition of a three-dimensional structure including several linked amino acids or monosaccharides. Therefore in order to mimic more closely the structure which a surface fucosaccharide would present to a receptor, we needed saccharides of known structure containing at least 4–5 sugar residues. Because SSEA-1 was known to be present, saccharides related to Gal β(1–4) [Fuc α(1–3)] GlcNAc were logical candidates for use in these inhibition studies. One of the best sources of such lactose and lactosamine based sugars are the A, B, H, blood group related saccharides from human milk (Kobata, 1972; Watkins, 1980). We were fortunate to obtain a series of these related fucosylated and unfucosylated oligosaccharides (a gift of Professor W. Watkins, CRC, Harrow, UK, and Professor V. Ginsburg, NIH, Bethesa, USA), including lacto-N-fucopentaose-III (LNFP-III; Kobata & Ginsburg, 1969) which carries the SSEA-1 determinant. We used these sugars at a range of concentrations from 0.5 to 10 mg/ml in our hapten-inhibition assay (Bird & Kimber, 1984). In all cases, embryos recompacted in the saccharide containing medium. However, towards the end of the 8-cell stage, those embryos which were incubated in LNFP-III, or a mixture of LNFP-II (composed of the same monosaccharides as LNFP-III but with α(1–4) linked fucose) and LNFP-III, started to decompact. All other fucosylated and unfucosylated saccharides were ineffective in perturbing compaction except 3-fucosyl-lactose which caused some decompaction (Fig. 14.5). In the absence of close cell–cell adhesion, embryos did not cavitate and form blastocysts. However, cell division continued, although at a slower rate, and the effects of the saccharide were reversible as long as embryos were washed and removed to fresh medium lacking saccharide before the time when control embryos started to cavitate (Fig. 14.6; Bird & Kimber, 1984). Those embryos which were 'rescued' from LNFP-III recompacted and cavitated but were not transferred to foster mothers to assess their post-implantation developmental potential.

Is LNFP-III acting on the cell surface to block a receptor which is normally active in binding SSEA-1 on adjacent cells? We can not be sure about the site of action, but an effect on metabolism seems unlikely since the sugar only reduced uptake of ^{3}H-leucine by 10% and had no effect on its incorporation into protein (Bird & Kimber, 1984). The possibility that LNFP-III binding has a non-specific effect is also unlikely since compaction is only reversed at a specific stage in development, at the end of the 8-cell stage and beginning of the 16-cell stage (Fig. 14.7). Even when embryos were incubated in the LNFP-III/II mixture starting at different times during the 8-cell stage, the timing of decompaction was relatively unaffected (Fig. 14.7).

The percentage of decompacted embryos was also reduced when embryos were incubated in LNFP-III/II starting progressively later in the fourth cell cycle, and 2–40% of embryos incubated from late in the 8-cell stage formed blastocysts. Apparently the saccharide acts at a specific period in development and thereafter embryos became resistant to its decompacting effect. The evidence indicates that the events interfered with by LNFP-III are involved in maintenance of the already apposed cell membranes since the initial mechanism of cell spreading and adhesion is unaltered by the sugar. Mono- and poly-clonal antibodies which recognise LNFP-II do not bind to the cell surface of mouse embryos (Fenderson *et al.*, 1986; Kimber, unpublished observation), so saccharide chains containing the $\alpha(1\text{–}4)$ linked fucose do not appear to be present. Thus the decompaction caused by the LNFP-II and LNFP-III mixture is presumably due to LNFP-III.

Evidence for the role of fucose in cell interactions from the use of lectins

Substrate bound lectins

Although we are dealing with 8, apparently equivalent cells (Kelly, 1975;

Fig. 14.5. Histogram showing the effect of incubating 8-cell embryos from 72 to 74 h post-hCG in 5 mg/ml of various oligosaccharides. Figures above histogram bars refer to the total number of embryos examined and figures in brackets to the number of replicates. Bars = percentage standard error between replicates. Abbreviations: Con, control medium; LNFP II/III, mixture of Lacto-N-fucopentaose-II and Lacto-N-fucopentaose-III; LNFP-III, Lacto-N-fucopentaose-III; LDFH I, Lacto-N-difucohexaose-I; 3FL, 3-fucosyllactose; LNFP I, Lacto-N-fucopentaose-I; LNT, Lacto-N-neotetraose; Fuco BSA, fucosyl bovine serum albumen (taken from Bird & Kimber (1984). *Devl. Biol.*, **104**, 449–60).

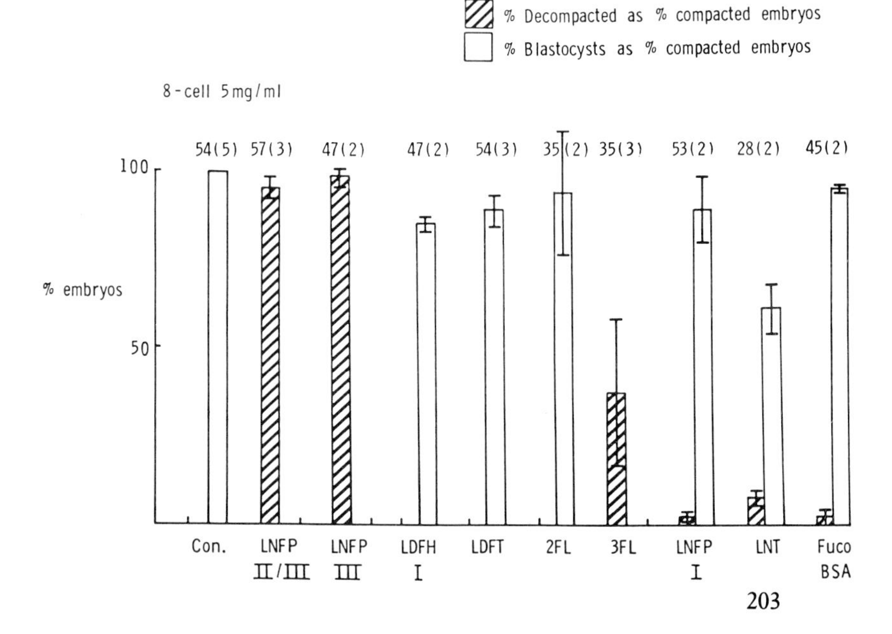

Johnson, 1985), the interaction of one cell with its neighbours is a complex process. It includes the response of each cell to a favourable substrate, encompassing a number of linked cell-surface and cytoskeletal events, and the provision by each cell of signals evoking a response in neighbouring cells. To simplify the system and study the response of a cell to surface-bound sugar receptors, we used the interaction of single cells from 8-cell embryos (1/8) with lectin conjugated agarose beads. Cells were paired with beads of approximately the same size placed in contact under standard culture

Fig. 14.6. a, Control 8-cell stage embryo showing complete recompaction after treatment with medium lacking Ca^{2+} and Mg^{2+}; b, normal blastocyst formed after decompaction in medium lacking Ca^{2+} and Mg^{2+} at the early 8-cell stage and return to control medium containing Ca^{2+} and Mg^{2+}; c, completely decompacted 8-cell embryo (enclosed within the zona pellucida) after culture in medium containing 5 mg/ml LNFP-III; d, embryos which had decompacted in LNFP-III, and after 14 h culture were returned to control medium, recompacted and formed fluid filled cavities which later expanded to form blastocyst like structures. Magnification × 1100 (modified from Bird & Kimber (1984). *Devl. Biol.*, **104**, 449–60).

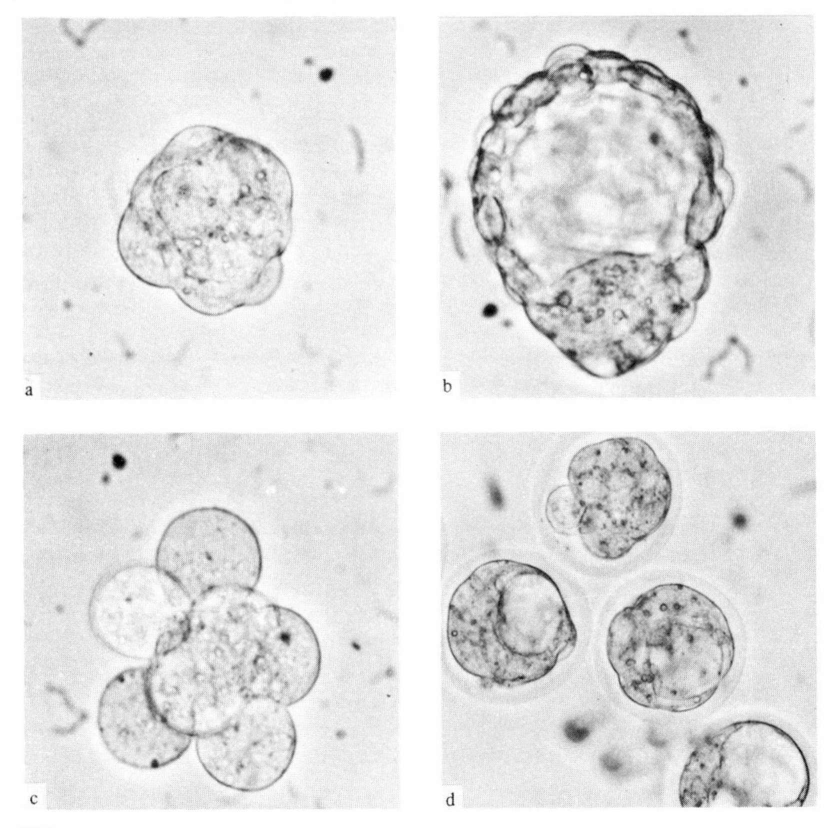

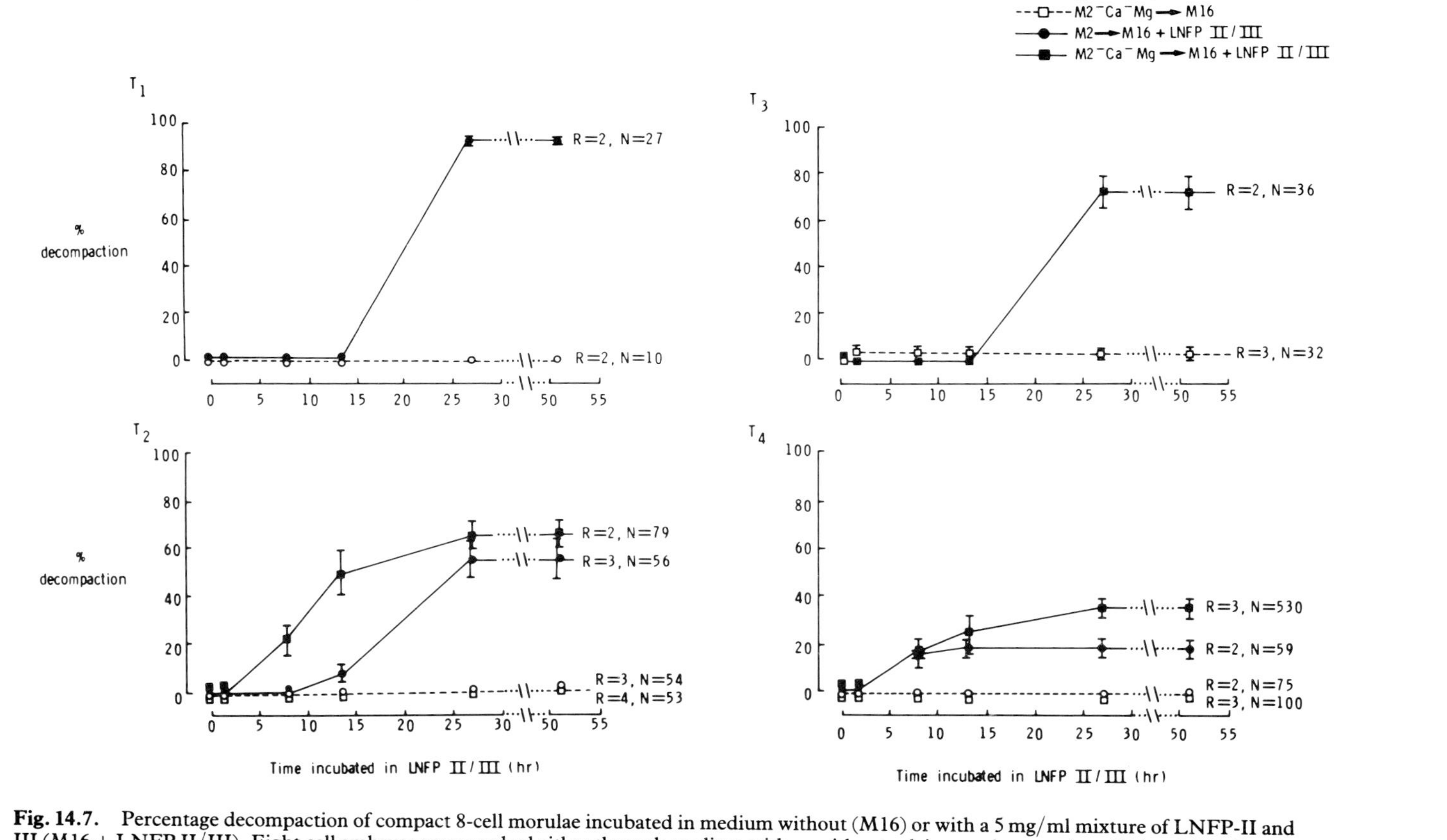

Fig. 14.7. Percentage decompaction of compact 8-cell morulae incubated in medium without (M16) or with a 5 mg/ml mixture of LNFP-II and III (M16 + LNFP II/III). Eight cell embryos were washed either through medium with or without calcium and magnesium (M2 and M2–Ca–Mg respectively) before transfer to control medium or medium with LNFP II/III at times T_1 to T_4 (T_1, 63 h post-hCG; T_2, 67–69 h post-hCG; T_3, 72–74 h post-hCG; T_4, 76–78 h post-hCG). R = number of replicates, N = number of embryos (from Bird & Kimber (1984). *Devl. Biol.*, **104**, 449–60).

conditions. They adhered to, and spread on, beads coated with wheatgerm agglutinin (WGA, affinity for $[B(1\text{–}4\text{-D-GlcNAC}]_2)$, peanut agglutinin (PNA, affinity for B-D-Gal) and Concanavalin A (Con A, affinity for α-D-man and α-D-Glc) but not beads coated with Soybean agglutinin (SBA) *Ricinus communis* agglutinin (RCA-I), *Limulus polyphemus* agglutinin (LPA) or *Lens culinaris* agglutinin (LCA) (Kimber & Surani, 1982, 1983). Using a semi-quantitative measure of spreading, we showed that spreading on WGA and PNA-beads (but non Con-A beads), was specifically related to the lectin mediated binding of surface glycoconjugates since it was inhibited only by sugars for which the respective lectins have high affinity.

Fucose was included among the control sugars used in testing for specificity of the lectin-cell interaction and, surprisingly it was found that this sugar (but no other) actually promoted spreading on WGA beads by 40%. Fucose itself is likely to be the active agent since it is an end product of metabolism (Kaufman & Ginsburg, 1968, Stoddart, 1984). Its action may be at the cell surface, for instance acting as a substrate for ecto-fucosyl transferase (see Roseman, 1970; Roth, McGuire & Roseman, 1971) and thereby modifying surface oligosaccharide chains which can then interact with WGA, or instigating activation of the cytoskeleton. Certain EC cell lines (F9) have been shown to possess high levels of fucosyl transferase activity resulting in the fucosylation of embryoglycans and this activity decreases as the cells differentiate (Muramatsu & Muramatsu, 1983). Alternatively, fucose may be taken up by the cells and used in the synthesis of glycoconjugates (Iwakura, 1983; Kimber, MacQueen & Bagley, in press and p. 211).

Effects of free lectin

In some preliminary experiments, we looked at the effect on compaction of a range of lectins from plant and animal sources with different sugar specificities encompassing most of the monosaccharides commonly present in the glycoconjugates of animal cells. If cell surface oligosaccharides functional in compaction were bound by any of these lectins one might expect the process to be perturbed. Embryos were cultured without the zona pellucida in groups of approximately 10 from the early 8-cell stage, in medium containing 20–200 μg/ml of the lectins *Griffonia simplifolia* lectin-I and -II (GSL-I and -II), *Maclura pomifera* agglutinin (MPA), *Ulex europaeus*-I agglutinin (UEA-I), DBA, SBA, RCA-1, Con-A, PNA, SBA, WGA and *Lotus tetragonolobus* agglutinin (LTA). Some lectins promoted agglutination of embryos, for instance DBA (α-D-GalNAC) SBA (α/β-D-GalNAC) and to a lesser extent WGA. Con-A caused a slight reduction in the percentage of embryos which compacted (70% of control embryos). However, Con-A proved toxic even at 20 μg/ml in our experience, although this latter effect was not reported by Reeve (1982) in similar experiments. In the presence of the fucose binding protein from *Lotus tetragonolobus* (LTA) at 23 μg/ml embryos compacted but subsequently 30% of them decompacted, when control embryos were between the 16-cell and 32-cell stage (Bird & Kimber, 1984). The effect of this lectin was difficult to interpret because the decompaction proved irreversible; embryos explanted to fresh, lectin-free, medium failed to recompact and form blastocysts. All the same, considered in the context of our other data it did suggest that binding of cell surface fucose by an exogenous agent was incompatible with continued cell–cell adhesion.

FITC-lectin binding

Even though our results from incubating embryos with LTA were open to some doubt, the evidence from hapten-inhibition experiments using saccharides indicated that it would be useful to look at the relative levels of exposed fucose on the cell surface during the 8- to 16-cell stage. Therefore, the level of binding of fluorescein isothiocyanate (FITC) conjugated LTA was examined over this period (Fig. 14.8) in parallel with the binding of other lectins with different sugar affinity. The effect of proteases on both lectin receptors and on compaction were also assessed (Kimber & Bird, 1985) and specificity of FITC–lectin binding was ensured by the abolition of fluorescence in the presence of the appropriate inhibitory sugar. At the 4-cell stage, binding of LTA was low, and even at the early 8-cell stage, FITC-LTA fluorescence was faint. It was noticeable though that binding was strongest opposite the point of cell–cell contact, as reported for other lectins (Handyside, 1980; Ziomek & Johnson, 1980), and also in the contact region (Fig. 14.9). Between the end of the 8-cell stage and the 16-cell fully

Fig. 14.8. Average photometer readings (arbitrary units) for intensity of fluorescence emitted from the cell surface of embryos labelled with FITC conjugated LTA (E-Y laboratories, San Mateo, USA) ○---○ control ○——○ after treatment of early 8-cell embryos with 1 mg/ml proteinase K for 30 min at 37 °C. The enzyme treatment delays compaction (after complete decompaction in calcium and magnesium-free medium) by 5–18 h. Embryos were stained with the minimum concentration of FITC-LTA which gave maximum binding to early 8–cell embryos (100 μg/ml) immediately after enzyme treatment (0 h), and 8–10 h or 17–18 h later. Embryos were late 8-cell and compact 16-cell morulae respectively for the last two time points. Specificity of binding was checked by lack of fluorescence when staining was undertaken in the presence of 200 mM fucose. Bars = standard error. Control: number of embryos = 330, number of replicates = 3–5. Proteinase K treated: number of embryos = 186; number of replicates = 2–4 (modified from Kimber & Bird (1985)). *Roux's Arch. Devl. Biol.*, **194**, 470–9.

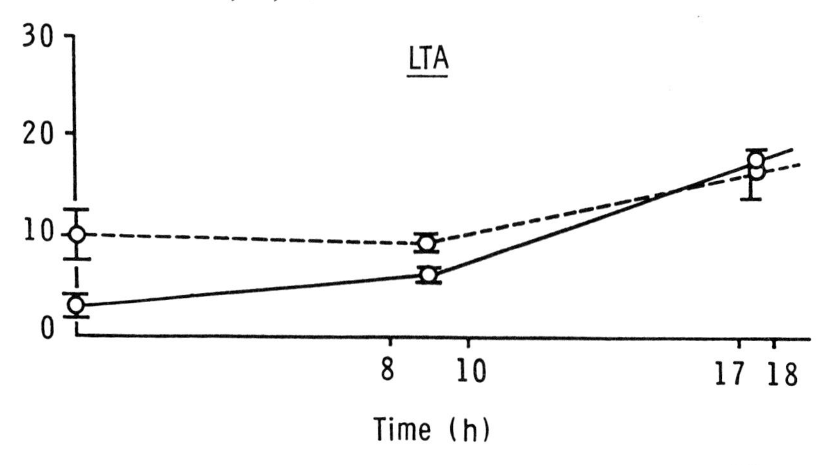

compacted morula, there was a considerable increase in LTA binding (Figs. 14.8 and 14.9). Thus the amount of cell-surface fucose available for binding to LTA appears to increase during this period, that is after the initial events of compaction, and around the time when LNFP-III is effective in reversing compaction. This is in agreement with the greater susceptibility of morulae, compared to 8-cell embryos, to complement-mediated lysis with the anti-SSEA-1 monoclonal (Solter & Knowles, 1978). The reduction in LTA-FITC fluorescence after treatment with proteases such as Proteinase k (Fig. 14.8) suggests that, at least at the early 8-cell stage a large proportion of all receptors for LTA (which will not necessarily include all surface fucosaccharides) are carried by protein. Receptors for the lectins BSL-II, MPA and RCA-1 with affinity for Gal and GlcNAC residues also increased between

Fig. 14.9. a, two cells from an early 8-cell embryo stained with FITC-LTA (100 μg/ml). Note the low level of fluorescence concentrated along, and opposite to, the region of cell–cell contact. b, 16-cell morula stained with FITC-LTA showing intense fluorescence at this stage.

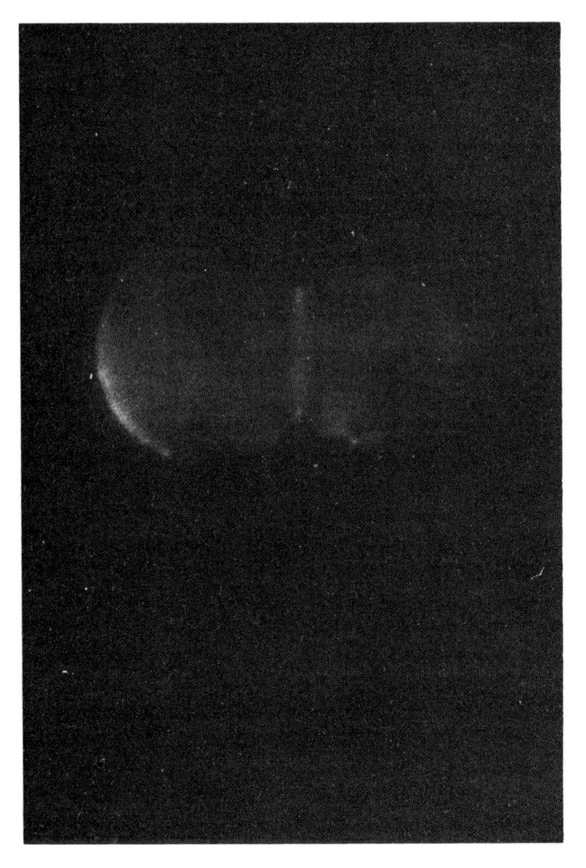

a

the early 8-cell and 16-cell stage while receptors for DBA (-D-GalNAC) decreased. Thus there appear to be overall changes in exposed surface saccharides during the period of compaction.

Incorporation of radio-labelled fucose into pre-implantation mouse embryos

In the light of the observations above, it seemed logical to examine the destination of fucose incorporated into macromolecules by mouse embryos, both biochemically and at the cellular level.

Embryos were incubated from the late 2-cell/early 4-cell stage in medium containing 200 μCi/ml. L-[6-^{3}H] or L-[1-^{14}C] fucose. Then they were either fixed and processed for plastic embedding immediately (Kimber, 1981) or cultured further in medium containing 10 mM unlabelled fucose. Analysis of the grain density in autoradiographs of semi-thin (1 μm) sections of embryos which had been 'chased' in non-radioactive medium for 0, 1.5,

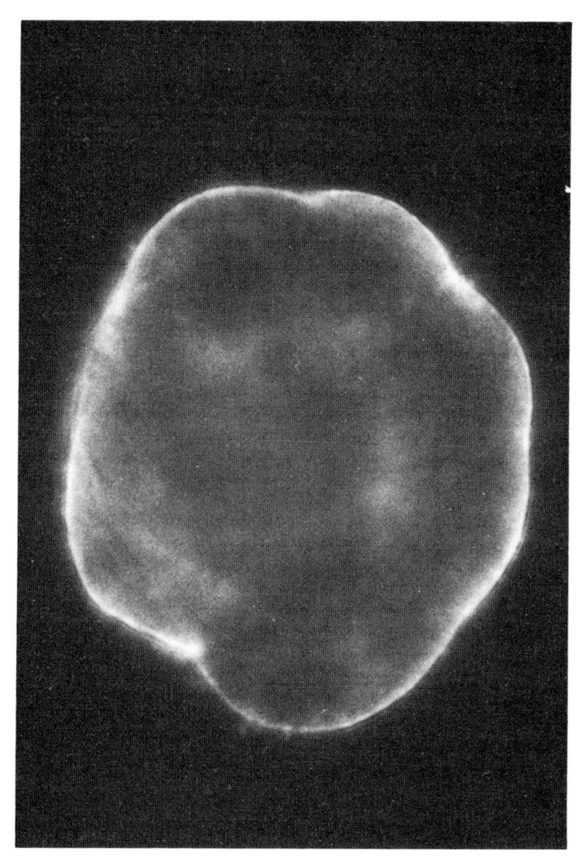

b

3, 24 and 48 h, was carried out (Kimber & Bagley, 1987). There was an enrichment of silver grains over the entire cell surface relative to the cytoplasm. However, immediately at the end of the radioactive incubation the density of silver grains was 40% higher over the apposed cell surfaces than over the outer microvillous regions. An autoradiograph from an embryo which had been 'chased' for 1.5 h is shown in Fig. 14.10. The grain density over the cell surface was high initially, dropped by about 20% during the first 3 h chase and then remained stable until between 24 h and 48 h when it decreased again. By 48 h there was no statistical difference in the density

Fig. 14.10. Autoradiograph of a 1 μm plastic section of an 8-cell embryo incubated from the 2-cell stage in medium containing 200 μCi/ml [^{3}H] fucose followed by 1.5 h in medium containing 10 mM unlabelled fucose. Embryos were fixed in 2% formaldehyde followed by 2.5% glutaraldehyde and 1% osmium tetroxide in phosphate buffer and processed as for electron microscopy. Sections were coated with L4 liquid nuclear emulsion (Ilford, UK), exposed for 5–7 days at 4 °C and developed in D19b developer (Kodak, UK). Note the concentration of silver grains over cell surfaces. Scale bar = 10 μm.

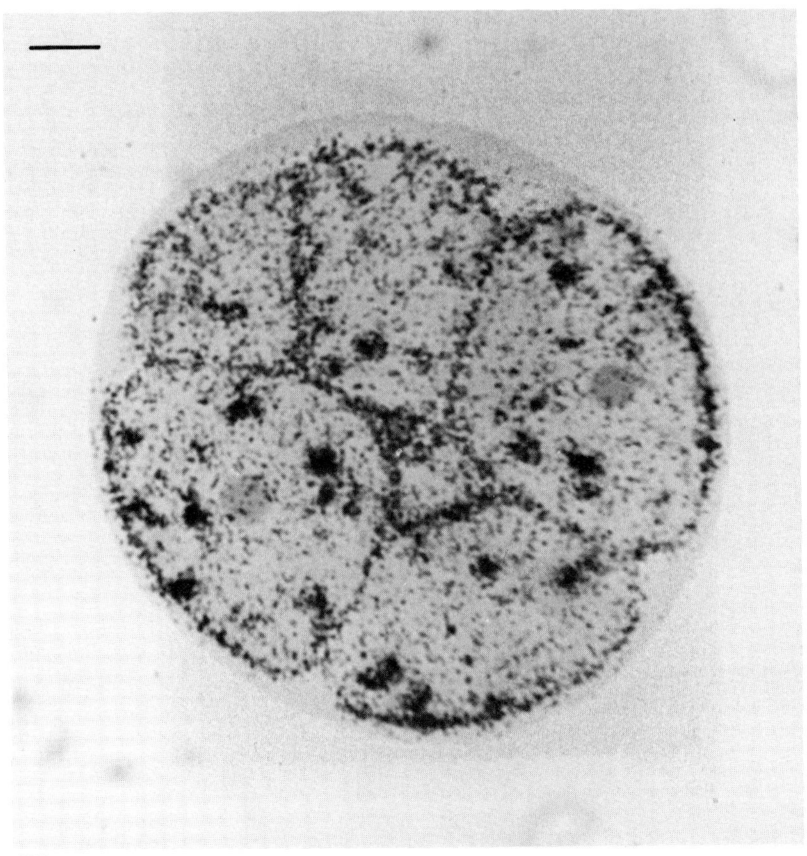

Table 14.1. *Incorporation of L-[6-^{3}H] into trichloroacetic acid precipitable material (APC) and lipid by preimplantation mouse embryos after 16–20 h incubation in ^{3}H-fucose containing medium*

Stage at harvest	Chase time, h	No. embryos (No. replicates)	Sample* treatment	DPM per embryo ± SE
Late 4 cell	0	173 (2)	APC	46 ± 9
6–8 cell	0	704 (6)	APC	62 ± 10
Compact cell 8–16 cell	6–9	579 (3)	APC	37 ± 7
Compact morulae/early blastocysts	24	262 (4)	APC	28 ± 7
Compact morulae	0	166 (2)	APC	74 ± 15
6–8 cell	0	722 (2)	Folch aqueous	311 ± 27
6–8 cell	0	1,362 (3)	Fucolipid	32 ± 4

* Embryos were washed free of radioactive medium and sonicated in phosphate buffered saline, containing 25 mg/ml bovine serum albumen as carrier for APC, or brain extract for the lipid extraction.
1. For APC, embryonic material was precipitated in trichloroacetic acid, the pellet dissolved in 1 M NaOH and placed in scintillation vials with Beckman Ready-Solv TM. Samples were counted on an LKB 1211 Minibeta liquid scintillation counter.
2. Lipids were extracted according to Folch *et al.* (1951). Polar lipids were further fractionated using the method of Hakomori *et al.* (1984). All the fucolipid was found in the first aqueous layer after the Hakomori method.

of silver grains over cell surfaces and cytoplasm which was reduced to 29% and 10% respectively of the levels at 0 h 'chase'. So fucose is indeed incorporated into cell-surface components particularly in the regions of cell–cell contact, but the resolution of this technique at the light microscope level does not allow us to distinguish if most of the label is in membrane fucoconjugates, extrinsic surface molecules or sub-membranous components.

Incorporation of fucose into protein and lipid in the embryo

While it is generally simple to obtain 10^6 or more cells from cell lines for biochemical analysis, for a similar study of early mouse embryos one must be content with 10^3 cells or less. Hence the biochemistry of these embryos is still in its infancy.

We examined the incorporation of the ^{3}H-fucose into trichloroacetic acid precipitable material at various stages during development (Table 14.1, Kimber, MacQueen & Bagley, in press). Incorporation of ^{3}H-fucose by 4-cell embryos was low but had increased by 35% by the time embryos had reached

Table 14.2. *Evidence for a role of fucosylated glycoconjugates in compaction of the mouse embryo*

1. Binding of SSEA-1 (specificity for Gal β(1–4) [Fuc α(1–3)] GlcNAc) to surface of mouse embryos from 8-cell stage (Solter & Knowles, 1978).
2. Gal β(1–4) [Fuc α(1–3)] GlcNAc, present in LNFP III, reverses compaction.
3. The fucose binding lectin from Lotus tetragonolobus (LTA) reverses compaction with similar time course to LNFP III.
4. Receptors for *Lotus tetragonolobus* agglutinin, absent on 4-cell embryos, increase between the early 8-cell stage and the compacted morula.
5. Fucoidin, a sulphated polymer of fucose causes partial decompaction of embryos.
6. Fucose stimulates spreading of single cells from 8-cell embryos on agarose beads conjugated to the lectin WGA.
7. Fucosylated glycoproteins and glycolipids are synthesised from late 4-cell/early 8-cell stage.
8. Incorporation of ^{3}H-fucose in 8-cell embryos is greater into cell surfaces, particularly apposed cell surfaces, than into the cytoplasm.
9. Fucan/mannan specific carbohydrate binding proteins implicated in cell adhesion of EC cells (Grabel *et al.*, 1979, 1981).

the 6- to 8-cell stage and by another 21% between the 6–8 cell stage and the compacted morula. The acid precipitable material, containing glycoproteins and glycolipids (Kaufman & Ginsburg, 1969; Iwakura, 1983; Stoddart, 1984), represented about 20% of the total ^{3}H-fucose taken up by embryos over this period of development. Those fucoconjugates known to be present at the 8-cell and 16-cell stage such as the 100 kD glucose regulated protein (McCormick & Babiarz, 1984), uvomorulin/cadherin (op. cit.) laminin (Hogan, 1980; Engvall *et al.*, 1983; Cooper & MacQueen, 1983) and embryoglycans (Muramatsu *et al.*, op. cit.) major carriers of the SSEA-1 determinant in EC-cells (Ozawa *et al.*, 1985), consist of protein-bound saccharides. Nothing is known about fucolipids in pre-implantation mouse embryos. Therefore, we were surprised to discover that the amount of fucose ending up in fucolipids in 6–8 cell mouse embryos, was approximately half the total incorporation of the fucose into fucoconjugates (Table 14.2). This indicates that almost equal amounts of fucoprotein and fucolipid are synthesised at this stage. SSEA-1 is carried by glycolipids in various cell types (Hakomori *et al.*, 1981; Watkins, 1980; Feizi, 1982) and our data provides evidence that it may also be carried by lipids in mouse embryos as previously suggested (Solter & Knowles, 1978). We need to investigate the nature of these fucolipids further and particularly whether they are involved in cell–cell interactions.

Separation of fucoproteins on 10% SDS-polyacrylamide gels (Kimber, MacQueen & Bagley, in press), revealed a number of bands at the 8-cell and later stages. Changes in components around 62–69 kD and 44–52 kD between the 8-cell stage and late morula were detected (Fig. 14.11) and most bands appear to be tunicamycin sensitive, although one band of M_r *c.* 62 kD remained after Tunicamycin treatment and may contain *o*-linked fuco-saccharide chains. When we carried out 'chase' experiments, incubating embryos in medium containing 10 mM unlabelled fucose, we found that most bands were relatively stable and to a similar extent although their intensity did decrease during the 24 h 'chase' incubation. This reflects the reduction by 40% within the first 6–9 h 'chase' in the amount of ^{3}H-fucose remaining in acid precipitable material (Table 14.1) and 20–24% in the density of silver grains over cell surfaces on autoradiographs after the first 3 h 'chase'. Alterations in bands in the region 62–69 kD occurred; one band, at *c.* 69 kD, had disappeared after 24 h chase. This band, which is present in the 8-cell embryo and compacted morula may have a shorter half life than others synthesised during the 4- to 8-cell stage. The fucoproteins

Fig. 14.11. Autofluorograph of glycoproteins from mouse pre-implantation embryos separated on a 10% SDS-polyacrylamide gel. a, precompact 8-cell embryos (86). b, compact 8-cell embryos (229); c, late morulae (128); d, early blastocysts (129). Embryos were incubated for 16–18 h in 200 μCi/ml [^{14}C]fucose, washed and solubilised in buffer containing 8% glycerol, 5% B-mercaptoethanol, 2.3% SDS and 0.076% TRIS base pH 6.8. Samples were stored at −20 °C with 2 mM phenylmethyl sulphonyl fluoride PMSF until required. Pre-stained molecular weight standards (BRL) were run in parallel (myosin H-chain, 200 kD; phosphorylase B, 92.5 kD; bovine serum albumen 68.0 kD; ovalbumen, 43.0 kD; α-chymotrypsin, 25.6 kD; B-lactoglobulin, 18.4 kD; cytochrome *c*, 12.3 kD). The Gel was exposed at −70 °C to pre-flashed (Bonner & Laskey, 1974) LKB Ultrafilm for 7 months.

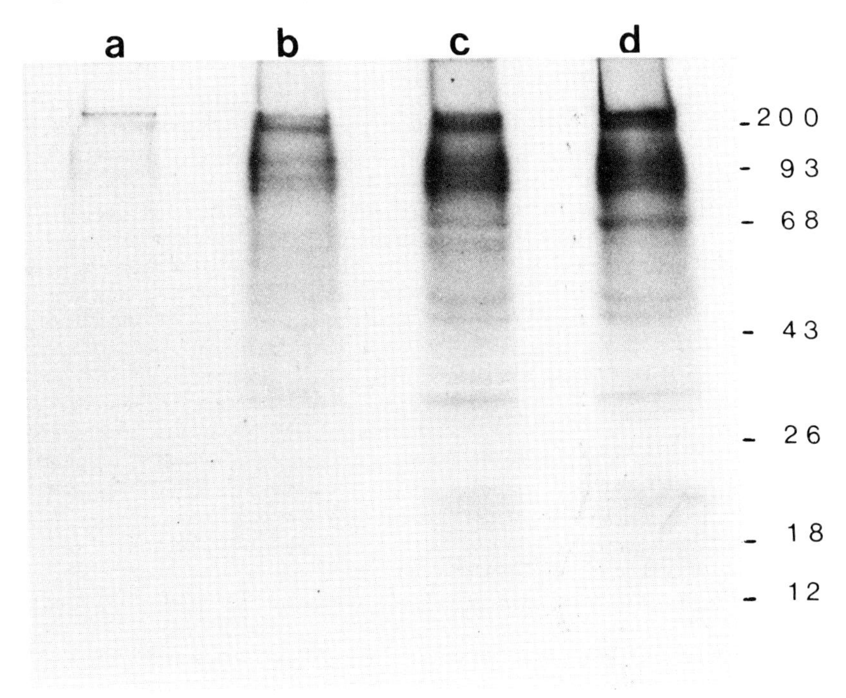

synthesised during this period of development seem to represent distinct polypeptides since no new bands appeared during the chase period.

Conclusions

The major data contributing to the idea that fucosylated glycoconjugates may be involved in maintaining cell-cell interaction in the compacted mouse embryo are shown in Table 14.2. The most pressing evidence comes from the specific reversal of compaction by LNFP-III, carrying the SSEA-I determinant. This is supported by the results of Fenderson, Zehavi & Hakomori (1984) showing that an LNFP-III conjugate of lysyllysine (containing 3 moles of oligosaccharide per mole of lysyllysine) reverses compaction with a similar time course to that which we observed for the monomer. Although with the strain of mice they used the oligosaccharides was not effective. The evidence outlined above has led the author to propose a three-step model for compaction (Bird & Kimber, 1984; Kimber & Bird, 1985) which is no doubt a simplification of the process but provides a useful basic skeleton which can be tested experimentally, and altered or added to as new data present themselves.

Phase 1 Initial calcium dependent spreading involving activation of microfilaments and the cell-surface protein uvomorulin/cadherin.

Phase 2 Stabilisation of interacting membranes by a mechanism including the binding of glycoconjugates carrying SSEA-I by surface receptors.

Phase 3 The subsequent formation of tight junctions and desmosomes leading to more substantial intercellular 'locking' between adjacent membranes.

These phases must overlap, and each succeeding phase gradually dominate over that preceding it. It must, however, be said that there are some pieces of evidence which do not, at first sight, support this model. The observed lack of decompaction when mouse embryos are cultured with anti-SSEA-I is somewhat at odds with the results reported above. Perhaps membranes are tightly apposed by the time the SSEA-I dependent mechanism comes into force so that the bulky IgM molecules cannot effectively penetrate and compete with natural receptors. Mutant PSA-1 and F9 EC cell lines have also been reported which show similar cell interaction behaviour to their parental lines but have reduced (or lack) SSEA-I (Rosenstraus, 1983; Buckalew, Sterman & Rosenstraus, 1985). However, it is quite probable that interaction of alternative receptors with different but appropriate surface oligosaccharides may be functional in cell interaction in these mutants. In any case no EC cell lines are completely identical to the 8- to 16-cell embryo or even the ICM/early egg cylinder cells that they resemble most closely (Rossant & Papaioannou, 1984). Indeed, the relative content of GlcNAC and fucose in the high molecular weight polysaccharide chains characteristic of embryonic cells has been shown to be different in F9 EC cells from compacted morulae (Iwakura, 1983), and there are also differences in the carbohydrate composition of low molecular weight saccharides and complex type glycopeptides.

Overall, the evidence seems to favour some cell interaction-related function for fucoconjugates in the early mouse embryo. Whether this is indeed a receptor–SSEA-I interaction acting in a trans fashion between cells to help maintain close membrane–membrane contact or some other process, perhaps linked to stabilisation of the cytoskeletal organisation necessary for the maintenance of the spread cell shape, remains to be determined. The critical time period for decompaction in LNFP-III at the beginning of the 16-cell stage may indicate that when cells round up for cell division LNFP-III bound to exposed membrane receptors prevents cells from spreading on one another again after the completion of mitosis. Whether this effect reflects the initiation of this mechanism of cell interaction at this particular stage or the new exposure of receptors for SSEA-1/LNFP-III on the poorly adhering cells at mitosis is unclear. The hypothesised fucosyltransferase suggested to be responsible for the appearance of SSEA-1 (Gooi *et al.*, 1981; Knowles *et al.*, 1982) may also regulate 'phase 2' of cell adhesion. The idea that surface glycosyltransferases might be involved in cell interactions is not a new one (Roseman, 1970; Roth *et al.*, 1971) and various cell surface glycosyltransferases have been demonstrated using appropriate controls (e.g. Patt *et al.*, 1976; Matsui *et al.*, 1983; Toporowicz & Reisner, 1986). In fact an ectogalactosyltransferase has been suggested to promote cell–cell adhesion by acting as a receptor for cell surface saccharides in F9 EC cells (Shur, 1982, 1983) and the enzyme activity has been shown to increase when these cells differentiate (Nakhasi, Nagarajan & Anderson, 1984). If this reflects cell adhesion processes in the embryo it probably mirrors mechanisms occurring later than the stages considered here, in the ICM and egg cylinder. Fucosyl transferase, if present, probably does not act as a 'lectin' in the compacting mouse embryo, but rather is responsible for the appearance, modification or exposure of components which are involved in cell interactions between these early undifferentiated embryonic cells. In keeping with this in the finding of decreased fucosyltransferase activity on differentiation of EC cells to parietal endoderm (Muramatsu & Muramatsu, 1983). In view of the evidence that the appearance of SSEA-1 in Chinese hamster ovary cells is controlled by the *de novo* expression of specific N-acetylglucosaminide α(1–3) fucosyl transferase activities (Campbell & Stanley, 1983) effort should now be directed to investigation of such enzymes in the mouse embryo.

Acknowledgements

I am most grateful to Mr P. Bagley and Mr J. Bird for technical help during the course of this work, to Dr M. A. H. Surani and Dr H. MacQueen for valuable discussions and collaborations as well as to Dr A. Handyside, Dr K. Ahuja and Ms K. Hardy for reading the manuscript and to Miss P. Sayers for typing it.

References

Andrews, P. W., Knowles, B. B., Cossu, G. & Solter, B. (1982). Teratocarcinoma and mouse embryo cell surface antigens: characterization of the molecule(s) carrying the SSEA-1 antigenic determinant. In *Teratocarcinoma and Embryonic Cell Interactions*, ed T. Muramatsu, G. Gachelin, A. A. Moscoma & Y. Iwakura, p. 103–19. New York: Academic press

Atienza-Samols, S. B., Pine, P. R. & Sherman, M. I. (1980). Effect of tunicamycin upon glycoprotein synthesis and development of early mouse embryos. *Dev. Biol.*, **79**, 19–31

Barondes, S. H. (1981). Lectins: Their multiple endogenous cellular functions. *Ann. Rev. Biochem.*, **5**, 207–31

Behrens, J., Birchmeier, W., Goodman, S. L. & Imhof, B. A. (1985). Dissociation of Madin-Darby canine kidney epithelial cells by the monoclonal antibody anti-Arc-1: mechanistic aspects and identification of the antigen as a component related to uvomorulin. *J. Cell Biol.*, **101**, 1307–15

Bird, J. M. & Kimber, S. J. (1984). Oligosaccharides containing fucose linked α (1–3) and α (1–4) to N-acetylglucosamine cause decompaction of mouse morulae. *Dev. Biol.*, **104**, 449–60

Boller, K., Vestweber, D. & Kemler, R. (1985). Cell-adhesion molecule uvomorulin is localized in the intermediate junctions of adult intestinal epithelial cells. *J. Cell Biol.*, **100**, 327–32

Bonner, W. M. & Laskey, R. A. (1974). A film detection method for tritium-labelled proteins and nucleic acids in polyacrylamide gels. *Eur. J. Biochem.*, **46**, 83–8

Bozzaro, S. (1985). Cell surface carbohydrates and cell recognition in Dictyostelium. *Cell Diff.*, **17**, 67–82

Brackenbury, R., Rutishauser, U. & Edelman, G. M. (1981). Distinct calcium-independent and calcium-dependent adhesion systems of chicken embryo cells. *Proc. Natl Acad. Sci. USA*, **78**, 387–91

Buckalew, J. J., Sterman, B., Rosenstraus, M. (1985). Variant embryonal carcinoma cells lacking SSEA-1 and Forsmann antigens remain developmentally pluripotent. *Dev. Biol.*, **107**, 134–41

Campbell, C. & Stanley, P. (1983). Regulatory mutations in CHO cells induce expression of the mouse embryonic antigen SSEA-1. *Cell*, **35**, 303–9

Childs, R. A., Pennington, J., Uemura, K. I., Scudder, P., Goodfellow, P. N., Evans, M. J. & Feizi, T. (1983). High-molecular-weight glycoproteins are the major carriers of the carbohydrate differentiation antigens I, i and SSEA-1 of mouse teratocarcinoma cells. *Biochem. J.*, **213**, 491–503

Cooper, A. R. & MacQueen, H. A. (1983). Subunits of laminin are differentially synthesised in mouse eggs and early embryos. *Dev. Biol.*, **96**, 467–71

Damjanov, I., Damjanov, A. & Damsky, C. H. (1986). Developmentally regulated expression of the cell-cell adhesion glycoprotein cell-CAM 120/80 in pre-implantation mouse embryos and extraembryonic membranes. *Dev. Biol.*, **116**, 194–202

Damsky, C. H., Richa, J., Solter, D., Knudsen, K. & Buck, C. A. (1983). Identification and purification of a cell surface glycoprotein mediating intercellular adhesion in embryonic and adult tissue. *Cell*, **34,** 455–66

Ducibella, T. (1980). Divalent antibodies to mouse embryonal carcinoma cells inhibit compaction in the mouse embryo. *Dev. Biol.*, **79**, 356–66

Ducibella, T. (1982). Depolymerization of microtubules prior to compaction. Development of cell polarity and cell spreading are not inhibited. *Exp. Cell Res.*, **138**, 31–8

Ducibella, T. & Anderson, E. (1975). Cell shape and membrane changes in the eight-cell mouse embryo: pre-requisite for morphogenesis of the blastocyst. *Dev. Biol.*, **47**, 45–58
Ducibella, T. & Anderson, E. (1979). The effects of calcium deficiency on the formation of the zonula occludens and blastostoel in the mouse embryo. *Dev. Biol.*, **73**, 46–58
Ducibella, T., Albertini, D. F., Anderson, E. & Biggers, J. D. (1975). The pre-implantation mammalian embryo: characterisation of intercellular junctions and their appearance during development. *Dev. Biol.*, **45**, 231-50
Edelman, G. M. (1986). Cell adhesion molecules in the regulation of animal form and tissue pattern. *Ann. Rev. Cell Biol.*, **2**, 81–116
Edelman, G. M., Hoffman, S., Chuong, C. M. & Cunningham, B. A. (1985). The molecular bases and dynamics of cell adhesion in embryo-genesis. In *Molecular Determinants of Animal Form.* UCLA Symp. Molec. Cell Biol. New Ser., Vol. 31, p. 195–221. New York: Alan R. Liss Inc
Ekblom, P., Vestweber, D. & Kemler, R. (1986). Cell-matrix interactions and cell adhesion during development. *Ann. Rev. Cell Biol.*, **2**, 27–47
Engvall, E., Krusius, T., Wewer, U. & Ruoslahti, E. (1983). Laminin from rat yolk sac tumour: isolation, partial characterization, and comparison with mouse laminin. *Arch. Biochem. Biophys.*, **222**, 649–56
Feizi, T. (1982). The antigens Ii, SSEA-1 and ABH are an interrelated system of carbohydrate differentiation antigens expressed on glycosphingolipids and glycoproteins. In *New Vistas in Glycolipid Research*, Proc. Biwako Symp. on Glycolipids, Otsu, Japan, 1981, eds A. Makita, S. Handa, T. Taketomi & Y. Nagai. *Adv. in Exp. Med. & Biol.*, **152**, 167–72
Fenderson, B. A., Hahnel, A. C. & Eddy, E. M. (1983). Immuno-histochemical localisation of two monoclonal antibody defined carbohydrate antigens during early murine embryogenesis. *Dev. Biol.*, **100**, 318–27
Fenderson, B. A., Holmes, E. H., Fukushi, Y. & Hakomori, S-I. (1986). Coordinate expression of X and Y haptens during murine embryogenesis. *Devl. Biol.*, **114**, 12–21
Fenderson, B. A., Zehavi, U. & Hakomori, S-I. (1984). A multivalent lacto-N-fucopentaose III–lysyllysine conjugate decompacts pre-implantation mouse embryos, while the free oligosaccharide is ineffective. *J. Exp. Med.*, **160**, 1591–6
Folch, J., Ascoli, I., Lees, M., Meath, J. A. & Le Baron, F. N. (1951). Preparation of lipide extracts from brain tissue. *J. Biol. Chem.*, **91**, 833–41
Fox, N., Damjanov, I., Martinez-Hernadez, A., Knowles, B. B. & Solter, D. (1981). Immuno-histochemical localisation of the early embryonic antigen (SSEA-1) in post implantation mouse embryos and fetal and adult tissues. *Devl Biol.*, **83**, 391–8
Fox, N., Shevinsky, L., Knowles, B. B., Solter, D. & Damjanov, I. (1982). Distribution of murine stage-specific embryonic antigens in the kidneys of these rodent species. *Exp. Cell Res.*, **140**, 331–9
Gallin, W. J., Edelman, G. M. & Cunningham, B. A. (1983). Characterisation of L-CAM, a major cell adhesion molecule from embryonic liver cells. *Proc. Natl Acad. Sci. USA*, **80**, 1038–42
Goodall, H. & Johnson, M. H. (1982). Use of carboxylfluorescein diacetate

to study formation of permeable channels between mouse blastomeres. *Nature (Lond.)*, **295**, 524–6

Goodall, H. & Johnson, M. H. (1984). The nature of intercellular coupling within the pre-implantation mouse embryo. *J. Embryol. Exp. Morph.*, **79**, 53–76

Gooi, H. C., Feizi, T., Kapadia, A., Knowles, B. B., Solter, D. & Evans, M. J. (1981). Stage-specific embryonic antigen involves α 1–3 fucosylated type 2 blood group chains. *Nature (Lond.)*, **292**, 156–8

Grabel, L. B. (1984). Isolation of a putative cell adhesion mediating lectin from teratocarcinoma stem cells and its possible role in differentiation. *Cell Diff.*, **15**, 121–4

Grabel, L. B., Glabe, C. G., Singer, M. S., Martin, G. R. & Rosen, S. D. (1981). A fucan specific lectin on teratocarcinoma stem cells. *Biochem. Biophys. Res. Commun.*, **102**, 1165–71

Grabel, L. B., Rosen, S. D. & Martin, G. R. (1979). Teratocarcinoma stem cells have a cell surface carbohydrate-binding component implicated in cell–cell adhesion. *Cell*, **17**, 477–84

Grabel, L. B., Singer, M. S., Martin, G. R., Rosen, S. D. (1983). Teratocarcinoma stem cell adhesion: the role of divalent cations and a cell surface lectin. *J. Cell Biol.*, **96**, 1532–7

Grunwald, G. B., Geller, R. L. & Lilien, J. (1980). Enzymatic dissection of embryonic cell adhesion mechanisms. *J. Cell Biol.*, **85**, 766–76

Gumbiner, G. & Simons, K. (1986). A functional assay for proteins involved in establishing an epithelial occluding barrier: identification of a uvomorulin-like polypeptide. *J. Cell Biol.*, **102**, 457–8

Hakomori, S-I., Nudelman, E., Levery, S., Solter, D. & Knowles, B. B. (1981). The hapten structure of a developmentally regulated glycolipid antigen (SSEA-1) isolated from human erythrocytes and adenocarcinoma: a preliminary note. *Biochem. Biophys. Res. Commun.*, **100**, 1578–86

Hakomori, S-I., Nudelman, E., Levery, S. B. & Kannagi, R. (1984). Novel fucolipids accumulating in human adenocarcinoma. *J. Biol. Chem.*, **259**, 4672–80

Handyside, A. H. (1980). Distribution of antibody- and lectin-binding sites on dissociated blastomeres from mouse morulae: evidence for polarisation at compaction. *J. Embryol. Exp. Morph.*, **60**, 99–116

Harrison, F. L. & Chesterton, C. J. (1980). Factors mediating cell-cell recognition and adhesion. *FEBS Lett.*, **122**, 157–65

Hogan, B. L. M. (1980). High molecular weight extracellular proteins synthesised by endoderm cells derived from mouse teratocarcinoma cells and normal extraembryonic membranes. *Devl. Biol.*, **76**, 275–85

Hyafil, F., Babinet, C. & Jacob, F. (1981). Cell–cell interactions in early embryogenesis: a molecular approach to the role of calcium. *Cell*, **26**, 447–54

Hyafil, F., Morello, D., Babinet, C. & Jacob, F. (1980). A cell surface glycoprotein involved in the compaction of embryonal carcinoma cells and cleavage stage embryos. *Cell*, **21**, 927–34

Iwakura, Y. (1983). Comparison of polysaccharide synthesis between pre-implantation stage mouse embryos and F9 embryonal carcinoma cells. *Exp. Cell Res.*, **146**, 329–38

Johnson, M. H., (1985). Three types of cell interaction regulate the generation of cell diversity in the mouse blastocyst. In The cell in contact. Adhesions and junctions as morphogenetic determinants, ed G. M. Edelman & J-P. Thiery, pp. 27–48, New York: John Wiley & Sons

Johnson, M. H., Chakraborty, J., Handyside, A. H., Willison, K. & Stern, P. (1979). The effect of prolonged decompaction on the development of the pre-implantation mouse embryo. *J. Embryol. Exp. Morphol.*, **54**, 241–61

Johnson, M. H. & Ziomek, C. A. (1981). The foundation of two distinct cell lineages within the mouse morula. *Cell*, **24**, 71–80

Kameyama, T., Oshi, K. & Aida, K. (1979). Stereochemical structure recognised by the L-fucose-specific hemagglutinin produced by *Streptomyces* sp. *Biochem. Biophys. Acta*, **587**, 407–14

Kannagi, R., Levery, S. B., Ishigami, F., Hakomori, S–I., Shevinsky, L. H., Knowles, B. B. & Solter, D. (1983). New globoseries glycosphingolipids in human teratocarcinoma reactive with the monoclonal antibody directed to a developmentally regulated antigen, stage-specific embryonic antigen 3. *J. Biol. Chem.*, **258**, 8934–42

Kaufman, R. L. & Ginsburg, V. (1968). The metabolism of L-fucose by HeLa cells. *Exp. Cell Res.*, **50**, 127–32

Kelly, S. J. (1975). Studies of the potency of the early cleavage blastomeres of the mouse. In *The Early Development of Mammals*. Brit. Soc. Devl. Biol. Symp. *2*, ed M. Balls & A. E. Wild, pp. 97–105. Cambridge: Cambridge University Press

Kemler, R., Babinet, D., Eisen, H., Jacob, F. (1977). Surface antigen in early differentiation. *Proc. Natl Acad. Sci. USA*, **74**, 4449–52

Kemler, R., Morello, D. & Jacob, F. (1979). Properties of some monoclonal antibodies raised against mouse embryonal carcinoma cells. In *Stem Cells and Cell Determination*, INSERM Symp. No. 10, ed N. Le Douarin, pp. 101–13, Amsterdam: Elsever/North Holland Biomedical Press

Kidder, G. M. & McLachlin, J. R. (1985). Timing of transcription and protein synthesis underlying morphogenesis in pre-implantation mouse embryos. *Dev. Biol.*, **112**, 265–75

Kimber, S. J. (1981). The secretion of the eggshell of *Schistocerca gregaria*. Analysis of the kinetics of secretion *in vitro* by light and electron microscope autoradiography. *J. Cell Sci.*, **50**, 225–43

Kimber, S. J. & Bagley, P. R. (1987). Cell surface enrichment of fucosylated glycoconjugates in the 8- to 16-cell mouse embryo: an autoradiographic study. *Roux's Arch. Devl. Biol.*, **196**, 492–8.

Kimber, S. J., MacQueen, H. A. & Bagley, P. R. (1987). Fucosylated glycoconjugates in mouse pre-implantation embryos. *J. Exp. Zool.*, in press

Kimber, S. J. & Bird, J. M. (1985). Cell surface changes in pre-implantation mouse embryos during compaction investigated using FITC conjugated lectins after proteolytic enzyme treatment. *Roux's Arch. Dev. Biol.*, **194**, 470–9

Kimber, S. J. & Surani, M. A. H. (1981). Morphogenetic analysis of changing cell association following release of 2-cell and 4-cell mouse embryos from cleavage arrest. *J. Embryol. Exp. Morph.*, **61**, 331–5

Kimber, S. J. & Surani, M. A. H. (1982). Spreading of blastomeres from eight-cell mouse embryos on lectin coated beads. *J. Cell Sci.*, **56**, 191–206

Kimber, S. J. & Surani, M. A. H. (1983). Lectin coated beads stimulate single cells from mouse embryos to spread. In *Lectins*, Vol. III, p. 63–72, eds T. C. Bøg-Hansen & G. A. Spengler, Berlin: Walter de Gruyter & Co.

Kimber, S. J., Surani, M. A. H. & Barton, S. C. (1982). Interactions of blastomeres suggest changes in cell surface adhesiveness during the formation of inner cell mass and trophectoderm in the pre-implantation mouse embryo. *J. Embryol. Exp. Morph.*, **71**, 133–52

Knowles, B. B., Rappaport, J. & Solter, D. (1982). Murine embryonic antigen (SSEA-1) is expressed on human cells and structurally related human blood group antigen I is expressed on mouse embryos. *Dev. Biol.*, **93**, 54–8

Kobata, A. (1972). Isolation of oligosaccharides from human milk. *Meth. Enzymol.*, **28**, 262–71

Kobata, A. & Ginsburg, V. (1969). Oligosaccharides of human milk II. Isolation and characterisation of a new pentasaccharide, Lacto-N-fucopentaose III. *J. Biol. Chem.*, **244**, 5496–502

Levy, J. B., Johnson, M. H., Goodall, H. & Maro, B. (1986). The timing of compaction: control of a major developmental transition in mouse early embryogenesis. *J. Embryol. Exp. Morph.*, **95**, 213–37

Lo, C. W. & Gilula, N. B. (1979). Gap junctional communication in the pre-implantation mouse embryo. *Cell*, **18**, 399–409

McCormick, P. J. & Babiarz, B. (1984). Expression of a glucose regulated cell surface protein in early mouse embryos. *Dev. Biol.*, **105**, 530–4

McCormick, P. J., Keys, B. J., Pucci, C. & Millis, A. J. T. (1979). Human fibroblast-conditioned medium contains a 100 k Dalton glucose-regulated cell surface protein. *Cell*, **18**, 173–82

McLachlin, J. R. & Kidder, G. M. (1986). Intercellular junctional coupling in pre-implantation mouse embryos: effect of blocking transcription or translation. *Dev. Biol.*, **117**, 146–55

McLachlin, J. R., Caveney, S. & Kidder, G. M. (1983). Control of gap junction formation in early mouse embryos. *Dev. Biol*, **98**, 155–64

Magnani, J. L., Thomas, W. A., Steinberg, M. S. (1981). Two distinct adhesion mechanisms in embryonic neural retina cells. I. A kinetic analysis. *Dev. Biol.*, **81**, 96–105

Magnuson, T., Demsey, A. & Stackpole, C. W. (1977). Characterisation of intercellular junctions in the pre-implantation mouse embryo by freeze-fracture and thin-section electron microscopy. *Dev. Biol.*, **61**, 252–61

Maro, B., Johnson, M. H., Pickering, S. J. & Louvard, D. (1985). Changes in the distribution of membranous organelles during mouse early development. *J. Embryol. Exp. Morph.*, **90**, 287–309

Maro, B. & Pickering, S. J. (1984). Microtubules influence compaction in pre-implantation mouse embryos. *J. Embryol. Exp. Morph.*, **84**, 2317–32

Marticorena, P., Hogan, B., DiMeo, A., Artzt, K. & Bennett, D. (1983). Carbohydrate changes in the pre- and peri-implantation mouse embryos as detected by a monoclonal antibody. *Cell Diff.*, **12**, 1–10

Matsui, Y., Lombard, D., Hoflack, B., Harth, S., Massarelli, R., Mandel, P. & Dreyfus, H. (1983). Ectoglycosyltransferase activities at the surface of cultured neurons. *Biochem. Biophys. Res. Commun.*, **113**, 446–53

Mentzer, S. J., Burakoff, S. J. & Faller, D. V. (1986). Adhesion of T Lymphocytes to human endothelial cells is regulated by the LFA-1 membrane molecule. *J. Cell Physiol.*, **126**, 285–90

Misevic, G. N. & Burger, M. M. (1987). Multiple low-affinity carbohydrates as a basis for cell recognition in the sponge *Microciona prolifera*, this volume

Muramatsu, T., Condamine, H., Gachelin, G. & Jacob, F. (1980). Changes in fucosyl-glycopeptides during early post-implantation embryogenesis in the mouse. *J. Embryol. Exp. Morph.*, **57**, 25–36

Muramatsu, T., Gachelin, G., Nicolas, J. F., Condamine, H., Jakob, H. & Jacob, F. (1978). Carbohydrate structure and cell differentiation: unique properties of fucosyl-glycopeptides isolated from embryonal carcinoma cells. *Proc. Natl Acad. Sci. USA*, **75**, 2315–19

Muramatsu, H., Ishihara, H., Miyauchi, T., Gachelin, G., Fujisaki, T., Tejima, S. & Muramatsu, T. (1983). Glycoprotein-bound large carbohydrates of early embryonic cells: structural characteristics of the glycan isolated from F9 embryonal carcinoma cells. *J. Biochem.*, **94**, 799–810

Muramatsu, H. & Muramatsu, T. (1983). A fucosyltransferase in teratocarcinoma stem cells. *FEBS Lett.*, **163**, 181–4

Nakhasi, H. L., Nagarajan, L. & Anderson, W. B. (1984). Increase in cell-surface N-acetylglucosaminide B(1-4) galactosyltransferase activity with retinoic acid-induced differentiation of F9 embryonal carcinoma cells. *FEBS Lett.*, **168**, 222–6

Ocklind, C. & Öbrink, B. (1982). Intercellular adhesion of rat hepatocytes. Identification of a cell surface glycoprotein involved in the initial adhesion process. *J. Biol. Chem.*, **257**, 6788–95

Ogou, S-I., Okada, T. S. & Takeichi, M. (1982). Cleavage stage mouse embryos share a common cell adhesion system with teratocarcinoma cells. *Dev. Biol.*, **92**, 521–28

Ogou, S-I., Yoshida-Noro, C., Takeichi, M. (1983). Calcium-dependent cell-cell adhesion molecules common to hepatocytes and teratocarcinoma stem cells. *J. Cell Biol.*, **97**, 944–8

Ozawa, M., Muramatsu, T. & Solter, D. (1985). SSEA-1, a stage-specific embryonic antigen of the mouse, is carried by the glycoprotein-bound large carbohydrate in embryonal carcinoma cells. *Cell Differ.*, **16**, 169–173

Papaioannou, V. E. & Rossant, J. (1983). Effects of the embryonic environment on proliferation and differentiation of embryonal carcinoma cells. *Cancer Surv.*, **2**, 165–83

Patt, L. M., Endres, R. O., Lucas, D. O. & Grimes, W. S. (1976). Ectogalactosyltransferase studies in fibroblasts and Concanavelin A stimulated lymphocytes. *J. Cell Biol.*, **68**, 799–802

Pennington, J. E., Rastan, S., Roelcke, D. & Feizi, T. (1985). Saccharide structures of the mouse embryo during the first eight days of development. *J. Embryol. Exp. Morph.*, **90**, 335–61

Peyriéras, N., Hyafil, F., Louvard, D., Ploegh, H. L. & Jacob, F. (1983). Uvomorulin: a nonintegral membrane protein of early mouse embryo. *Proc. Natl Acad. Sci. USA*, **80**, 6274–7

Peyriéras, N., Louvard, D. & Jacob, F. (1985). Characterisation of antigens recognised by monoclonal and polyclonal antibodies directed against uvomorulin. *Proc. Natl. Acad. Sci. USA*, **82**, 8067–71

Pratt, H. P. M., Chakraborty, J. & Surani, M. A. H. (1981). Molecular and morphological differentiation of the mouse blastocyst after manipulation of compaction with Cytochalasin D. *Cell*, **26**, 279–2

Pratt, H. P. M., Ziomek, C. A., Reeve, W. J. D. & Johnson, M. H. (1982). Compaction of the mouse embryo: an analysis of its components. *J. Embryol. Exp. Morph.*, **70**, 113–32

Reeve, W. J. D. (1982). Effect of Concanavelin A on the formation of the mouse blastocyst. *J. Reprod. Immunol.*, **4**, 53–64

Richa, J., Damsky, C. H., Buck, C. A., Knowles, B. B. & Solter, D. (1985). Cell surface glycoproteins mediate compaction trophoblast attachment and endoderm formation during early mouse development. *Dev. Biol.*, **108**, 513–21

Roseman, S. (1970). The synthesis of complex carbohydrates by multiglycosyltransferase systems and their potential function in intercellular adhesion. *Chem. Phys. Lipids*, **5**, 270–9

Rosenstraus, M. J. (1983). Isolation and characterisation of an embryonal carcinoma cell lacking SSEA-1 antigen. *Dev. Biol.*, **99**, 318–23

Rossant, J. & Papaioannou, V. E. (1984). The relationship between embryonic, embryonal carcinoma and embryo-derived stem cells. *Cell Diff.*, **15**, 144–61

Roth, S., McGuire, E. J. & Roseman, S. (1971). Evidence for cell surface glycosyltransferases. *J. Cell Biol.*, **51**, 536–47

Sato, M., Muramatsu, T. & Berger, E. G. (1984). Immunological detection of cell surface galactosyltransferase in pre-implantation mouse embryos. *Dev. Biol.*, **102**, 514–18

Schuh, R., Vestweber, D., Riede, I., Ringwald, M., Rosenberg, U. B., Jackle, H. & Kemler, R. (1986). Molecular cloning of the mouse cell adhesion molecule uvomorulin: CDNA contains a B1-related sequence. *Proc. Natl Acad. Sci. USA*, **83**, 1364–8

Shevinsky, L. H., Knowles, B. B., Damjanov, I. & Solter, D. (1982). Monoclonal antibody to murine embryos defines a stage-specific embryonic antigen expressed on mouse embryos and human teratocarcinoma cells. *Cell*, **30**, 697–705

Shirayoshi, Y., Okada, T. S. & Takeichi, M. (1983). The calcium-dependent cell–cell adhesion system regulates inner cell mass formation and cell surface polarisation in early mouse development. *Cell*, **35**, 631–8

Shur, B. D. (1982). Evidence that galactosyltransferase is a surface receptor for poly (N)-acetyllactosamine glyconjugates on embryonal carcinoma cells. *J. Biol. Chem.*, **257**, 6871–8

Shur, B. D. (1983). Embryonal carcinoma cell adhesion: the role of surface galactosyltransferase and its 90 k lactosaminoglycan substrate. *Dev. Biol.*, **99**, 360–72

Solter, D. & Knowles, B. B. (1978). Monoclonal antibody defining a stage specific mouse embryonic antigen (SSEA-1). *Proc. Natl Acad. Sci. USA*, **75**, 5565–9

Springer, T. A., Sastre, L. & Anderson, D. C. (1985). The LFA–1, Mac-1

leucocyte adhesion glycoprotein family and its deficiency in a heritable human disease. *Biochem. Soc. Trans.*, **13**, 3–6
Stahl, P. & Gordon, S. (1982). Expression of a mannosylfucosyl receptor for endocytosis on cultured primary macrophages and their hybrids. *J. Cell Biol.*, **93**, 49–56
Stoddart, R. W. (1984). *The Biosynthesis of Polysaccharides*. London: Croom Helm. p. 354
Surani, M. A. H. (1979). Glycoprotein synthesis and inhibition of glycosylation by tunicamycin in pre-implantation mouse embryos: compaction and trophoblast adhesion. *Cell*, **18**, 217–27
Surani, M. A. H., Barton, S. C. & Burling, A. (1980). Differentiation of 2-cell and 8-cell mouse embryos arrested by cytoskeletal inhibitors. *Exp. Cell Res.*, **125**, 275–86
Surani, M. A. H., Kimber, S. J. & Handyside, A. H. (1981). Synthesis and role of cell surface glycoproteins in pre-implantation mouse development. *Exp. Cell Res.*, **133**, 331–9
Surani, M. A. H., Kimber, S. J. & Osborn, J. O. (1983). Mevalonate reverses the developmental arrest of pre-implantation mouse embryos by Compactin, an inhibitor of HMG COA reductase. *J. Embrol. Exp. Morph.*, **75**, 205–23
Sutherland, A. E. & Calarco-Gillam, P. G. (1983). Analysis of compaction in the pre-implantation mouse embryo. *Devl. Biol.*, **100**, 328–38
Takatsuki, A., Arima, A. & Tamura, G. (1971). Tunicamycin, a new antibiotic. I isolation ahd characterisation of tunicamycin. *J. Antibiotics*, **24**, 215–23
Takeichi, M., Atsumi, T., Yoshida, C. & Ogou, S-I. (1982). Molecular approaches to cell–cell recognition mechanisms in mammalian embryos. In *Teratocarcinoma and Embryonic Cell Interactions*, ed. T. Muramatsu, G. Gachelin, A. A. Moscona & Y. Iwakura, pp. 283–93. New York: Academic Press
Takeichi, M., Ozaki, H. S., Tokunaga, K., Okada, T. S. (1979). Experimental manipulation of cell surface to affect cellular recognition mechanisms. *Dev. Biol.*, **70**, 195–205
Toporowicz, A. & Reisner, Y. (1986). Changes in sialytransferase activity during murine T cell differentiation. *Cell. Immunol.*, **100**, 10–19
Townsend, R. & Stahl, P. (1981). Isolation and characterisation of a mannose/Nacetylglucosamine/fucose-binding protein from rat liver. *Biochem. J.*, **194**, 209–14
Vestweber, D. & Kemler, R. (1984*a*). Some structural and functional aspects of the cell adhesion molecule uvomorulin. *Cell Diff.*, **15**, 269–73
Vestweber, D. & Kemler, R. (1984*b*). Rabbit antiserum against a purified surface glycoprotein decompacts mouse pre-implantation embryos and reacts with specific adult tissues. *Exp. Cell Res.*, **152**, 169–78
Vestweber, D., Kemler, R. (1985). Identification of a putative cell adhesion domain of uvomorulin. *EMBO J.*, **4**, 3393–8
Vestweber, D., Kemler, R. & Ekblom, P. (1985). Cell-adhesion molecule uvomorulin during kidney development. *Dev. Biol.*, **112**, 213–21
Vestweber, D., Ocklind, C. Gossler, A., Odin, P., Öbrink, B. & Kemler,

R. (1985). Comparison of two cell-adhesion molecules, uvomorulin and cell-CAM 105. *Exp. Cell Res.*, **157**, 451–61

Watkins, W. M. (1980). Biochemistry and genetics of the ABO Lewis and P blood group systems. In *Advances in Human Genetics*. Vol. **10**, pp. 1–36, ed H. Harris & K. Hirschhorn. London: Plenum Press

Willison, K. R., Karol, R. A., Suzuki, A., Kundu, S. K. & Marcus, D. M. (1982). Neutral glycolipid antigens as developmental markers of mouse teratocarcinoma and early embryos: an immunologic and chemical analysis. *J. Immunol.*, **129**, 603–9

Willison, K. R. & Stern, P. L. (1978). Expression of a Forssman antigenic specificity in the pre-implantation embryo. *Cell*, **14**, 785–93

Yamamoto, M., Boyer, A. M. & Schwarting, G. A. (1985). Fucose-containing glycolipids are stage- and region-specific antigens in developing embryonic brain of rodents. *Proc. Natl Acad. Sci. USA*, **82**, 3045–9

Yednock, T. A., Stoolman, L. M. & Rosen, S. D. (1987). Phosphomannosyl-derivatised beads detect a receptor involved in lymphocyte homing. *J. Cell Biol.*, **104**, 713–23

Yoshida, C. & Takeichi, M. (1982). Teratocarcinoma cell adhesion: identification of a cell-surface protein involved in calcium-dependent cell aggregation. *Cell*, **28**, 217–24

Yoshida-Noro, C., Suzuki, N. & Takeichi, M. (1984). Molecular nature of the calcium-dependent cell–cell adhesion system in mouse teratocarcinoma and embryonic cells studied with a monoclonal antibody. *Dev. Biol.*, **101**, 19–27

Ziomek, C. A. & Johnson, M. H. (1980). Cell surface interaction induces polarization of mouse 8-cell blastomeres at compaction. *Cell*, **21**, 435–42

15

Behaviour of mouse and human trophoblast cells during adhesion to and penetration of the endometrial epithelium

Paul V. Holmes and Svend Lindenberg

Departments of Obstetrics & Gynaecology,
Sahlgrenska Sjukhuset and Rigshospitalet,
Universities of Gothenburg,
Sweden and Copenhagen, Denmark

Abstract

Substances from the embryo lying free in the uterine lumen seem to be involved in signalling the maternal host to prepare the site of impending implantation (decidualisation), and for extending and increasing progesterone production from the corpus luteum. These embryonic signals must be extracellular biochemical compounds which are membrane-bound or in a free form (Heap, Flint & Gadsby, 1979). This comparative, morphological study is concerned with investigating the existence of signalling in three experimental models using differential-interference light microscopy, plus transmission and scanning electron microscopy.

The first experimental model involves observations made on diapausing mouse blastocysts, where subcutaneous estradiol or intrauterine-cAMP is used to induce blastocyst implantation. The second model involves observations made on human blastocysts from an IVF/ET programme, where the blastocysts were cultured from fertilisation to the stage of activated, hatched blastocysts in medium containing maternal serum. The third model involves observations made on a human *in vitro* implantation system, where human blastocysts are permitted to attach to, and invade, a monolayer of human endometrium epithelial cells.

The photomicrographic evidence suggests a similarity between human and mouse blastocysts in their induction of intracellular vesicle production, which are apparently exuded from the activated trophoblast cells by exocytosis in a regionally organised manner. This occurs coincident with increased trophoblast cell adhesiveness and outgrowth. At present, the nature of the contents of these exocytotic vesicles is unknown and it seems that exuded material may not be membrane-bound in the normal physiological situation. During *in vitro* implantation, the human trophoblast cells appear to induce the human uterine epithelial cells to move laterally away from the impending implantation site without causing their destruction, possibly by an intercellular process of contact inhibition. Moreover, in the near-contact regions between the two cell types, the epithelial cells exhibit large numbers of pinocytotic vesicles which develop in those of their membranes that oppose trophoblast cells. These observations suggest trophoblast cell secretion and epithelial cell uptake, further confirming that signal substances pass from the conceptus to the maternal host just prior to implantation.

Introduction

Blastocyst implantation in the uterus can be induced in normal mice undergoing lactational delay by removing the litter (McLaren, 1968), and in ovariectomised, delayed-implantation mice by injection of oestrogen (McLaren, 1971). Oestrogen-induced activation of diapausing blastocysts is characterised by a trophoblast cell transformation (Dickson, 1963), including various cytological (Dickson, 1969), cell surface (Bergström, 1972; Holmes & Dickson, 1973; Holmes & Bergström, 1975, 1976) and enzyme (Wong & Dickson, 1969; Holmes & Dickson, 1973) changes. Blastocyst activation also appears to be dependent on a rapid increase in RNA, protein and DNA synthesis (Weitlauf & Greenwald, 1965; Prasad, Dass & Mohla, 1969; Unger & Dickson, 1969; Holmes & Dickson, 1975).

Oestrogen-induced changes in the rat uterus appear to involve stimulation of adenyl cyclase activity (Rosenfeld & O'Malley, 1970). Szego & Davis (1973) found an increase in the uterine levels of cyclic AMP after oestrogen administration. Furthermore, not only oestrogen but also cAMP has been shown to induce blastocyst implantation in diapausing mice (Holmes & Bergström, 1973, 1975; Webb, 1975) and to induce surface morphological changes in the early trophoblast cells (Holmes & Bergström, 1976).

Very early in pregnancy, when a blastocyst is still lying free in the uterine lumen, interactions must occur between the uterine endometrium and the pre-implantation blastocyst in order to stimulate the endometrium to undergo local decidualisation for the reception of the blastocyst, to activate the blastocyst for the adhesive and invasive phases of implantation, and finally to maintain the corpus luteum function. These interactions or signals must be extracellular macromolecules, either membrane-bound or in a free form (Heap, Flint & Gadsby, 1979). Rather strong evidence exists today that pre-implantation blastocysts are able to synthesise and secrete glycoproteins (Saxena, Hasan, Haour & Schmidt-Gollwitzer, 1974; Bebing, Cedarqvist & Fuchs, 1976), in particular gonadotrophins, as well as steroids and other macromolecules (Knobil, 1973; Perry, Heap & Amoroso, 1973; Wiley, 1974; Dickman, Dey & Gupta, 1976; VanBlerkom, Chavez & Bell, 1979.) Moreover, it is apparent that the preimplantation blastocysts have the capability of both exocytosis and secretory functions (Enders, 1971).

In the present work the possibility of signal transfer between the preimplantation blastocyst and the endometrium was investigated utilising morphological techniques. Because of the inherent problem with artefacts in morphological investigations, more than one preparative technique was investigated. In addition, although the bulk of the investigations were done with a diapausing mouse embryo system, human blastocysts from an *in vitro* fertilisation/embryo transfer (IVF-ET) program were also investigated morphologically for comparison with the mouse blastocysts. Furthermore, an *in vitro* human implantation model was used to monitor the uterine epithelial cell reaction to implanting trophoblast cells.

Materials and methods

Randomly bred Swiss Webster albino mice were maintained under a controlled ten-hour night regime, centred on midnight, and were fed a standard diet *ad libitum*. Female mice weighing 25 to 35 grams were mated with fertile males and bilaterally ovariectomised under sodium pentobarbital anaesthesia on day 3 of gestation, a vaginal plug having been found on day 1. At ovariectomy 1.0 mg medroxyprogesterone acetate (Depo Provera, Upjohn Company) in 0.02 ml of sterile water was given subcutaneously to produce delayed implantation (Dickson, 1969) and an identical injection was repeated every 5 days where necessary.

Oestrogen-activated blastocysts were flushed from the uterine horns of diapausing mice on day 8, 24 to 36 hours after subcutaneous injection of 0.5 μg oestradiol benzoate in corn oil. Control blastocysts were collected from the diapausing mice after the injection of the corn oil vehicle only. The carefully excised uterine horns were flushed either with fixative in preparation for scanning and transmission electron microscopy or with physiological saline to obtain blastocysts for differential-interference microscopy (DIM).

In a further investigation, diapausing mice at laparotomy received 10 μl of sterile buffered saline solution containing dibutyryl adenosine 3,5-cyclic monophosphate (8 mM, pH 7.4; Sigma Chemical Company, St Louis, Mo.). This was instilled into the lumen of each uterine horn by microsyringe. The cervix was ligated to prevent hormone leakage and blastocyst loss. Control mice received installation of only the saline vehicle.

For transmission electron microscopy (TEM), the mouse blastocysts were fixed for 1 hour in 2.5% glutaraldehyde. They were repeatedly rinsed in 0.1 M sodium cacodylate buffer at pH 7.3, post fixed for 1 hour in 1% buffered osmium tetraoxide, rinsed in distilled water, stained with uranyl acetate for 30 minutes, rinsed twice and set in 1 mm blocks of 4% agar. These blocks were dehydrated, embedded in Epon, and sectioned for ultrastructural examination in a Philips transmission electron microscope. For scanning electron microscopy (SEM), the mouse blastocysts were fixed for 1 hour in 2.5% glutaraldehyde, rinsed in sodium cacodylate buffer at pH 7.3, post-fixed for 1 hour in 1% buffered osmium tetraoxide and then rinsed several times in redistilled water. These blastocysts subsequently were freeze dried at high vacuum and at liquid nitrogen temperature, then coated with evaporated gold–palladium for observation in a Cambridge Stereoscan electron microscope.

In order to be aware of possible artefacts introduced into the TEM and SEM methods above, the presence of vesicles and some cell surface structures were observed and recorded using differential-interference microscopy (DIM). Using this technique, direct observations could also be made of living blastocysts flushed with buffered physiological saline from similar experimental and control groups of mice. The blastocysts were placed in shallow chambers on microscope slides and observed on the temperature-controlled stage of a Zeiss Photoscope III equipped with DIM optics.

Human blastocysts were cultured from oocytes recovered at ultrasound-guided, percutaneously, transvesical aspiration of ovarian follicles. The patients ovulatory cycles were controlled with clomiphene citrate and hCG, as described previously (Lindenberg, Lauritsen & Lenz, 1985).

The oocytes were matured, inseminated and cultured to blastocysts in Earle's balanced salt solution (EBS) supplemented with the individual patient's serum (8% serum for maturation and insemination and 15% for the remaining culture period). The oocytes were washed in follicular fluid and matured for 4–6 hours in 0.2 ml culture medium under oil before insemination with approximately 50,000 motile sperm per ml. After 18 hours, those ova containing 2 pronuclei were transferred to a medium with 15% serum for cell division.

Blastocysts exhibiting signs of hatching from their zona pellucida were placed over and allowed to sink on to monolayer cultures of human epithelial cells. Normal endometrial specimens were obtained by curettage and the monolayer cultures were prepared under sterile conditions, as previously described in detail (Lindenberg, Lauritsen, Nielsen & Larsen, 1984). The co-cultures (human blastocysts on human endometrial cells) were followed on a daily basis via an inverted phase contrast microscope to observe implantations.

The human blastocyst implantation sites were fixed directly in their culture vessels for 20 minutes at 4 °C with 70% Karnowsky fixative. After repeated rinsing with 0.1 M cacodylate buffer at pH 7.3, the cultures were postfixed for 2 hours in 1% osmium tetraoxide in 0.1 M cacodylate buffer at 4 °C. For TEM, these implanation sites were dehydrated in graded alcohols and propylene dioxide followed by embedding in Epon. Ultra-thin sections were obtained after 5–10 semi-thin sections. The semi-thin sections were stained with toluidine blue, while the ultra-thin sections were contrasted with lead citrate.

Results

Mouse blastocyst implantations

Over 200 blastocysts were flushed from 42 female mice for investigation of the hormone-induced morphogenesis which seems to be necessary for implantation. The morphological characteristics of the blastocysts appeared to be the same whether they were activated for implantation by estrogen or cyclic AMP. Hormone induction resulted in hatching and expansion of the diapausing blastocysts, and activation of the trophoblast cells in a regionally orientated manner (Fig. 15.1). With blastocyst expansion the trophoblast cells became sticky on their surfaces and large vesicles containing unknown substances began developing, first in the most abembryonic cells (Fig. 15.1). This intracellular vesicle development was quite considerable and it could be visualised easily with DIM (Fig. 15.1) and TEM (Fig. 15.2). The activation and transformation process proliferated from the abembryonic pole over all trophoblast cells to the embryonic pole. A transition zone

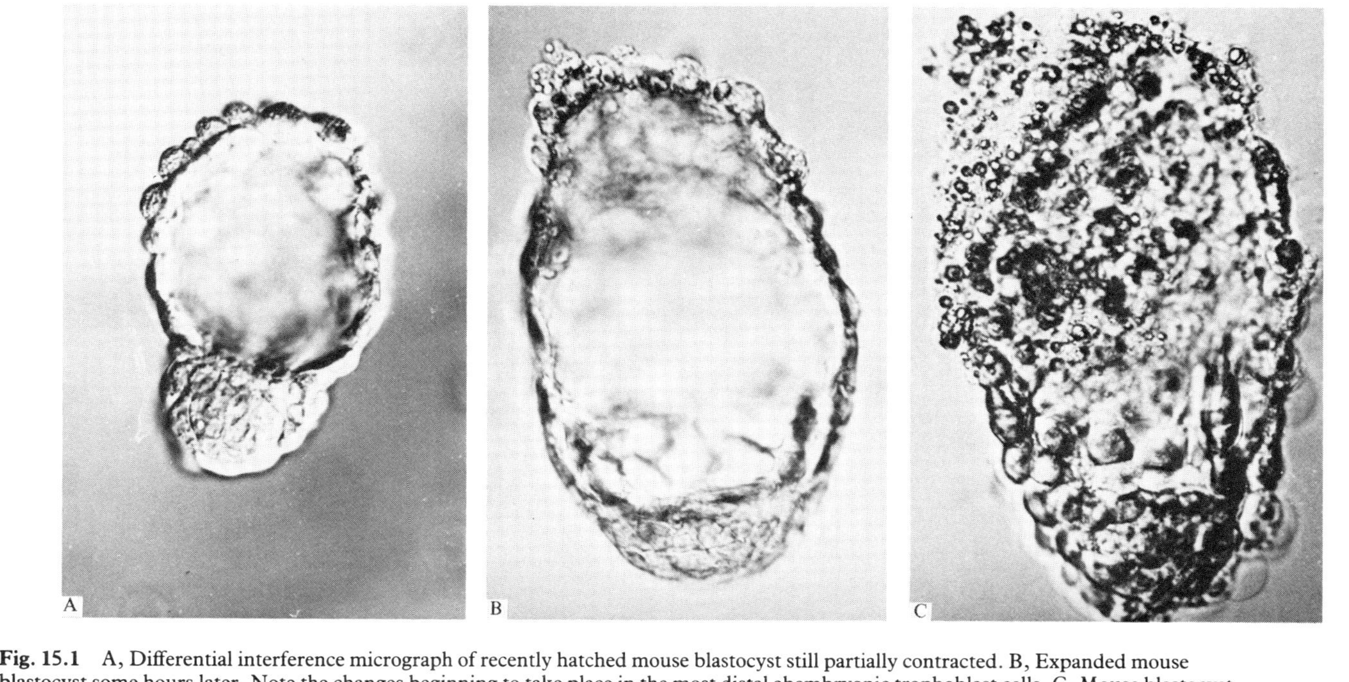

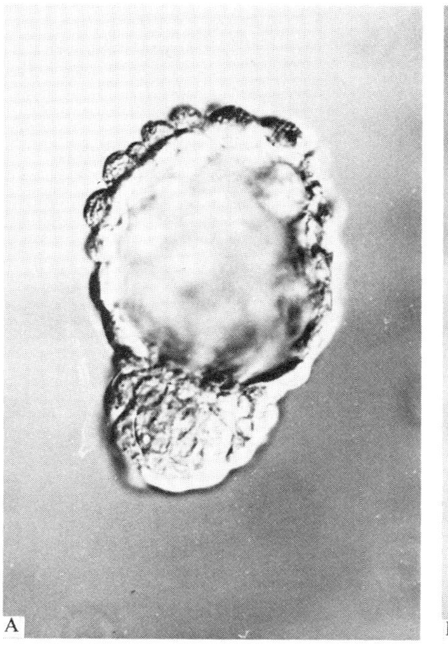

Fig. 15.1 A, Differential interference micrograph of recently hatched mouse blastocyst still partially contracted. B, Expanded mouse blastocyst some hours later. Note the changes beginning to take place in the most distal abembryonic trophoblast cells. C, Mouse blastocyst at the stage when implantation begins. The activation and transformation of the trophoblast cells has now progressed over the abembryonic half of the blastocyst, a process which eventually takes place in all trophoblast cells. Note the vesicles developing in the abembryonic cells and the fluid droplets being released from the embryonic and mural trophoblast cells. These morphological events can be temporally and spatially related to enzyme and cell surface changes previously described (Holmes & Dickson, 1973) and to an increased adhesiveness of the transforming cells (Böving, 1966).

Fig. 15.2. TEM illustration of the abembryonic trophoblastic vesicles containing electron-dense material which is secreted from the external surface of the cells. The arrow indicates one such vesicle in the process of secretion.

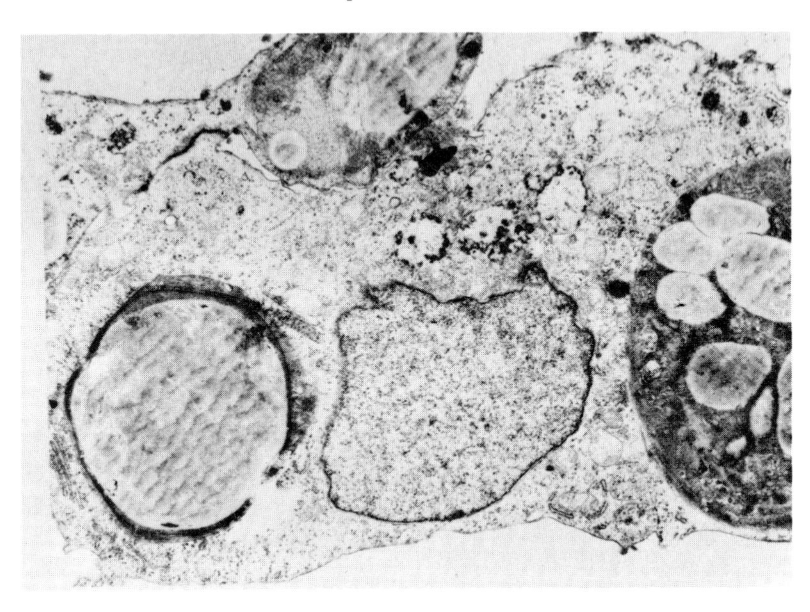

Fig. 15.3. DIM tangential view of equatorial polygonial trophoblast cells of a mouse blastocyst, illustrating their progressive activation and transformation with the production of secretory vesicles.

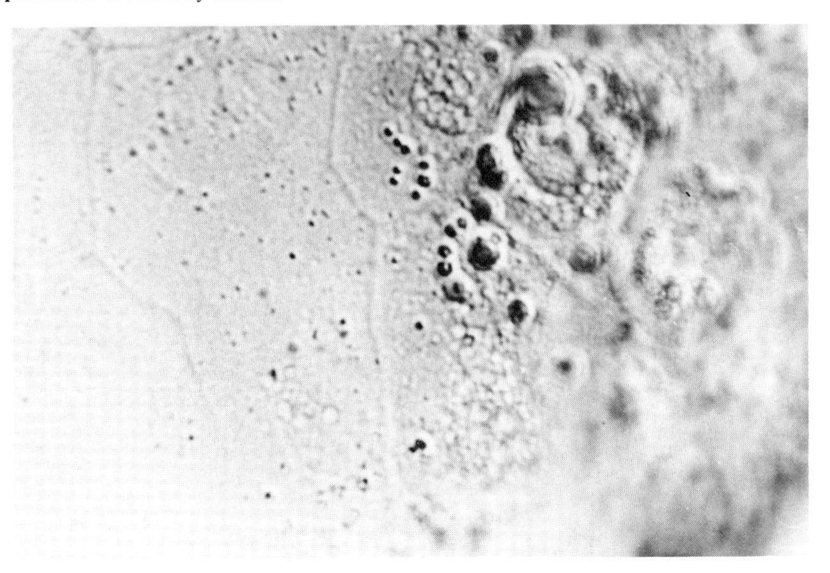

between transformed and untransformed trophoblast cells can be seen in Fig. 15.3, in the tangential equatorial DIM photomicrograph of mural trophoblast cells.

The high magnification SEM micrograph (Fig. 15.4) provides evidence of two secretory events occurring, one at each pole of the blastocyst, differing morphologically. Approximately 1 hour later in the implantation process, the secretory activity characterising the abembryonic trophoblast also envelopes the embryonic trophoblast cells. Furthermore, the abembryonic pole is usually torn open during the flushing procedure in these late-implantation blastocysts, which itself provides evidence of a better developed, regional attachment. The blastocyst activation initiated by intrauterine cAMP stimulation could not be differentiated from that initiated by subcutaneous oestrogen stimulation. In addition, no trophoblast cell transformation or secretory activity could be observed in the oestrogen-vehicle and cAMP-vehicle control blastocysts, although they did hatch, and the zona pellucida did dissolve, with time. These control blastocysts also failed to exhibit any trophoblast cell stickiness and, therefore, presented no practical manipulation problem.

Human blastocysts

Despite the restriction in the availability of blastocysts from IVF/ET programmes, 12 could be observed with Köhler and Normaski optics. These blastocysts were not diapausing and received no specific stimulation. However, they were cultured to the blastocyst stage in medium containing 15% serum collected from the donor just prior to the impending ovulation, at

Fig. 15.4. SEM micrograph of an activating, half-transformed mouse blastocyst, illustrating morphologically different secretory events occurring at the embryonic and abembryonic poles.

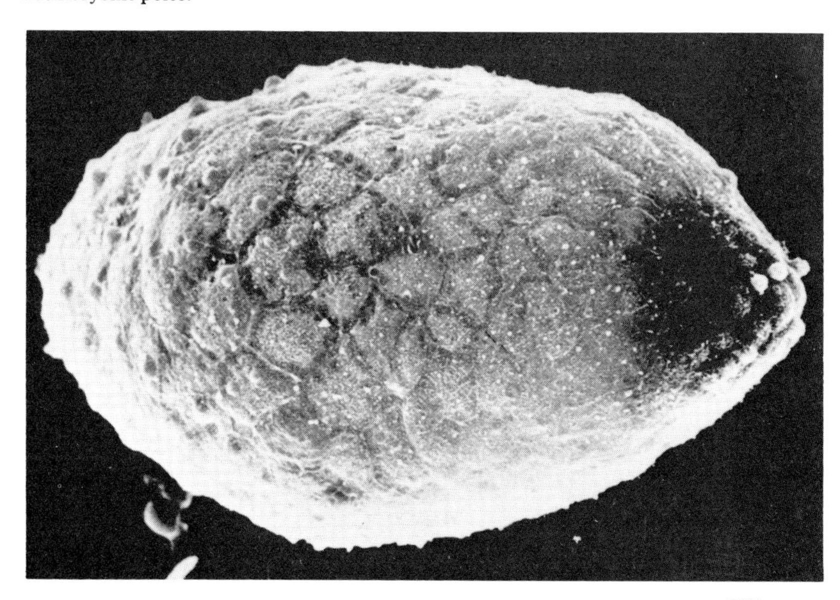

a time of high oestrogen levels in the peripheral blood. Consequently, these cultured blastocysts will have been influenced directly by oestrogen and, in fact, do hatch and expand spontaneously *in vitro*. After blastocyst expansion, the human trophoblast cells also undergo some form of transformation and become firmly adherent either to the culture vessel (Fig. 15.5) or to the epithelium monolayer (Fig. 15.7). Furthermore, secretory vesicles similar to those released from mouse blastocysts can be observed bulging from the surfaces of the trophoblast cells (Fig. 15.5, 15.6). However, it could not be discerned whether the transformation began regionally in these human blastocysts, since their firmly attached orientation did not permit precise location of the embryonic pole. Human blastocysts have previously been observed to adhere first at the embryonic pole, this being the opposite of what happens in the mouse blastocysts.

As seen by TEM, the mural trophoblast cells had an abundance of microvilli on their outer surfaces, while their inner surfaces facing the blastocoele were relatively smooth (Fig. 15.6). Their cytoplasm often contained membrane-bound vacuoles, while membrane-bound vesicles or granules of lightly electron dense material were present in the superficial or cortical cytoplasm (Fig. 15.6).

Human* in vitro *implantations

Of the four implantation sites investigated with both DIM and TEM, none showed any signs of stromal cell growth and the epithelial cells made up a confluent monolayer. Microvilli and occasional cilia formation were present

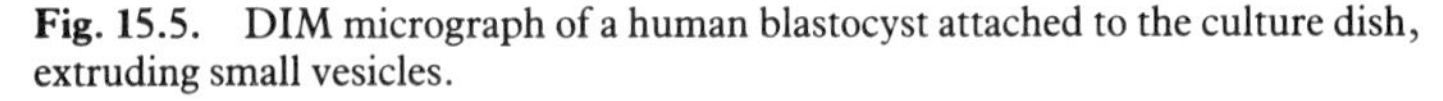

Fig. 15.5. DIM micrograph of a human blastocyst attached to the culture dish, extruding small vesicles.

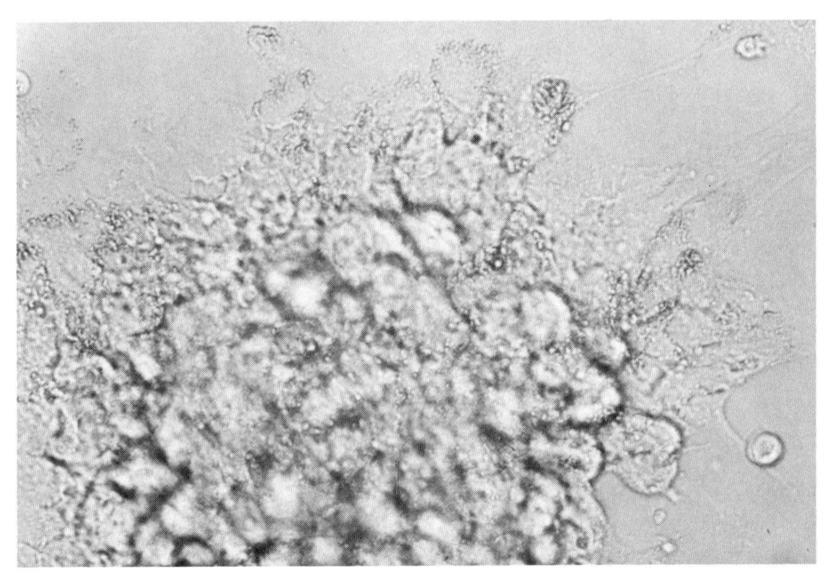

on this monolayer of flattened endometrium epithelial cells. In addition, desmosomes developed between these cells.

The adhesion phase of implantation was considered to have been achieved when the blastocyst no longer moved relative to the epithelial monolayer when the culture vessel was gently tapped. Initial adhesion was seen at day 8–9 after *in vitro* fertilisation.

After fixation, sectioning and staining, no invasive phase with destruction of the epithelial cells was seen. Instead, some form of specialised cell–cell interaction occurred, possibly due to a contact inhibition. When the outgrowing cytotrophoblast cells came into near-contact with the monolayer, either

Fig. 15.6. TEM micrograph of a human mural trophoblast, visualising membrane-bound vesicles extruded from the cell surface.

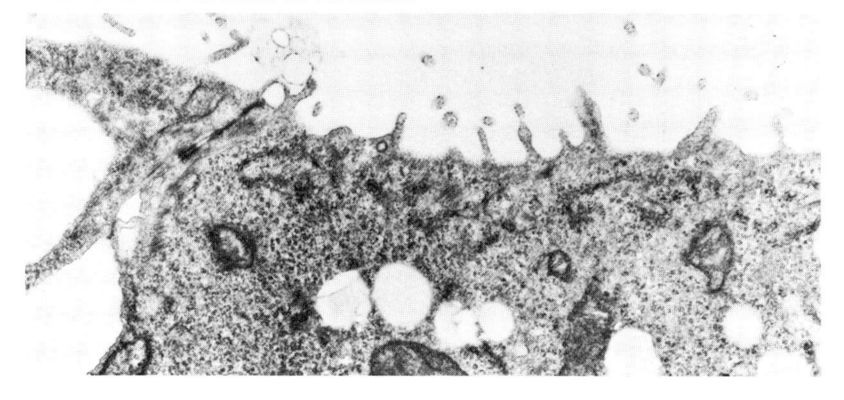

Fig. 15.7. TEM micrograph showing trophoblasts intruding between two endometrial epithelial cells. The plasmalemma of endometrial cells apposing the trophoblast have small (100 μm), possibly pinocytotic, vesicles.

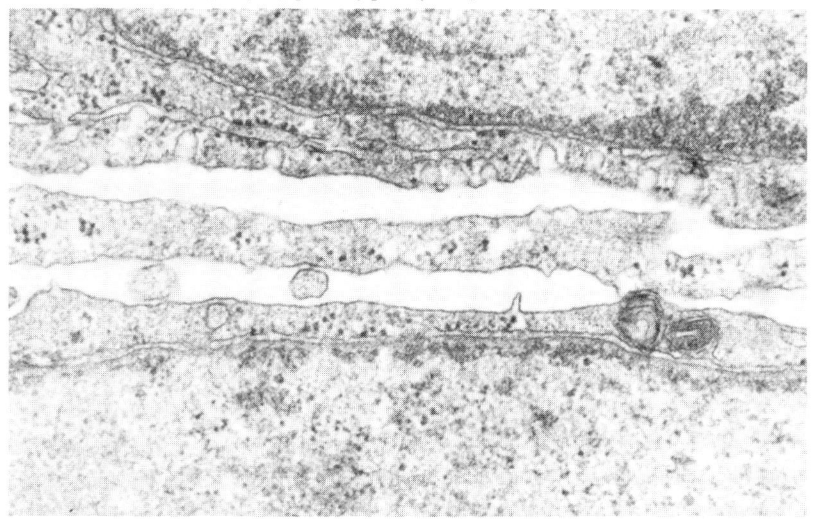

the epithelial cells moved laterally away from the impending implantation site or the outgrowing cytotrophoblast cells displaced the epithelial cells laterally, thus permitting the cytotrophoblasts to establish contact with the culture vessel.

In the contact zone between these two cell types, the endometrial epithelium cells became stacked into a multilayer of 3–4 cells, while the cytotrophoblasts formed long slender ectoplasmic protrusions which penetrated into the intercellular space between the adjacent endometrical cells (Fig. 15.7). Locally, the cytotrophoblasts form close apposition with the epithelial cells, although no junctions between the two cell types were observed. However, in the near contact region between these two cell types the endometrial cell plasmalemma, facing the cytotrophoblast cell membrane had fewer microvilli and exhibited large numbers of membrane invaginations (100 nm in diameter) (Fig. 15.7). This epithelial cell membrane activity is observed nowhere else in the monolayer cell culture, but only at the implantation site where these epithelial membranes appose cytotrophoblast membranes. In addition, similar membrane activity has been reported to represent endocytosis in the endometrium during conceptus attachment in the cow by Guillomot *et al.* (1986).

Discussion

In the foregoing, specific experiments, conducted in the area of mouse and human implantation, are described which provide morphological evidence for signalling between allograft conceptus cells and maternal-host endometrium epithelial cells. In contrast to mouse blastocyst implantation, no evidence exists in the literature (to the authors' knowledge) of an oestrogen stimulation or of a trophoblast cell activation necessary for human blastocyst adhesion and implantation. All culture media, used for human blastocysts in this work, contained 15% maternal serum, the serum itself containing steroid hormones which could be responsible for the trophoblast cell activation. It is our experience from two successfully established IVF/ET programs that human blastocysts have a negligible survival rate without the addition of maternal serum to their culture medium. Thus, if the human conceptus should have an absolute requirement for oestrogen for development to the blastocyst stage, it is provided in the serum. Additionally, in the mouse implantation model, it is clear that cyclic AMP can activate trophoblast cells and induce implantation (Holmes & Bergstrøm, 1973; Webb, 1975), this second messenger possibly being involved in the oestrogen induction mechanism.

It seems obvious from the present morphological findings, that both mouse and human blastocysts must undergo a phase of trophoblast cell activation before implantation can take place. Furthermore, this activation has similar morphological characteristics in the mouse and human, both clear and electron dense vesicles developing in and being released from the trophoblast cells. The blastocysts appear to be able to communicate with their surroundings by releasing signal substances, mainly of unknown character but also via hCG secretion from the human blastocysts, as reported previously (Fishel, Edwards & Evans, 1984).

The transformation of the mouse trophoblast commences abembryonically, as seen in the present work, while the orientation of the human trophoblast activation could not be clearly determined. The abembryonic activation in the mouse coincides with the initial abembryonic adhesion in this species and, furthermore, also coincides temporally and regionally with oestrogen-induced surface coat and enzyme changes prior to implantation (for references see Holmes & Dickson, 1973, 1975).

The character of the human implantation studied here *in vitro* was definitely not invasive as no destruction and phagocytosis of the endometrial cells by trophoblast was observed. The redistribution of the epithelial cells during cytotrophoblast cell outgrowth was more characteristic of an intrusive implantation, with a co-operation between the two cell types. Moreover, the endocytotic activity of the endometrial epithelial cells and the secretory activity of the trophoblast cells provided morphological evidence for conceptus-maternal signals, both prior to and during intrusive implantation.

References

Bebing, C. G., Cedarqvist, L. L. & Fuchs, F. (1976). Demonstration of gonadotrophin during the second half of the cycle in women using intrauterine contraception. *Am. J. Obstet. Gynecol.*, **125**, 855–8

Bergström, S. (1972). Scanning electron microscopy of ovoimplantation. *Arch. Gynßk.*, **212**, 285–307

Böving, B. G. (1966). Some mechanical aspects of trophoblast penetration of the uterine epithelium of the rabbit. In *Egg Implantation*, eds G. E. W. Wolstenholme & M. O'Connor, pp. 72–93. London: J. & A. Churchill

Dickmann, Z., Dey, S. K. & Gupta, J. S. (1976). A new concept: control of early pregnancy by steroid hormones originating in the preimplanting embryo. *Vitam. Horm.*, **34**, 215–42

Dickson, A. D. (1963). Trophoblastic giant cell transformation of mouse blastocysts. *J. Reprod. Fert.*, **6**, 465–6

Dickson, A. D. (1969). Cytoplasmic changes during the trophoblastic giant cell transformation of blastocysts from normal and ovariectomized mice. *J. Anat.*, **105**, 371–80

Enders, A. C. (1971). The fine structure of the blastocyst. In: Blandau, R. T (ed.) *The Biology of the Blastocyst.* Chicago: University of Chicago Press, pp. 71–94

Fishel, S. B., Edwards, R. G. & Evans, C. J. (1984). Human chorionic gonadotrophin secreted by preimplantation embryos cultured *in vitro*. *Science*, **233**, 816–18

Guillomot, M., Betteridge, K. J., Harvey, D. & Goff, A. K. (1986). Endocytotic activity in the endometrium during conceptus attachment in the cow. *J. Reprod. Fert.*, **78**, 27–36

Heap, R. B., Flint, A. P. & Gadsby, J. E. (1979). Role of embryonic signals in the establishment of pregnancy. *Br. Med. Bull.*, **35**, 129–35

Holmes, P. V. & Bergström, S. (1973). Estrogen and cyclic AMP induced changes in the surfaces morphology of delayed implantation mouse blastocysts. *Proc. Can. Fed. Biol. Soc.*, **16**, 116

Holmes, P. V. & Bergström, S. (1975). Induction of blastocyst implantation in the mouse by cyclic AMP. *J. Reprod. Fert.*, **43**, 329–32
Holmes, P. V. & Bergström, S. (1976). Cyclic AMP-induced changes in the surface morphology of diapausing blastocysts and the effect on implantation. *Am. J. Obstet. Gynec.*, **124**, 301–8
Holmes, P. V. & Dickson, A. D. (1973). Estrogen induced surface coat and enzyme changes in implanting mouse blastocyst. *J. Embryol Exp. Morph.*, **29**, 639–45
Holmes, P. V. & Dickson, A. D. (1975). Temporal and spatial aspects of estrogen-induced RNA, protein and DNA synthesis in the delayed implantation mouse blastocysts. *J. Anat.*, **119**, 453–9
Knobil, E. (1973). On the regulation of the primate corpus luteum. *Biol. Reprod.*, **8**, 246–58
Knoth, M. & Larsen, J. F. (1972). Ultrastructure of a human implantation site. *Acta Obstet. Gynecol. Scand.*, **51**, 385–93
Lenz, S. (1985). Percutaneous oocyte recovery using ultrasound. *Clin. Obstet. Gynecol.*, **12**, 785–98
Lindenberg, S., Lauritsen, J. G., Nielsen, M. H. & Larsen, J. F. (1984). Isolation and culture of human endometrial cells. *Fertil. Steril.*, **41**, 650–2
Lindenberg, S., Nielsen, M. H. & Lenz, S. (1985). *In vitro* studies of the human blastocyst implantation. *Ann. NY Acad. Sci.*, **442**, 368–74
Lindenberg, S., Hyttel, P., Lenz, S. & Holmes, P. (1986). Ultrastructure of the early human implantation *in-vitro*. *J. Hum. Reprod. Embryol.*, **1(8)**, 533–8
McLaren, A. (1968). A study of blastocysts during delay and subsequent implantation in lactating mice. *J. Endocrinol.*, **42**, 453–63
McLaren, A. (1971). Blastocysts in the mouse uterus: the effect of ovariectomy, progesterone and estrogen. *J. Endocrinol.*, **50**, 515–26
Perry, F. S., Heap, R. B. & Amoroso, E. C. (1973). Steroid hormone production by pig blastocysts. *Nature* (*Lond.*) **245**, 45–7
Prasad, M. R. N., Dass, C. M. S. & Mohla, S. (1969). Time sequences of action of estrogen on nucleic acid and protein synthesis in the uterus and blastocyst during delayed implantation in the rat. *Endocrinology* **85**, 528–36
Rosenfeld, M. G. & O'Malley, B. W. (1970). Steroid hormones: effect on adenyl cyclase activity and adenosine 3,5-monophosphate in target tissue. *Science NY*, **468**, 253–5
Saxena, B. B., Hasan, S. H., Haour, F. & Schnidt-Gollwitzer, M. (1974). Radioreceptor assay of human chorionic gonadotrophins: detection of early pregnancy. *Science*, **184**, 793–5
Szego, C. M. & Davis, S. (1967). Adenosine 3,5-monophosphate in rat uterus: acute elevation by oestrogen. *Proc. Natl Acad. Sci. USA*, **58**, 1711–18
Unger, B. & Dickson, A. D. (1971). Effect of cycloheximide ant actomycin D on the mouse blastocyst undergoing the giant cell transformation. *J. Anat.*, **108**, 519–25
VanBlerkom, J., Chavez, D. J. & Bell, H. (1979). Molecular and cellular aspects of facultative delayed implantation in the mouse. *Ciba Foundation Symposium*, **64**, 141–72

Webb, F. T. G. (1975). Implantation in ovariectomized mice treated with dibutyryl adenosin 3,5-monophosphate. *J. Reprod. Fert.*, **42**, 511–17
Weitlauf, H. M. & Greenwald, G. S. (1965). A comparison of 35-S-methionine incorporation by the blastocysts of normal and delayed implanting mice. *J. Reprod. Fert.*, **10**, 203–8
Wiley, L. D. (1974). Presence of a gonadotrophin on the surface of preimplanting mouse embryos. *Nature (Lond.)*, **252**, 715–16
Wong, Y. C. & Dickson, A. D. (1969). A histochemical study of ovo-implantation in the mouse. *J. Anat.*, **105**, 547–55

PART IV

Recognition between organisms

16 Recognition processes in lettuce downy mildew disease

J. W. Mansfield, A. M. Woods, P. F. S. Street and P. M. Rowell

Wye College (University of London)

Abstract

Lettuce downy mildew disease is a model system for studies of the recognition processes underlying race specific resistance to plant pathogens. A gene for gene relationship exists between the obligately biotrophic fungus *Bremia lactucae* Regel and its host *Lactuca sativa* L. At least 18 specific resistance factors matched by 18 complementary virulence factors control the relationship. Microscopical studies have shown that resistance conferred by resistance genes *Dm*1–*Dm*10 is expressed as the hypersensitive reaction (HR) of penetrated epidermal cells and that restriction of fungal growth follows the death of the plant cell. Lettuce cells undergoing the HR become brown and easily distinguished under the light microscope.

Experiments with mixed inocula have demonstrated that *B. lactucae* does not produce suppressors of the HR. Attempts to isolate elicitors of the HR from various fungal structures including isolated primary and secondary vesicles, and haustoria will be described. During the development of methods to assay putative elicitors it was found that responses characteristic of race-specific resistance are expressed by lettuce cells in suspension culture but not by protoplasts prior to wall regeneration.

Introduction

Downy mildew disease of lettuce is caused by the obligate fungal parasite *Bremia lactucae* Regel. The fungus is biotrophic, that is it absorbs nutrients from living lettuce cells (Ingram, Sargent & Tommerup, 1976). Successful infection by *B. lactucae* requires 1) germination of conidiosporangia on the leaf surface, 2) formation of a germ-tube and appressorium, 3) direct penetration of the underlying epidermal cell, 4) invagination of the host plasmalemma during expansion of a primary vesicle within the epidermal cell, 5) development of a secondary vesicle, 6) growth of intercellular hyphae from the secondary vesicle, and 7) formation of haustoria in cells adjacent to intercellular hyphae. In susceptible cultivars the first penetrated epidermal cells remain alive at least until hyphae grow from the secondary vesicle into the intercellular space; cells containing haustoria remain alive for at least 7 days, when the fungus begins to sporulate. Resistance to the downy mildew disease is expressed by the premature death of penetrated cells. Death of plant cells is thought to prevent nutrient uptake by the biotrophic fungus; restriction of fungal growth may subsequently be caused by starvation (Maclean & Tommerup, 1979). The response of resistant lettuce cells to penetration by *B. lactucae* is an example of a hypersensitive reaction (HR) to infection (Stakman, 1915; Mansfield, 1984, 1986). Lettuce cells undergoing the HR become brown and easily recognised under the light microscope.

A gene for gene relationship (Flor, 1956; Ellingboe, 1982) exists between the pathogen and its host. At least 18 dominant genes for resistance in lettuce are matched by 18 complementary virulence factors in *B. lactucae* (Crute & Johnson, 1976; Crute, I. R. pers. comm). Norwood & Crute (1984) have demonstrated that avirulence appears to be dominant in the fungus. The downy mildew disease is therefore a model system for studies of the recognition processes underlying race-specific resistance. The aim of our work is to determine the biochemical mechanisms controlling resistance to *B. lactucae* and in particular to characterise the molecules involved in the recognition event which initiates the hypersensitive reaction. Recent progress in the research is summarised in this article.

Characterisation of the hypersensitive reaction

Timing of plant cell death

Although detailed light microscope studies of incompatible interactions have been made, it is not clear from the results of Maclean (1974) and Maclean & Tommerup (1979) at which stage of fungal development the HR is induced in penetrated cells. Identification of the time of the earliest response of cells to infection will allow attention to be focussed on the critical stages of fungal development at which regulatory molecules may be produced.

In recent experiments, failure to plasmolyse in 0.85 M KNO_3 solution has been used as a physiological probe for the initiation of irreversible membrane damage (IMD) which occurs during the very early stages of the HR and perhaps chronologically close to the recognition event. Development of plasmolytic methods proved difficult because of physical damage to epidermal cells during preparation for microscopy. However, use of 2–3 cell-thick paradermal sections of petiolar epidermis and cortex allowed recovery of viable layers of cells plasmolysing in a range of solutes introduced into the tissue by brief vacuum infiltration. Results obtained for the interaction between cv. Mildura (resistance genes *Dm*1 and *Dm*3) and the Vo/11 isolate of *B. lactucae* are presented in Fig. 16.1. In this incompatible interaction IMD occurs following the production of a secondary vesicle 5–15 μm in length. Until IMD, growth of the fungus in cv. Mildura is the same as in the susceptible cv. Cobham Green. Preliminary experiments with other cultivars have shown that the timing of IMD varies depending on the resistance gene being expessed. Thus in cv. Valmaine with resistance gene *Dm* 5, IMD occurred during expansion of the primary vesicle, that is within

1.5 h of penetration. The plasmolytic method developed to examine permeability changes has the advantage of being a physiological probe which reveals effects at the cellular level. We are currently using the plasmolytic approach to examine the effects of inhibitors of protein synthesis and other metabolic activities on the timing of IMD during the HR.

Suppression or elicitation of the HR?

Bushnell & Rowell (1981) have presented a hypothesis describing the possible role of suppressor molecules in controlling race-specific resistance. They argue that virulent isolates produce molecules which suppress resistance reactions by blocking receptor sites for elicitors of the HR. Diffusible suppressors of processes of resistance have recently been isolated from rust fungi and *Phytophthora infestans*; in the latter example, suppressors possessed a degree of race specificity (Doke & Tomiyama, 1980a; Heath, 1981). In both cases evidence for the production of suppressor molecules was obtained from dual inoculation experiments in which prior inoculation with a virulent race suppressed resistance to a subsequent challenge with a non-pathogen or avirulent race (Tomiyama, 1966).

We have used dual inoculation experiments to examine the production of diffusible suppressors by *B. lactucae* (Crucefix, Mansfield & Wade, 1984). The principal requirement for studies on induced susceptibility is that the two isolates used must be clearly distinguished within infected tissue. Two approaches were adopted to allow certain identification of isolates; spatial

Fig. 16.1. Stages of fungal development and occurrence of irreversible membrane damage in epidermal cells of cv. Mildura (expressing resistance genes *Dm*1 and *Dm*3) penetrated by isolate Vo/11 of *Bremia lactucae*. Ability to plasmolyse in 0.85 M KNO_3 was used as a test for membrane integrity; ○ = plasmolysing cell, ● = cell failing to plasmolyse, ◑ = incomplete plasmolysis.

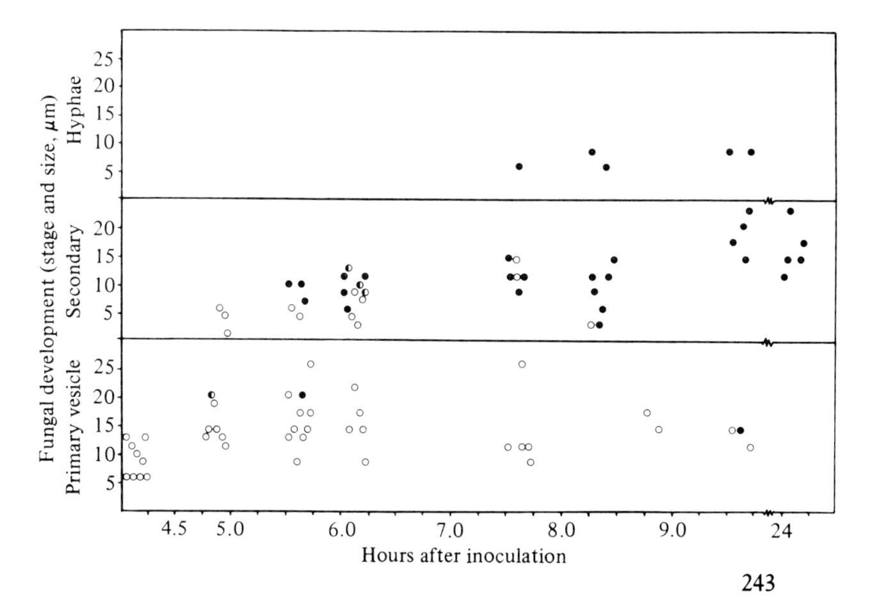

separation of inocula and labelling conidiosporangia of one isolate with the fluorescent brightener tinopal. The latter was found to have no adverse effect on fungal development.

In the first series of experiments, isolate TV (virulent) was inoculated onto the adaxial surface of cv. Mildura cotyledons and after incubation for 4 days to allow extensive colonisation, isolate IM43 (avirulent on Mildura) was inoculated onto the abaxial surface. Development of IM43 was recorded 1 and 2 days after inoculation. In controls uninfected by isolate TV, growth of IM43 was restricted to the production of a primary vesicle and, frequently, a short hypha within the penetrated epidermal cell. The hypha occasionally extended into the intercellular space. Prior colonisation by TV did not suppress resistance. The HR was expressed in all cells penetrated by isolate IM43 and no infections of IM43 developed expanded secondary vesicles. Resistance to the avirulent isolate was expressed by cells adjacent to compatible intercellular hyphae or even those cells which contained compatible haustoria. Use of the fluorescent marker allowed the two isolates to be inoculated at the same site. Conidiosporangia of TV were stained with tinopal and inoculated 24 h before inoculation of unstained IM43. Most infections by the virulent race had reached the expanded secondary vesicle stage at the time of challenge with IM43. Growth of IM43 within penetrated cells or cells adjacent to those penetrated by TV was recorded 1 and 2 days after inoculation (Table 16.1). Incompatible hyphae were again restricted to penetrated cells; their growth was similar to that in the absence of the virulent isolate.

The elicitation of the HR in lettuce cells containing compatible haustoria or infection structures indicates that *B. lactucae* is most unlikely to produce diffusible suppressors of race-specific resistance. In some cases compatible and incompatible fungal structures appeared to be in contact within lettuce cells but resistance was still expressed. These findings suggest that if suppressor molecules are involved in race specificity in lettuce downy mildew disease their activity must be highly localised, perhaps to the area of contact between the fungal cell wall and host plasmalemma.

Elicitors of the HR

Detection of diffusible elicitors

The failure to detect suppression of resistance to *B. lactucae* indicates that the HR in lettuce may result from direct interaction between products of avirulence genes in the fungus and the products of resistance genes in the plant. In studies of plant/fungus interactions, although a number of non-specific elicitors of the processes of resistance have been isolated from fungal mycelia and culture filtrates produced *in vitro*, molecules with race or cultivar specific activity have proved particularly elusive (Yoshikawa, 1983; De Wit, 1986). The production of race-specific elicitors by the tomato leaf mould fungus, *Cladosporium fulvum, in vivo*, was recently demonstrated by De Wit & Spikman (1982). They found that intercellular fluids obtained from infected tomato leaves contained race-specific elicitors of the hypersensitive reaction (HR), the characteristic response of tomato to challenge by avirulent isolates of *C. fulvum*.

Table 16.1. *Development of the avirulent isolate IM43 in cotyledons previously inoculated with the virulent islolate TV*

Time after inoculation with isolate IM43 (h)	Pretreatment of host[a]	Extent of fungal development (% of total infections)[b]			
		Restricted to penetrated cell	Hypha < 150 μm long	Hypha > 150 μm long	Single haustorium
24	Inoculated with isolate TV	92.5 (15.0)[c]	7.5	–	–
	Control (inoculated with water)	83.3	12.3	2.4	–
48	Inoculated with isolate TV	82.5 (7.5)	17.5 (2.5)	–	–
	Control (inoculated with water)	85.4	14.6		

[a] Inoculation with suspensions of isolate TV or water alone were made 24 h before challenge with isolate IM43.
[b] Forty incompatible infections were scored at each time.
[c] Figures in parentheses give the percentage of infections in which both isolates had penetrated the same epidermal cell.
Data from Crucefix *et al*. (1984).

We recovered intercellular fluids from lettuce cotyledons infected with *B. lactucae* and injected the fluids into cotyledons in order to detect elicitors of the HR as described by De Wit & Spikman (1982). Irrespective of the isolate of *B. lactucae* or lettuce cultivar employed intercellular fluids did not elicit the HR in lettuce (Crucefix *et al.*, 1984).

An alternative approach adopted to detect diffusible elicitors of the HR was to use colonised donor leaf transplants to examine the growth of mycelium from susceptible tissue into resistant recipient cotyledons. This experiment was designed to maximise the exposure of resistant tissue to compounds diffusing from vigorous intercellular hyphae of *B. lactucae*. Following transplantation, hyphae grew rapidly into resistant tissue but no symptoms were observed in cells in contact with incompatible hyphae unless there were obvious signs of penetration and the development of rudimentary haustoria. Individual penetrated cells underwent the HR (Crucefix *et al.*, 1984).

The failure to detect elicitors in intercellular fluids, or widespread necrosis following tissue transplantation, indicate that no diffusible elicitors of race-specific resistance are produced by *B. lactucae*. It may be concluded that recognition in lettuce downy mildew disease involves highly localised interactions probably occurring at the plant cell membrane/fungal wall interface.

Elicitors from fungal infection structures

Crucefix *et al.* (1987) described attempts to isolate race specific elicitors of the HR from germinating conidiosproangia and from various infection structures produced by *B. lactucae* including isolated mycelium, primary and secondary vesicles and haustoria recovered from infected tissue. Potential elicitors were assayed by injection into lettuce cotyledons and also following their addition to lettuce protoplasts. The protoplast assay was designed to ensure contact between elicitors and putative receptor sites in the plant cell membrane. Scanning electron microscopy showed that enzymic digestion of cotyledons for 12 h released protoplasts free of residual wall fragments and that the wall reformed slowly following protoplast isolation.

Conidiosporangia germinated poorly in the protoplast osmoticum. Protoplast viability (assessed with fluorescein diacetate, FDA) was reduced by 20–45% in the presence of germinating spores but no cultivar or isolate specific effects were observed. Prior germination of conidiosporangia in water or use of protoplasts recovered following enzymic digestion for 3, 5, 7 or 9 h did not affect the response of protoplasts. Frozen, thawed and ultrasonicated suspensions of conidiosporangia did not kill protoplasts rapidly but caused non-specific agglutination.

Mycelium of *B. lactucae* was isolated following enzymic digestion of infected cotyledons. Intercellular hyphae and haustoria recovered appeared to be packed with cytoplasm but were dead as indicated by their failure to exclude Evans blue stain. Primary and secondary vesicles were recovered by homogenisation of epidermal strips and purified by combinations of sieving and application to sucrose density gradients. Very few infection structures were viable following extraction and all were dead within 10 min of isolation. The presence of cultivar specific elicitors of the HR was indicated in certain assays of primary vesicles of isolate NW3 recovered 6 h after inoculation of the susceptible cv. Cobham Green. Repeated experiments summarised in Table 16.2 showed that the occurrence of differential activity was very inconsistent and may have been associated with the presence of many viable vesicles in the active extracts. No suppressors of the HR were detected in extracts of infection structures following their injection into cotyledons prior to inoculation.

The inconclusive nature of the results obtained highlights difficulties in the direct approach to isolate elicitors of the HR. Failure to detect activity following injection of extracts into the intercellular spaces was not unexpected in view of the potential role of the plant cell wall as a barrier between elicitors and their proposed receptor sites in the plant symplast. A protoplast assay appeared likely to overcome this problem and has proved useful in studies of other plant/pathogen interactions (Earle *et al.*, 1978; Doke & Tomiyama, 1980*b*) but in the absence of a clear positive result it is impossible to know if protoplasts were competent to respond to putative elicitors. An alternative assay attempted was to inject extracts into the paramural space using techniques applied to studies on stomatal physiology as developed by Edwards & Meidner (1979). Problems associated with the movement of materials across cell walls, removal of receptor sites or cognon/cognor binding interference by compounds in the bathing solution surrounding protoplasts should have been avoided by direct injection into epidermal cells. Unfortunately, despite repeated attempts to perfect the injection system, the method proved insufficiently reliable to allow meaningful results to be obtained. Too often, injection alone caused irreversible membrane damage and particulate extracts blocked the very narrow capillary needles pushed into epidermal cells.

In vitro interactions

Infection of lettuce cells in suspension culture

Following our lack of success with protoplasts we investigated the possibility of using lettuce cells in suspension culture for assays of elicitors of the HR. The responses of isolated cells and cell suspension cultures to plant pathogens have frequently been reported to differ from those observed in whole plants. In particular, the hypersensitive reaction (HR) characteristic of race specific resistance is usually not expressed *in vitro* (Tomiyama *et al.*, 1958; Ingram, 1967; Otsuki *et al.*, 1972; Fett & Zacharius, 1983; Helgeson & Deverall, 1983). It was, therefore, essential to test if cultured cells expressed race specific resistance to *Bremia lactucae*. Details of the methods used to prepare cell suspensions and challenge cells with *B. lactucae* are given in Street *et al.* (1986).

Conidiosporangia germinated rapidly in cell suspension cultures. Germ-tubes which made contact with lettuce cells differentiated appressoria, penetrated and formed primary vesicles. Figure 16.2 shows the rapid expansion of the primary vesicle following penetration. In cells from susceptible cultivars, expanded secondary vesicles and hyphae usually developed and many hyphae had grown out of penetrated susceptible cells by 36 h after inoculation. In some cases the hyphae penetrated additional cells forming structures resembling the haustoria found in cotyledon mesophyll. Essentially, fungal development was very similar to that observed in the intact cotyledons.

Table 16.2. *Effect of infection vesicles on protoplast viability*

Source of vesicles	Isolate	Time after inoculation (h)	Reduction in percentage viability compared with no vesicle controls[a]	
			cv. Cobham Green (susceptible)	cv. Mildura (resistant)
cv. Cobham Green	NW3	6	13[c], +0.4[c], +1, 14, (9)[b]	67[c], 51[c], +1, 16, (11)
	IM43	6	2, 11, 5, 8, (5)	15, 9, +5, +12, (+9)
cv. Mildura	NW3	6	6, 3, 16, (1, 19)	7, 26, 20, (36, 5)
		22	34, 35	31, 23

[a] Results from separate experiments with each vesicle preparation are given. All data are means from two replicates in each experiment, + indicates higher viability in treated protoplasts. Viability was assessed after incubation for 18 h, approx. 250 protoplasts were examined in each replicate.

[b] Figures in parentheses are from tests with vesicles homogenised in a glass Griffiths tube.

[c] Experiments in which cultivar specific activity was detected.

Data from Crucefix *et al*. (1987).

Fig. 16.2. Cultured cell of cv. Cobham Green infected by the virulent isolate TV. Photographs b, c and d were taken 10, 30 and 60 min after location of a recently penetrated cell (Fig. 16.2a). Note the rapid expansion of the primary vesicle (arrowed) and movement of cytoplasm from the attached appressorium (a) into the vesicle (compare b and c). Bar = 40 μm.

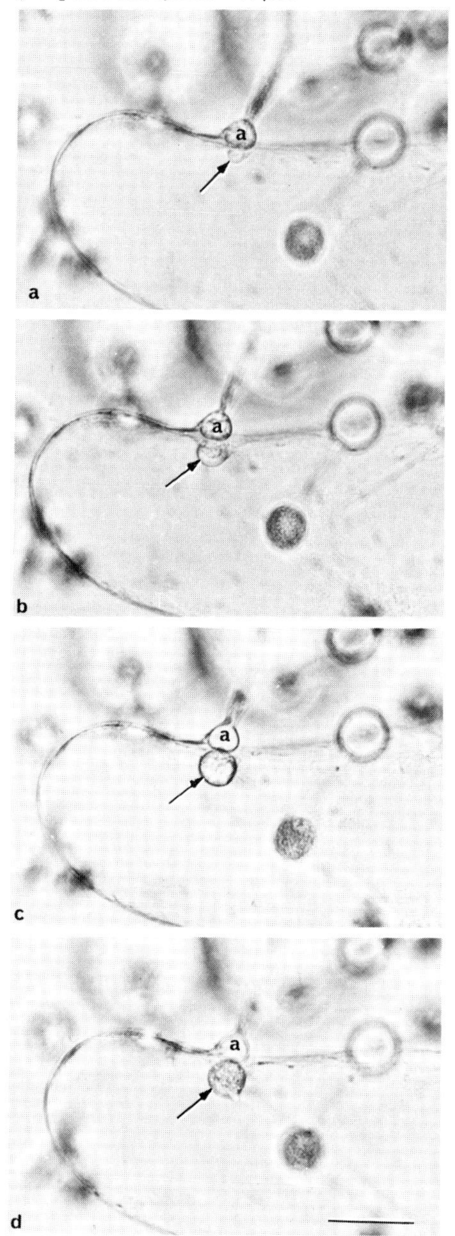

Primary vesicles were also formed in cells from the resistant cv. Mildura but secondary vesicles were rarely produced by the avirulent isolates NW3 or IM43. Instead, the fungus developed single narrow hyphae or several small vesicle-like swellings. The distinction between hyphae and poorly formed secondary vesicles was not always clear but fungal growth was usually limited to the production of the small intracellular structures. Restricted fungal development followed death of penetrated plant cells as indicated by their failure to stain with FDA.

Quantitative studies of plant cell death in cultured cells of cvs. Cobham Green and Mildura inoculated with three isolates of *B. lactucae* are summarised in Fig. 16.3. The majority of penetrated cells remained alive in the compatible interactions (cv. Cobham Green with all isolates, cv. Mildura with isolate TV) at least until the growth of hyphae out of the first penetrated cell. By contrast, most penetrated cells were dead 12 h after inoculation during incompatible interactions (cv. Mildura with isolates IM43 and NW3).

Isolated cultured cells appeared to undergo the HR as expressed in cotyledon epidermal cells. It should be stressed, however, that the differential response of cultured cells to avirulent and virulent isolates was not as clear as that observed in the intact plant. Thus, fewer cultured cells were killed

Fig. 16.3. Viability of penetrated cells in suspension culture. Conidiosporangia of isolates IM43, NW3 and TV were incubated with cultured cells of cvs. Cobham Green (open columns) or Mildura (closed columns) as described in Street *et al.* (1986). The cv. Cobham Green is susceptible to all isolates, cv. Mildura is resistant to isolates IM43 and NW3 but susceptible to TV. Cell viability was determined by staining with FDA. Data for each time after inoculation are from duplicate samples from which a total of *c.* 50 penetrated cells were examined.

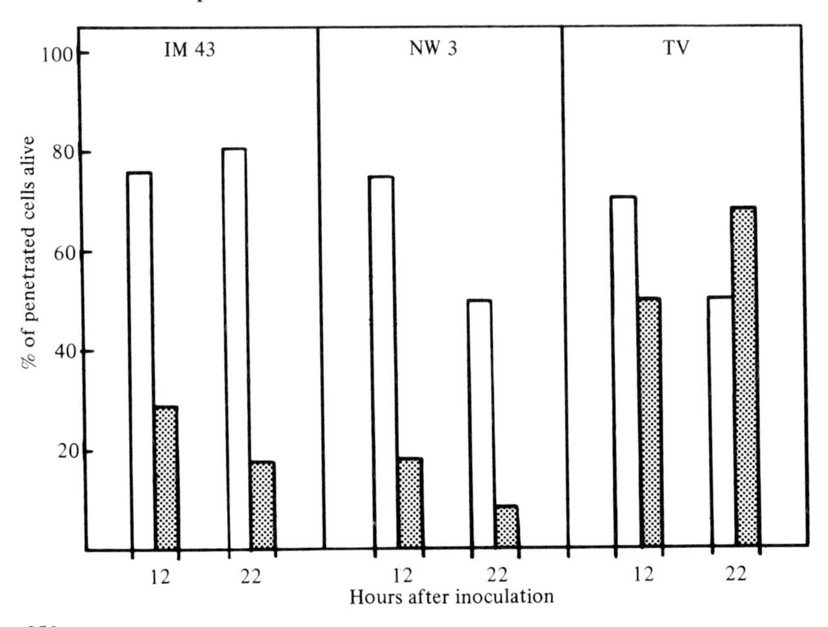

during the early stages of incompatible interactions and more cells died in culture than in the intact cotyledon following penetration by virulent isolates. Nevertheless, race specific resistance to *B. lactucae* does appear to be expressed by isolated cultured plant cells. The HR occurred only in cells penetrated by *B. lactucae*. No evidence was obtained for the release of diffusible elicitors of the HR by germinating conidiosporangia or by cells undergoing necrosis. These observations again suggest that the molecules controlling the HR may be highly localised.

Phytoalexin accumulation has frequently been implicated as the cause of the inhibition of fungal growth during the HR. Experiments with small groups of dissected cells of French bean and potato have demonstrated that a critical mass of tissue is required for the expression of the HR and inhibition of fungal growth by avirulent isolates of the facultative parasites *Colletotrichum lindemuthianum* and *Phytophthora infestans* respectively (Skipp & Deverall, 1972; Bailey, 1982). Live tissues in contact with the cell undergoing the HR are thought to be the sites of biosynthesis of phytoalexins which restrict fungal development (Mansfield *et al.*, 1974; Bailey, 1982). The restricted development of *B. lactucae* in isolated lettuce cells in suspension cultures may not be caused by the accumulation of antifungal compounds but may simply reflect starvation due to the failure of the obligate parasite to absorb nutrients from isolated, killed cells.

Responses of protoplasts with regenerating walls

The failure of protoplasts and contrasting success of cells in suspension culture to differentiate between avirulent and virulent races of *Bremia*, prompted an investigation of the responses of protoplasts with regenerating walls.

Protoplasts were incubated in regeneration medium (Nagata & Takebe, 1970) and samples tested at daily intervals for their viability and reaction to conidiosporangia of *B. lactucae*. Results obtained are presented in Fig. 16.4. Protoplasts from resistant and susceptible cultivars, allowed to regenerate walls for 3–4 days, were found to respond differentially to isolates of *B. lactucae*. Protoplasts from the resistant cv. Mildura died following exposure to isolate NW3, in effect reproducing the HR observed in whole plants.

Our finding that protoplasts with partially regenerated walls respond differentially to avirulent and virulent isolates of *B. lactucae* raises important questions concerning the significance of the cell wall for plant responses. For example, is the HR triggered by a factor released from the plant cell wall during attempted penetration or does the presence of the wall induce differentiation and the production of cultivar specific elicitors of the HR by the fungus? A major difficulty with regenerating protoplasts is the high background of dead protoplasts which tends to mask differential reactions. If long term viability can be improved by optimising protoplast preparation and regeneration media, the *in vitro* system should prove valuable for studies on recognition in the downy mildew disease.

Concluding remarks

Speculation abounds on the molecules involved in the recognition processes controlling the HR in race-specific disease resistance in plants. Attempts to isolate from bacteria or fungi race (isolate) and variety-specific elicitors of the HR and/or phytoalexin accumulation have proved remarkably unsuccessful, perhaps because insufficient attention has been paid to the biology of plant/pathogen interactions (Darvill & Albersheim, 1984; Mansfield, 1984). Recent developments with broad host range cloning vectors have

Fig. 16.4. Viability of protoplasts with regenerating walls incubated with conidiosporangia of isolate NW3. Protoplasts were incubated in wells in microtitre plates in regeneration medium (Nagata & Takebe, 1970). At intervals, individual wells were inoculated with suspensions of conidiosporangia in osmoticum or osmoticum alone (controls), and protoplast viability determined after 18 h by staining with FDA (Crucefix *et al.*, 1987), *c.* 250 protoplasts were examined in each assay. Open symbols = controls, closed symbols = spores added to protoplasts.

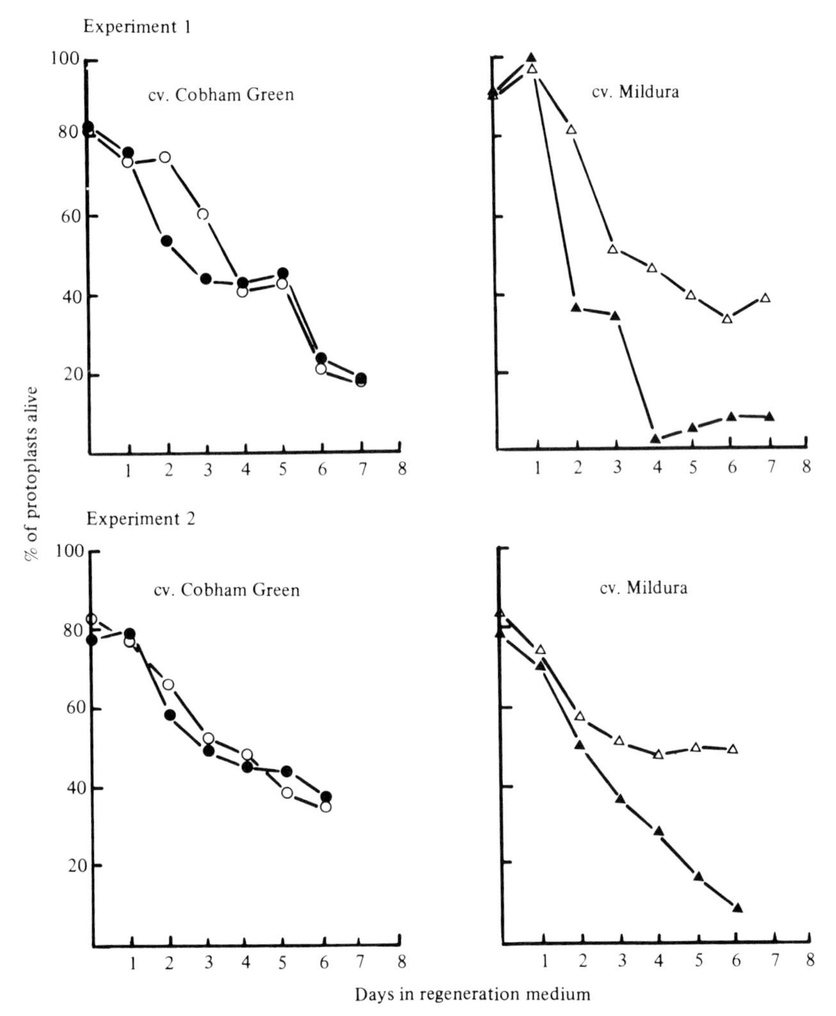

allowed alternative genetical approaches to be adopted. The new strategy is to isolate the genes in the parasite controlling infection development and subsequently to identify gene products. Recent work with bacterial pathogens has resulted in the cloning of race specific avirulence genes associated with induction of the HR by *Xanthomonas campestris* pv. *malvacearum* and *Pseudomonas syringae* pv. *glycinea* (Staskawicz *et al.*, 1984; Gabriel, 1985).

Bacterial pathogens are particularly amenable to analysis by molecular genetics, but rapid developments are being made in the application of equivalent technology to the filamentous fungi (Yoder *et al.*, 1986). Unfortunately, the obligate fungal parasites (which cause the greatest economic losses) as represented by the downy mildew fungus *Bremia lactucae*, are unsuitable for the application of molecular biology to the isolation of avirulence genes and thence their products which control the HR. Transformation systems currently available for fungi involving regeneration of fungal protoplasts, cannot be applied to the obligate parasites (Yoder *et al.*, 1986). We must therefore continue the direct physiological and biochemical search for factors controlling recognition in diseases caused by the fastidious fungi.

Many of the experiments described in this article have involved the interaction between lettuce cvs. Mildura (resistance genes *Dm*1 and *Dm*3) and isolates IM43, NW3 and TV of *B. lactucae*. The isolates possess the following 'virulence factors' corresponding to resistance genes in the host: IM43 virulence factors (5–10), NW3 (1, 2, 4, 5, 7–10) and TV (1–10). Thus cv. Mildura is susceptible to TV; resistance of the variety to IM43 is conferred by both *Dm*1 and *Dm*3 and to NW3 by resistance gene *Dm*3 alone. It is tempting to speculate that if the molecules controlling recognition, as observed by the expression of resistance governed by one gene, for example *Dm*3, can be identified and characterised, then molecules associated with other gene interactions (e.g. *Dm*1 or *Dm*5) will prove to be closely related. Such a system of modified but essentially similar recognition molecules may prove useful targets for the development of rationally designed disease control measures. It is however, equally probable that the biochemical processes controlling each proposed avirulence gene/resistance gene interaction are quite unrelated. Each interaction should be considered separately. Thus, although we have good evidence that *Dm*1 and *Dm*3 controlled resistance does not involve diffusible elicitors, *Dm*2, 4, 5, 6, 7, 8, 9, 10, 11, 12, 13, 14, 15, 16, 17 and 18 have yet to be investigated!

Acknowledgements

We wish to acknowledge support from the Agricultural and Food Research Council and valuable discussion with Dr D. N. Crucefix.

References

Bailey, J. A. (1982). Mechanisms of phytoalexin accumulation. In: *Phytoalexins*, eds J. A. Bailey & J. W. Mansfield, pp. 289–318. Glasgow: Blackie

Bushnell, W. R. & Rowell, J. A. (1981). Suppressors of defence reactions: A model for roles in specificity. *Phytopathology*, **71**, 1012–14

Crucefix, D. N., Mansfield, J. W. & Wade, M. (1984). Evidence that determinants of race specificity in lettuce downy mildew disease are highly localized. *Physiol Plant Pathol.*, **24**, 93–106
Crucefix, D. N., Rowell, P. M., Street, P. F. S. & Mansfield, J. W. (1987). A search for elicitors of the hypersensitive reaction in lettuce downy mildew disease. *Physiol. & Mol. Plant Pathol.*, **30**, 39–54
Crute, I. R. & Johnson, A. G. (1976). The genetic relationship between races of *Bremia lactucae* and cultivars of *Lactuca sativa*. *Ann. App. Biol.*, **83**, 125–37
Darvill, A. G. & Albersheim, P. (1984). Phytoalexins and their elicitors – a defense against microbial infection in plants. *Ann. Rev. of Plant Physiol.*, **35**, 243–75
de Wit, P. J. G. M. (1986). Elicitation of active resistance mechanisms. In *Biology and Molecular Biology of Plant–Pathogen Interactions* ed. J. A. Bailey pp. 149–69. Berlin: Springer Verlag
de Wit, P. J. G. M. & Spikman, G. (1982). Evidence for the occurrence of race and cultivar-specific elicitors of necrosis in intercellular fluids of compatible interactions of *Cladosporium fulvum* and tomato. *Physiol. Plant Pathol.*, **21**, 1–11
Doke, N. & Tomiyama, K. (1980*a*). Suppression of the hypersensitive response of potato tuber protoplasts to hyphal wall components by water soluble glucans isolated from *Phytophthora infestans*. *Physiol. Plant Pathol.*, **16**, 177–86
Doke, N. & Tomiyama, K. (1980*b*). Effect of hyphal wall components from *Phytophthora infestans* on protoplasts of potato tuber tissue. *Physiol. Plant Pathol.*, **16**, 169–76
Earle, E. D., Gracen, V. E., Yoder, O. C. & Gemmil, K. P. (1978). Cytoplasm specific effects of *Helminthosporium maydis* race T toxin on survival of corn mesophyll protoplasts. *Plant Physiol.*, **61**, 420–4
Edwards, M. & Meidner, H. (1979). Direct measurements of turgor pressure potentials. IV. Naturally occurring pressures in guard cells and their relation to solute and matric potentials in the epidermis. *J. Exp. Bot.*, **30**, 829–37
Ellingboe, A. H. (1982). Genetical aspects of active defense. In *Active Defense Mechanisms in Plants*, ed. R. K. S. Wood, NATO Advanced Study Institutes Series, Series A: Life Sciences Vol. 31. pp. 179–92. New York: Plenum Press
Fett, W. F. & Zacharius, R. M. (1983). Bacterial growth and phytoalexin elicitation in soybean cell suspension cultures inoculated with *Pseudomonas syringae* pathovars. *Physiol. Plant Pathol.*, **22**, 151–72
Flor, H. H. (1956). The complementary genetic systems in flax and flax rusts. *Adv. in Genetics*, **8**, 29–54
Gabriel, D. W. (1985). Molecular cloning of specific avirulence genes from *Xanthomonas malvacearum*. In *Advances in Molecular Genetics of the Bacteria–Plant Interaction*, eds A. A. Szalay and R. P. Legock, pp. 202–4. Cornell University, Ithaca, USA
Heath, M. C. (1981). Effects of infection by compatible species or injection of tissue extracts on the susceptibility of non-host plants to rust fungi. *Phytopathology*, **70**, 356–60

Helgeson, J. P. & Deverall, B. J. (eds) (1983). *Use of Tissue Culture and Protoplasts in Plant Pathology*. Sydney: Academic Press

Ingram, D. S. (1967). The expression of R-gene resistance to *Phytophthora infestans* in tissue cultures of *Solanum tuberosum*. *J. Gen. Microbiol.*, **49**, 99–108

Ingram, D. S. Sargent, J. A. & Tommerup, I. C. (1976) Structural aspects of infection by biotrophic fungi. In *Biochemical Aspects of Plant–Parasite Relationships*, ed. J. Friend & D. R. Threlfall pp. 43–78. London: Academic Press

Maclean, D. J., Sargent, J. A., Tommerup, I. C. & Ingram, D. S. (1974). Hypersensitivity as the primary event in resistance to fungal parasites. *Nature (Lond.)*, **249**, 186–7

Maclean, D. J. & Tommerup, I. C. (1979). Histology and physiology of compatibility and incompatibility between lettuce and the downy mildew fungus, *Bremia lactucae* Regel. *Physiol. Plant Pathol.*, **14**, 291–312

Mansfield, J. W. (1984). Plant cell death during infection by fungi. In *Cell Ageing and Cell Death*, ed. I. Davies & D. C. Sigee pp. 323–45, Cambridge: Cambridge University Press

Mansfield, J. W. (1986). Recognition, elicitors and the hypersensitive reaction. In *Recognition in Microbe-plant Symbiotic and Pathogenic Interactions* ed. B. Lugtenberg pp. 433–7, Berlin: Springer Verlag

Mansfield, J. W., Hargreaves, J. A. & Boyle, F. C. (1974). Phytoalexin production by live cells in broad bean leaves infected with *Botrytis cinerea*. *Nature (Lond.)*, **252**, 316–17

Nagata, T. & Takebe, I. (1970). Cell wall regeneration and cell division in isolated tobacco mesophyll protoplasts. *Planta* (Berl.), **92**, 301–8

Norwood, J. M. & Crute, I. R. (1984). The genetic control and expression of specificity in *Bremia lactucae* (lettuce downy mildew). *Plant Pathol.*, **33**, 385–400

Otsuki, Y., Shimomura, T. & Takebe, I. (1972). Tobacco mosaic virus multiplication and expression of the N gene in necrotic responding tobacco varieties. *Virology*, **50**, 45–50

Skipp, R. & Deverall, B. J. (1972). Relationships between fungal growth and host changes visible by light microscopy during infection of bean hypocotyls (*Phaseolus vulgaris*) susceptible and resistant to physiologic races of *Colletotrichum lindemuthianum*. *Physiol. Plant Pathol.*, **2**, 357–74

Stakman, E. C. (1915). Relations between *Puccinia graminis* and plants highly resistant to its attack. *J. Agricultural Res.*, **4**, 193–200

Staskawicz, B. J., Dahlbeck, D. & Keen, N. T. (1984). Cloned avirulence gene of *Pseudomonas syringae* pv. *glycinea* determines race-specific incompatibility on *Glycine max* (L) Merr. *Proc. Natl Acad. Sci., USA*, **81**, 6024–8

Street, P. F. S., Rowell, P. M., Crucefix, D. N., Didehvar, F. & Mansfield, J. W. (1986). Race specific resistance to *Bremia lactucae* is expressed by lettuce cells in suspension culture. In *Recognition in Microbe–plant Symbiotic and Pathogenic Interactions*, ed. B. Lugtenberg, pp. 243–51. Berlin: Springer Verlag

Tomiyama, K. (1966). Double infection by an incompatible race of *Phytoph-*

thora infestans of a potato plant cell which has previously been infected by a compatible race. *Ann. Phytopath. Soc. Japan*, **32**, 181–5

Tomiyama, K., Takakuwa, M. & Takase, N. (1958). The metabolic activity in healthy tissue neighbouring the infected cells in relation to resistance to *Phytophthora infestans* (Mont.) De Bary in potatoes. *Phytopath. Zeit.*, **31**, 237–50

Yoder, O. C., Weltring, K., Turgeon, B. G., Garber, R. C. & Van Etten, H. D. (1986). Technology for molecular cloning of fungal virulence genes. In *Biology and Molecular Biology of Plant–Pathogen Interactions* ed. J. A. Bailey, pp. 371–84. Berlin: Springer Verlag

Yoshikawa, M. (1983). Macromolecules recognition and the triggering of resistance. In *Biochemical Plant Pathology* ed. J. A. Callow pp. 267–98. Chichester: John Wiley & Son

17

Self-incompatibility in *Brassica oleracea*: a recognition system with characteristics in common with plant–pathogen interactions?

T. Hodgkin[1], H. G. Dickinson[2] and G. D. Lyon[1]

[1]Scottish Crop Research Institute,
Invergowrie, Dundee DD2 5DA, UK
[2]Department of Botany, Plant Science Laboratories,
University of Reading, Whiteknights,
Reading RG6 2AS, UK

Abstract

Self-incompatibility systems, which prevent self-fertilisation in flowering plants, involve highly effective mechanisms for self-recognition. Recent studies on the sporophytic self-incompatibility system of *Brassica oleracea* have provided fresh insights into recognition and response following incompatible pollinations. These results are described and compared with data from plant-pathogen interactions which also involve highly specific recognition reactions. The similarities and differences between the two systems are discussed and their significance for our understanding of the nature of recognition and response in flowering plants evaluated.

Introduction

Recognition reactions involving flowering plants are well established phenomena. They include both plant/plant interactions and interactions between plants and other organisms and vary greatly with respect to the degree of discrimination shown. The self-incompatibility (SI) system in flowering plants is a highly specific recognition reaction which prevents the production of zygotes following self-pollination (de Nettancourt, 1977). Such systems occur in most families of flowering plants and have been the subject of many investigations (for reviews see de Nettancourt, 1977; Clarke & Gleeson, 1981; Heslop-Harrison, 1983).

The diploid *Brassica* species *B. campestris*, *B. nigra* and *B. oleracea* possess a sporophytic SI system controlled by a single locus (S-) with a large number of alleles (Thompson, 1957; Ockendon, 1974; Mackay, 1977). The SI system in the brassicas has considerable agronomic significance as it is utilised in the production of F1 cultivars of a wide range of different crops of which those of *B. oleracea* are particularly important (Wills & North, 1978). However the production of such cultivars is not without problems. In particular, many hybrid seedlots contain an unacceptably high proportion of inbreds as a result of the partial failure of the SI system (Hodgkin, 1981).

As well as having agronomic importance, the SI system of *B. oleracea* provides an interesting model system, not only for studies of angiosperm pollen-pistil interactions, but also for more general investigations of recognition and response in flowering plants. Over the last five years a number of significant advances have been made in our understanding of the events following an SI pollination in *B. oleracea* (Dickinson & Elleman, 1985; Hodgkin & Lyon, 1985) and of the molecular genetics of the S-gene (Nasrallah *et al.*, 1985.

Plant host–pathogen (HP) interactions can also involve highly specific recognition reactions in which individual races of a particular pathogen may fail to infect particular host genotypes (i.e. the reaction is incompatible) while remaining highly pathogenic or compatible with others (for general reviews see Wood & Graniti, 1976; Horsefall & Cowling, 1980; Yoshikawa, 1983). Recently, several authors have drawn attention to the existence of a number of common features which are shared by the SI and HP interactions (Heslop-Harrison, Knox & Heslop-Harrison, 1974; Teasdale *et al.*, 1974). More detailed consideration was given to this by Lewis (1980), who drew attention to the role of boron in pollen pistil interactions and in the synthesis of phenolic phytoalexins and Hogenboom (1983) who emphasised the similarities between the genetics of intraspecific and interspecific incompatibility systems and that of HP interactions. We have also commented on similarities that appear to exist between the two systems (Hodgkin & Lyon, 1979, 1985; Roberts, Harrod & Dickinson, 1984).

In many respects, the recent work on the SI system of *B. oleracea* tends to support the possibility that the SI interaction has significant features in common with those HP interactions which conform to the gene-for-gene hypothesis and, typically, produce phytoalexins following an incompatible interaction. It is the object of this paper to summarise the major features of the SI response in *B. oleracea*, and explore the extent to which these are also characteristic of such HP interactions.

While we believe such a comparison to be informative, it is clear that it needs to be approached with some caution. There is a wealth of variation in both SI and HP interactions which preclude the development of a satisfactory model for either at present (see, for example, Ayers, Goodall & DeAngelis, 1985 & Heslop-Harrison, 1983) and, in this paper, we have selected those examples which we consider to be the most useful. Further, the SI reaction occurs only between pollen and pistil as part of a complex series of interactions in a highly specialised organ (the flower), the function of which is the production of seed. Many of the characteristic features of the flower, and particularly the pollen and pistil, act to promote pollen capture, germination, pollen-tube growth and penetration and, ultimately fertilisation. In this context SI prevents the occurrence of fertilisation by self pollen while fertilisation by cross or compatible pollen is unaffected. In contrast, HP interactions are not usually confined to a single organ or tissue and the host plant derives no apparent benefit from a compatible interaction.

Genetics of self-incompatibility

The SI system in *B. oleracea* is controlled by a single locus with multiple alleles (Thomson, 1957). The presence of the same active S-allele(s) in pollen and stigma confers incompatibility and, as with other sporophytically controlled SI systems, pollen specificity is determined by the paternal genome rather than by the pollen genome as in gametophytic self-incompatibility systems. The S-gene is expressed only in open flowers, whereas pistils from buds can accept self-pollen up to about 2 days before anthesis. The precise number of alleles is unknown but is likely to exceed 50 and they show complex non-linear dominance relationships which are often different in the stigma and pollen (Thompson & Taylor, 1966). Thus, S_x may be dominant to S_y in pollen while they act independently in the stigma and S_z may be recessive to S_x in both pollen and stigma. Despite this, it is possible to describe most S-alleles so far detected as recessive, intermediate or dominant (Ockendon, 1975).

The SI response may be modified by genes which confer full self-compatibility. Such genes may affect any S-allele but genes which affect specific alleles have also been reported (Thompson & Taylor, 1971). In both cases these appear to be unlinked to the S-locus. In addition, partial self-compatibility, the production of some seed from pollinations between plants with the same active S-allele(s), occurs frequently in a number of *B. oleracea* crops. This is often under genetic control (Hodgkin, 1978, 1980*a*, *b*) and it has been frequently suggested that it occurs commonly in plants with recessive S-alleles which are thereby classified as 'weak'. However the evidence for this is inconclusive (Hodgkin, 1980*b*).

Hogenboom (1983) discussed the similarities between pollen-pistil and HP interactions and drew attention to the differences between the genetics of intraspecific incompatibility and of interspecific incompatibility systems. Thus, he considered that intraspecific incompatibility resulted from the interaction of complementary products of a single gene while interspecific incompatibility involved an incomplete matching of genetic systems, and lack of congruity and suggested that both of these systems may occur in HP interactions.

HP interactions in which a gene-for-gene relationship has been established (Flor, 1956), seem genetically very similar to the SI interaction of *B. oleracea* in that both are genetically simple and are expressed as 'yes' and 'no' types of responses (Hogenboom, 1983). A major apparent difference is that in SI systems a single gene determines the specificity of both partners (although in *B. oleracea* there is as yet no evidence that the gene products in each are identical) and recognition is of 'self', while in HP systems different genes operate in host and pathogen and recognition is of 'non-self'. It should be noted, however, that although the race of pathogen recognised in such an interaction is non-self, it is not necessarily the case that the component recognised in the interaction is from the pathogen. Indeed, there is evidence to suggest that there are stages of the HP interaction which involve 'self' recognition (see below).

Other differences should also be noted; firstly, in the most studied HP interactions there are many resistance genes which are inherited independantly of each other (Day, 1974), although there is also evidence that resistance genes occur in linked clusters (Day, 1974). Further, most resistance genes can act independently of each other, correspondence at one plant and pathogen locus being sufficient to give incompatibility. In contrast, the brassicas have a single S-locus and, even where more than one S-locus has been detected (e.g. *Eruca sativa*, Verma, Malik & Dhir, 1977), there is inter-locus complementation. Secondly, although resistance genes with multiple alleles have been described (e.g. Flor, 1971), we are not aware that complex dominance relationships of the kind found in sporophytic SI systems have been reported. Thirdly, alteration in pathogen specificity and the detection of new resistance genes is well established but there have been no reports of mutation from one S-allele to a novel form, although changed specificities have been detected in *Lycopersicon peruvianum* following inbreeding (de Nettancourt *et al.*, 1971). This is not surprising since, presumably, changes in both pollen and stigma specificity determinants would have to occur simultaneously whilst the HP interaction is subject to quite different selection constraints.

An interpretation of events following SI pollinations from electron microscopic studies

The most complete description of the events following compatible and incompatible intraspecific pollination in plants with a sporophytic SI system has been provided by Dickinson and his colleagues both for *Raphanus sativus* (Dickinson & Lewis, 1973*a*, *b*) and for *B. oleracea* (Dickinson & Roberts, 1986; Dickinson & Elleman, 1985). In both species the sequence of events is broadly similar. Following a compatible pollination, pollen grains adhere positively to the stigmatic surface, hydrate to a variable extent, germinate and penetrate the stigmatic cuticle. These tubes do not grow through the walls of the papillae but appear to grow within the wall to the base of the papillae where they continue to the ovary, growing through the middle lamellae.

In incompatible pollinations, there is evidence that much of the pollen adheres with less tenacity to the stigma surface (Stead, Roberts & Dickinson, 1980) and that, in general, only slow and slight hydration occurs. In many pollinations some pollen will germinate but the tubes formed are generally short, contorted, and fail to penetrate the stigmatic papillae successfully (Ockendon, 1972; Dickinson & Lewis, 1973*a*). On a few occasions the tip of the pollen tube does achieve some dissolution of the outer cuticle of the papillae and, where this occurs, the formation of the characteristic callosic lenticule is stimulated at the point of penetration (Kanno & Hinata, 1969; Dickinson & Lewis, 1973*a*).

The pollen grain itself is invested by a highly specialised coating derived from the sporophytic tapetal cells which invested the pollen at its last stages of development in the anther (Dickinson & Lewis, 1973*b*). On the surface of this coating, often referred to as tryphine, there is a superficial layer which can be isolated, and has been shown to bind both lanthanum (a property of biological membranes) and alcian blue, indicating the presence of mucopolysaccharide (Elleman & Dickinson, 1986). The surface of the stigma papillae is also complex in structure; a rather granular and perhaps perforate cuticle (Dickinson & Roberts, 1986) is invested by a thin proteinaceous pellicle which possesses considerable esterase activity (Mattson *et al.*, 1974).

Thus, the initial contact between pollen and stigma is not between sporopollenin exine of the grain and the cuticle of the stigmatic papillae, but rather between two highly specialised layers which invest them. Following both compatible and incompatible intraspecific pollinations, the passage of water from the stigma to the pollen grain causes that area of the coating immediately in contact with the stigma to become converted into a highly structured electron-opaque form. Additional changes occur in the coating at the point at which the tube emerges from the colpus of the grain, and elements of the coating, generally in the form of vesicles and cisternae, accompany the tube as it travels to the stigma surface. In the case of a compatible pollination this sporophytic coating is lost from the tube as it enters the stigmatic cuticle and grows towards the ovary. At this level incompatible pollen can be differentiated from compatible only by its failure to develop beyond a particular stage (hydration, germination or penetration). In other respects it appears similar to compatible pollen.

Characteristically in most incompatible HP interactions, inhibition of the pathogen follows host cell wall degradation and host cell penetration (e.g. Sato, Kitazawa & Tomiyana, 1971; Jones, Graham & Ward, 1975; Coffey & Wilson, 1983). In contrast, the SI response need not involve any penetration of, or visible alteration to, the stigmatic surface. The hypersensitive HP response is also characterised by rapid host cell necrosis, browning and the accumulation of phytoalexins. In the SI interaction necrosis only occurs in pollen grains although stigma browning is a commonly observed feature of compatible pollinations.

There appear to be differences in the rate at which the HP and SI responses occur. In the former, differences between incompatible and compatible interactions may only become apparent after several hours whilst in *B. oleracea* the SI response appears to be extremely rapid (2–15 minutes). In HP interactions several hours can elapse before the pathogen germinates and this may account, in part, for the difference in the rate of the two responses. In fact, some host responses have been shown to occur extremely rapidly. Thus, increases in phytoalexin biosynthesis have been detected 10 min after cell damage (Sakai, Tomiyama & Doke, 1979) and gene transcription of pathogenesis-related proteins within 5 min of treatment with a phytoalexin elicitor (Somssich *et al.*, 1986).

In HP interactions cross-protection experiments have clearly demonstrated that inoculation of plants with incompatible pathogens can prevent their subsequent infection by compatible ones. (Ouchi, Oku & Hibino, 1976). Similarly, infection by compatible pathogens can prevent the subsequent development of an incompatible response following infection by an incompatible isolate (Ouchi *et al.*, 1976). In marked contrast, data from pollinations involving mixtures of compatible and incompatible pollen have shown that individual pollen grain/stigma papilla interactions must remain largely independent of each other; most compatible pollen is not inhibited by the presence of incompatible pollen (Ockendon & Currah, 1977) and incompatible pollen largely fails to grow in the presence of compatible pollen (Hodgkin, 1977).

Modificaion of the SI response

The SI response in *B. oleracea* is sensitive to a number of environmental and physiological influences. Increased temperature, relative humidity and plant or flower age all tend to weaken the SI response (van Marrewijk & Visser, 1975; Carter & McNeilly, 1975; Hodgkin, 1976) and result in the production of seed from SI pollinations. Similarly, many factors can be demonstrated to affect the infection process but few of them provide any direct information on the nature of the recognition response. With respect to SI, four areas of investigation may have a particular significance, i.e. the effect of relative humidity, carbon dioxide concentration, boron availability, and the role of metabolic inhibitors.

Relative humidity

Recently, Zuberi & Dickinson (1985) have re-examied the effect of relative humidity on the SI response. They reported that at high RH, germination of pollen on the stigmatic surface was enhanced. Such germination was particularly extensive if the plants used possessed a recessive S-allele, when it was often followed by penetration of a number of pollen tubes through the stigma (partial self-compatibility). It has been suggested that SI may depend to a large extent on the flow of water to the grain (Dickinson & Roberts, 1986) but as Zuberi & Dickinson (1985) observe, the fact that much of the incompatible pollen on the stigma surface is normally hydrated and still inhibited suggested that such a theory is not directly applicable to the SI response of brassicas.

Carbon dioxide

Raised levels of carbon dioxide (3–5%) have been shown to have a profound effect on the SI response (Nakanishi, Esashi & Hinata, 1969) resulting in extensive tube penetration and seed set following self-pollinations. The effect is so marked that the technique has been used successfully in the production of seed of parental lines of F1 cultivars (Taylor, 1982). The causes of the effect are not understood. It has been reported that CO_2 also increases the numbers of tubes penetrating the stigma in compatible pollinations and it may be that the effect is not directly related to the SI mechanism. We are not aware of any evidence that CO_2 levels comparable to those shown to prevent the expression of SI inhibit the incompatible HP response.

Boron

Hodgkin (1980*c*) reported that increasing the available boron to incubated flowers resulted in a marked decrease in the strength of the SI response and that tubes which penetrated the stigma appeared to contain increased amounts of callose. Boron is essential to brassica pollen germination and tube growth (Hodgkin, 1983) and has been implicated in the production of phenolic phytoalexins (Lewis, 1980) although there is little experimental evidence to support this. It has also been suggested that it may replace calcium (another essential element in *B. oleracea* pollen germination, Hodgkin, 1983) in the targetting of callose precursors in plant cell metabolism.

Metabolic inhibitors

The addition of cycloheximide to the pistil prior to pollination prevents the expression of the normal SI response. Self pollen grains germinate and pollen tubes penetrate in a manner that is indistinguishable from a compatible pollination (Roberts *et al.*, 1984). However, treatment of pistils with tunicamycin (an inhibitor of protein–carbohydrate complexing in the formation of glycoproteins) had no effect on the SI response (Roberts *et al.*, 1984). The conversion of incompatible to compatible pollination by cycloheximide can be induced by its introduction up to 5 hours after pollination, evidence that, at least for this period, the SI response can be overcome. These results suggest that the SI response requires protein synthesis at some early stage in the reaction (but see McMahon, 1975 for further consideration of the effect of cycloheximide).

Recent video studies by Sarkar (pers. comm.) have shown that hydration of compatible pollen grains is much more rapid on stigmas treated with cycloheximide (comparable to rates that occur in bud pollinations) than on untreated ones. It may be that during their maturation stigmatic papillae synthesise compounds which slow down the rate of hydration of pollen and that cycloheximide inhibits this synthesis. Interestingly, it has also been shown (Sarkar, pers. comm.) that the interspecific incompatibility response that occurs between *B. napus* pistils and *B. oleracea* pollen can also be overcome by cycloheximide, possible evidence that the reactions affected also include some which involve recognition of species.

Metabolic inhibitors have been used extensively to investigate the biosynthesis of those compounds considered to have a role in the incompatible HP interaction. Thus, at sufficient concentrations cycloheximide inhibits the accumulation of phytoalexins (e.g. Bailey, 1969). Further, from the work of Cramer *et al.* (1985), we now know that increased transcription of mRNA of the enzymes involved in the biosynthesis of phytoalexins appears to be characteristic of the incompatible HP response. It is not known whether this is the case in SI.

It should be noted that the experiments with cycloheximide also confirm that the effect of the SI reaction on self pollen is at first biostatic and that up to 4 hours after pollination they can still be induced to germinate and penetrate the stigma surface. It is well known that pathogens can be successfully cultured from many hypersenstive reactions even after several days (e.g. Harrison, 1984).

The biochemistry of SI

The biochemical basis of the SI response remains unknown, particularly with respect to the components that may be present in the pollen. However, in addition to what may be deduced from genetic, morphological or physiological experiments information is available on the product of the S-gene and the metabolic changes that occur following an SI pollination.

The S-specific glycoprotein

Using serological techniques Nasrallah & Wallace (1967) showed that antigenically distinct compounds could be detected in the stigmas of plants with different S-alleles but such compounds were not detected in the pollen. Sedgely (1974) confirmed this work but was unable to detect antigenic activity for the recessive S-alleles investigated. Glycoproteins specific to different S-alleles have also been detected in stigma extracts using isoelectric focussing and have been shown to cosegregate with S-alleles (Nishio & Hinata, 1977). While these glycoprotein bands are absent from *B. oleracea* buds which lack a functional SI system (Roberts *et al.*, 1979), glycoproteins similar to those found in SI material have been detected in self-compatible lines. However, glycoproteins have not been found for every S-allele investigated and in some cases 2, rather than 1, glycoprotein bands have been allocated to a specific S-allele (Nishio & Hinata, 1977).

The glycoproteins detected have been reported as having molecular weights of 40–50,000 and are heavily glycosylated (Nishio & Hinata, 1982). Recently, Nasrallah *et al.* (1985) have described a partial gene sequence for an S-specific glycoprotein. Although no S-specific products have been reported from pollen it has been reported that antibodies raised against the stigmatic glycoproteins react with mature pollen extracts and that reactivity may also be detected in the tapetum during the last stages of pollen development (Elleman, pers. comm.; Nasrallah, J., pers. comm.).

The role of the S-specific glycoproteins in the SI response is unknown. It has been suggested that extracts containing S-specific glycoprotein inhibit germination of pollen with the same S-allele in *in vitro* tests (Ferrari, Bruns & Wallace, 1981) but we have been unable to confirm this (Roberts *et al.*, 1983). We have also been unable to detect S-glycoproteins in stigma diffusates (as opposed to extracts) suggesting that either they are bound to the stigma surface or are absent from it.

Several glycoproteins associated with HP interactions have been demonstrated to possess various types of biological activity. However, no gene products equivalent to the S-glycoprotein have yet been described from HP interactions with the possible exception of a peptide from tomato plants infected with *Cladosporium fulvum* (De Wit *et al.*, 1985). Extracellular glycoproteins from culture filtrates of incompatible but not compatible races of *Phytophthora megasperma* var *glycinae* have been reported to cross-protect soybean plants from compatible races of the fungus (Wade & Albersheim, 1979) although unidentified confounding factors in these experiments have made further progress difficult (Desjardins *et al.*, 1982). Cell wall glycoprotein extracts from isolates of *P. megasperma* elicited the production of the soybean phytoalexin glyceollin with the same specificity as the fungal isolates from which they were obtained (Keen & Legrand, 1980).

Ziegler & Pontzen (1982) have also isolated a glycoprotein from *P. megasperma* var *glycinea* which inhibited the accumulation of the phytoalexin glyceollin in soybean cotyledons treated with a glucan elicitor from the fungus.

This inhibition was race specific and was considered to be due to the carbohydrate moeity. In addition, hydroxyproline-rich glycoproteins from plant cell walls have been reported to increase after infection by pathogens and to be associated with resistance although their specific role is not understood (Esquerré-Tugayé *et al.*, 1979).

Lectins

There has been considerable speculation that lectins may play a part as receptors in SI interactions (Anderson *et al.*, 1983) as in HP and legume/*Rhizobium* interactions (Schmidt & Bohlool, 1981). Gaude, Fumex & Dumas (1983) reported that lectins were present in *B. oleracea* pollen and lectin binding sites have been reported on *B. oleracea* stigma surfaces (Kerhoas *et al.*, 1983). These sites bound the lectin Concanavalin A, and following such treatment, the stigma papillae failed to produce callose when fractions of self pollen which normally induced callose production were applied. As noted below there is some doubt as to whether callose production can be considered diagnostic of the SI response in brassicas and the direct relevance of these observations to the SI reaction is therefore uncertain.

Enzymes

While there is evidence that enzymes such as esterase and acid phosphatase diffuse freely from both pollen and stigma surface in *B. oleracea* (Vithanage & Knox, 1976; Hodgkin, 1984), and *B. oleracea* pollen grains contain an active cutinase (Linskens & Heinen, 1962), no SI related role has been assigned to the enzymes detected. In contrast, in HP interactions, a number of enzymes have been isolated from both pathogen and host which elicit the production of phytoalexins from plant cells. Of particular interest have been those enzymes with a cell wall degrading function. There is now considerable evidence that such enzymes are present in both pathogen and plant and that the cell wall fragments produced by them (oligosaccharides) elicit phytoalexins in host cells (Lee & West, 1981*a*, *b*; Davis *et al.*, 1984; Yoshikawa *et al.*, 1981; Keen & Yoshikawa, 1983; Lyon & Albersheim, 1982).

Pollen germination inhibitors

The possibility that low molecular weight compounds similar to phytoalexins might be produced following an SI recognition reaction and play a role in inhibiting germination of SI pollen was first suggested by Hodgkin & Lyon (1979) following experiments which showed that the potato phytoalexin rishitin inhibited germination of *Solanum demissum* pollen. Subsequent experiments demonstrated that novel pollen germination inhibitors were produced following self pollination of *B. oleracea*, which could be separated and detected using a thin layer chromatographic bioassay. The inhibitor(s) attained a maximum quantity 2 hours after self-pollination and inhibited *Lilium lankongense* and *Petunia hybrida* pollen as well as *B. oleracea* pollen (Hodgkin & Lyon, 1984, 1986). Thus, as with the phytoalexins, the compounds detected were not involved in the specificity of the recognition reaction although Hodgkin & Lyon (1985) suggested that they played a part in the response. To date there has been no information on the nature of the compounds involved or whether they are produced by the pollen or stigma. However, recent evidence suggest that they are present in extremely small amounts (1–5 μg per gm fresh weight or 1 ng per stigma).

The accumulation of post-infection inhibitors such as phytoalexins by host cells is one of the characteristic features of the incompatible HP interaction and it has been shown that this is preceded by increases in the transcription of the mRNAs specific to the enzymes involved in their biosynthesis (Cramer *et al.*, 1985). Phytoalexins are non-specific in their action and the precise relationship of their accumulation to the HP recognition reaction has yet to be discovered.

Callose

Considerable attention has been given to the possibility that callose production may be a characteristic indicator of the SI response (e.g. Dumas & Knox, 1983). Certainly, callose is usually produced in *B. oleracea* and *R. sativus* stigma papillae when penetration is attempted by incompatible pollen tubes (Dickinson & Lewis, 1973*a*; Dickinson & Elleman, 1985). However, we have also observed that callose production may be slight or even absent in self-pollinated stigmas on which no pollen germination occurs. Similarly, many interspecific pollinations between different members of the Brassicaceae are characterised by large depositions of callose in the stigma whether or not they are successful.

Callose production is characteristic both of many HP interactions and of wounding responses in flowering plants and the similarity between these and the SI response with respect to callose production has been described by a number of workers (Lewis, 1980; Hinch & Clark, 1980; Dumas & Knox, 1983). It should be noted however that stigma cells possess a number of features common to other cells in which callose synthesis plays a prominant part. Thus, sieve elements contain large amounts of p-protein which closely resembles the proteinaceous bodies seen in stigmatic papillae in *Brassica* and *Raphanus*.

Discussion

Recognition reactions in biological organisms necessarily involve similar kinds of processes. There must be detection of the components of the interaction by at least one of the participants, discrimination involving a 'yes' or 'no' process and a response, the nature of which depends on the results of the discrimination. Compatible pollen and pathogens go through very similar processes to gain entry to the plant. They adhere to the plant surface, hydrate, germinate and penetrate the outer surface. This similarity is reflected both in the visual characteristics of the interactions and in their biochemistry (e.g. the presence of lectin binding sites and cuticle degrading enzymes). The key feature of the difference is that the compatible pollen–stigma interaction takes place in a complex of organs which maximise the possibility of success of that interaction while the HP one does not. Any comparison between SI and HP interactions must take account of all these basic features and the questions become less whether there are similarities between the two as what is the nature and extent of the similarities.

The genetic evidence shows that for the HP and SI interactions considered in this paper single loci are sufficient to provide the discriminatory power required by such recognition systems. The fact that in HP interactions a number of different loci can operate independently in the host, while in SI interactions complementation occurs between loci, may be evidence that in the former, different stages in the interaction can be involved whilst only one stage is involved in the latter. However, the occurrence of mutations to self-compatibility which involve loci unlinked to the S-locus, or affect particular S-alleles, suggests that such a picture may be too simple and the genetic data are not evidence that the SI reaction only involves a single stage.

In the SI interaction both pollen and stigma specificity are controlled by a single locus while in the HP interaction the loci involved are in different organisms. To this extent, recognition in SI systems is often described as 'self' recognition and contrasted with 'non-self' recognition as it occurs in HP interactions. The general characteristics of 'self' and 'non-self' recognition have been considered by Burnet (1971, 1976) but it should be noted that although the race of a pathogen recognised in the HP interaction is not self, the compounds involved in the interaction may not come from the pathogen.

From research on some HP interactions there is evidence that the compounds characteristic of the incompatible response can be elicited by plant cell wall framents ('self' components). Thus cell walls possess significant information carrying potential by virtue of the generation of numbers of structurally different fragments. Elegant proposals have been made as to how this might operate through the action of both pathogen and plant wall degrading enzymes (Darvill *et al.*, 1985; McNeil *et al.*, 1984; Albersheim *et al.*, 1981). Although similar enzymes to those implicated in HP interactions have been found in a number of pollens, as noted above, there is no evidence that penetration of the stigma surface occurs in the SI response or that the generation of wall fragments is involved.

The initial interaction between pollen and stigma occurs between two complex layers, the pollen tryphine and stigma pellicle. These layers have been shown to possess a wide range of enzymes and non-protein components. Much of the contents of the pollen tryphine are derived from the degenerating tapetum of the anther. It could be that compounds analogous to the cell wall fragments of the HP interaction, coming from the tapetum, interact with the stigmatic glycoprotein to initiate the incompatible response. This would account for the apparent rapidity of the SI response. In this sense SI could be considered as 'pre-programmed' in contrast to the HP interaction where the cell wall fragments have to be generated. However, pre-programming alone cannot account for all the experimental observations. As the experiments with cycloheximide demonstrate the SI reaction seems to involve some protein synthesis.

The independence of each pollen grain/stigma papilla interaction with respect to surrounding ones is a key feature of the difference from HP interactions where cross-over effects are common. Dickinson & Elleman (1985) have suggested that in *B. oleracea* the SI response is centred on the pollen grain itself rather than depending on inhibition by the stigma of the pollen.

This attractive suggestion would account for the individuality of pollen/stigma responses and the absence of cross protection in the SI response. It might also explain the variability observed with respect to the stage at which SI pollen development ceases.

The response to an SI pollination has striking similarities with the HP interactions discussed. The generation of pollen germination inhibitors following self-pollination closely resembles the accumulation of phytoalexins. The occurrence of callose is also characteristic of both. What is not clear is the extent to which these phenomena are prerequisites of inhibition or whether they are related features of a quite different mechanism. As in HP interactions there is no evidence of specificity residing in the germination inhibitors detected and this is unlike the situation found in gametophytic SI species such *Prunus* (Williams *et al.* 1982) where there is evidence that a stylar glycoprotein inhibits germination of pollen possessing an identical S-allele.

From the evidence presented above it seems most likely that the SI interaction in *B. oleracea* involves more than one stage. Dickinson & Elleman (1985) have suggested that a two-stage mechanism may be involved in which a primary recognition reaction triggers a secondary messenger which elicits the response. The identity of the secondary messenger is unknown and a detailed investigation is needed to estabish the link between the S-glycoproteins and the pollen germination inhibitors.

Conclusions

By comparing the SI interaction in *B. oleracea* with data from HP interactions it has been possible to explore the significance of notable similarities between the interactions and, more importantly, introduce possible explanations for particular features of difference. The evidence suggests that the SI system in *B. oleracea* can be viewed as functioning through a particular modification of more general defence systems which flowering plants can mobilise to protect themselves from attack by pathogens and also possibly from wounding.

It seems to us likely that there is a many staged 'dialogue' between pollen and stigma and that the recognition reaction itself is likely to be unique to the SI system but that the responses which incompatible combinations of any type initiate will appear very similar. Where differences are detected these are likely to be explicable in terms of particular characteristics of the interactants. Thus, for the SI reaction, the information (in terms of molecules which trigger a protective response) may be present from the start of the interaction and the response could be initially vested in the pollen. In contrast, in the HP interaction the recognition molecules must be released and the response is vested in the host.

Finally, although we are aware that much of the evidence remains inconclusive, we consider that the concepts proposed are of value in the development of our understanding of the significance of the experimental data, provide a framework for the interpretation of the observations and can be used to develop experimentally testable hypotheses.

Acknowledgements

We thank our collaborators, particularly C. Elleman, I. Roberts, R. Sarkar, A. Stead and E. Wiseman for their contributions to this research over the years.

References

Albersheim, P., Darvill, A. G., McNeil, M., Valent, B. S., Hahn, M. G., Lyon, G. D., Sharp, J. K., Desjardins, A. E., Spellman, M. W., Ross, L. M., Robertson, B. K., Aman, P. & Franzen, L. (1981). Structure and function of complex carbohydrates active in regulating plant–microbe interactions. *Pure & App. Chem.*, **53**, 79–88

Anderson M. A., Hoggart, R. D. & Clarke, A. E. (1983). The possible role of lectins in mediating plant cell–cell interactions. *Chemical Taxonomy, Molecular Biology, and Function of Plant Lectins*, pp. 143–61. Alan R. Liss., New York

Ayers, A. R., Goodell, J. J. & DeAngelis, P. L. (1985). Plant detection of pathogens. In *Chemically Mediated Interactions Between plants and Other Organisms*, ed. G. A. Cooper-Driver, T. Swain & E. E. Conn, *Rec. Adv. in Phytochem.*, **19**, 1–20

Bailey, J. A. (1969). Effects of antimetabolites on production of the phyto-alexin pisatin. *Phytochemistry*, **8**, 1393–5

Burnet, F. M. (1971). 'Self-recognition' in colonial marine forms and flowering plants in relation to evolution of immunity. *Nature (Lond.)*, **232**, 230–5

Burnet, F. M. (1976). The evolution of receptors and recognition in the immune system. In *Receptors and Recognition*, vol. 1A, ed. P. Cuatrecasas & M. F. Greaves, pp. 35–58. Chapman & Hall, London

Carter, A. L. & McNeilly, T. (1975). Effects of increased humidity on pollen tube growth and seed set following self-pollination in Brussels sprout (*Brassica oleracea* var. *gemmifera*). *Euphytica*, **24**, 805–13

Clarke, A. E. & Gleeson, P. A. (1981). Molecular aspects of recognition and response in the pollen–stigma interaction. In *The Phytochemistry of cell recognition and cell surface interactions, Recent Advances in Phytochemistry, Vol. 15* ed. F. A. Loewus & C. A. Ryan, pp. 161–211, New York: Plenum Press

Coffey, M. D. & Wilson, U. E. (1983). An ultrastructural study of the late-blight fungus *Phytophthora infestans* and its interaction with the foliage of two potato cultivars possessing different levels of general (field) resistance. *Can. J. Bot.*, **62**, 2669–85

Cramer, C. L., Ryder, T. B., Bell, J. N. & Lamb, C. J. (1985). Rapid switching of plant gene expression induced by fungal elicitor. *Science*, **227**, 1240–3

Darvill, A. G., Albersheim, P., McNeil, M., Lau, J. M., York, W. S., Stevenson, T. T., Thomas, J., Doares, S., Gollin, D. J., Chelf, P. & Davis, K. (1985). Structure and function of plant cell wall polysaccharides. *J. Cell Sci.* (Suppl.), **2**, 203–17

Davis, K. R., Lyon, G. D., Darvill, A. G. & Albersheim, P. (1984). Host-pathogen interactions. XXV. Endopolygalacturonic acid lyase from

Erwinia carotovora elicits phytoalexin accumulation by releasing plant cell wall fragments. *Plant Physiol.*, **74**, 52–60

Day, P. R. (1974). *Genetics of host–parasite interaction.* Freeman, W. H. and Company, San Francisco

Desjardins, A. E., Ross, L. M., Spellman, M. W. Darvill, A. G. & Albersheim, P. (1982). Host–pathogen interactions. XX. Biological variation in the protection of soybeans from infection by *Phytophthora megasperma* f. sp. *glycinea*. *Plant Physiol.*, **39**, 1046–50

De Wit, P. J. G. M., Hofman, A. E., Velthuis, G. C. M. & Kuc, J. A. (1985). Isolation and characterisation of an elicitor of necrosis isolated from intercellular fluids of compatible interactions of *Cladosporium fulvum* (syn. *Fulvia fulva*) and tomato. *Plant Physiol.*, **77**, 642–7

Dickinson, H. G. & Elleman, C. J. (1985). Structural changes in the pollen grain of *Brassica oleracea* during dehydration in the anther and development on the stigma as revealed by anhydrous fixation techniques. *Micron & Microscopica Acta*, **16**, 255–70

Dickinson, H. G. & Lewis, D. (1973*a*). Cytochemical and ultrastructural differences between intraspecific compatible and incompatible pollinations in *Raphanus*. *Proc. Roy. Soc. Lond., B.*, **183**, 21–38

Dickinson, H. G. & Lewis, D. (1973*b*). The formation of the tryphine coating the pollen grains of *Raphanus*, and its properties relating to the self-incompatibility system. *Proc. Roy. Soc., Lond., B.*, **184**, 149–65

Dickinson, H. G. & Roberts, I. (1986). Cell-surface receptors in the pollen–stigma interaction of *Brassica oleracea*. In, *Hormones, Receptors and Cellular Interactions in Plants*. ed. (Chadwick & Garrod) CUP pp. 255–81

Dumas, C. & Knox, R. B. (1983). Callose and determination of pistil viability and incompatibility. *Theoretical & Appl. Genetics*, **67**, 1–10

Elleman, C. J. & Dickinson, H. G. (1986). Pollen–stigma interactions in *Brassica*. IV. Structural reorganisation in the pollen grains during hydration. *J. Cell Sci.*, **80**, 141–57

Esquerré-Tugayé, M. T., Lafitte, C., Mazau, D., Toppan, A. & Touze, A. (1979). Cell surfaces in plant–microorganism interactions. II. Evidence for the accumulation of hydroxyproline-rich glycoproteins in the cell wall of diseased plants as a defense mechanism. *Plant Physiol.*, **64**, 320–6

Ferrari, T. E., Bruns, D. & Wallace, D. H. (1981). Isolation of a plant glycoprotein involved with control of intercellular recognition. *Plant Physiol.*, **67**, 270–7

Flor, H. H. (1956). The complementary genic systems in flax and flax rust. *Adv. Genetics*, **8**, 29–54

Flor, H. H. (1971). Current status of the gene-for-gene concept. *Ann. Rev. Phytopathol.*, **9**, 275–96

Gaude, T., Fumex, B. & Dumas, C. (1983). Are lectin-like compounds involved in stigma-pollen adhesion and/or recognition in *Populus* and *Brassica*? In *Pollen: Biology and Perspectives for Plant Breeding*, ed. D. L. Mulcahy & E. Ottaviano, pp. 265–72. Elsevier, New York, Amsterdam, Oxford

Harrison, J. G. (1984). *Botrytis cinerea* as an important cause of chocolate spot in field beans. *Trans. Br. Mycol. Soc.*, **83**, 631–7

Heslop-Harrison, J. (1983). Self-incompatibility: phenomenology and physiology. *Proc. Roy. Soc., Lond.*, **218**, 371–95

Heslop-Harrison, J., Knox, R. B. & Heslop-Harrison, Y. (1974). Pollen-wall proteins: Exine- held fractions associated with the incompatibility response in *Cruciferae. Theor. & App. Gen.*, **44**, 133–7

Hinch, J. M. & Clarke, A. E. (1980). Callose formation as a response to infection of *Zea mays* roots by *Phytophthora cinnamoni. Physiol. Plant Pathol.*, **16**, 303–8

Hodgkin, T. (1976). *Studies on the inheritance of pseudocompatibility in Brassica oleracea.* PhD Thesis, Dundee University, pp. 286

Hodgkin, T. (1977). Sib numbers in individual Brussels sprout siliquae. *Cruciferae Newsl.*, **2**, 25–6

Hodgkin, T. (1978). The inheritance of partial self-compatibility in *Brassica oleracea L.*: Results from a half diallel homozygous for a highly recessive S-allele. *Theor. & Appl. Genetics*, **53**, 81–7

Hodgkin, T. (1980*a*). The inheritance of partial self-compatibility in *Brassica oleracea L.* Results from a half diallel homozygous for a moderately recessive S-allele. *Euphytica*, **29**, 65–71

Hodgkin, T. (1980*b*). The inheritance of partial self-compatibility in *Brassica oleracea* L. inbreds homozygous for different S-alleles. *Theor. & Appl. Gen.*, **58**, 101–6

Hodgkin, T. (1980*c*). Boron increases pollen tube numbers in self-pollinated *Brassica oleracea* stigmas. *Cruciferae Newsl.*, **5**, 18–19

Hodgkin, T. (1981). Some aspects of sib production in F1 cultivars of *Brassica oleracea. Acta Hort.*, **111**, 17–24

Hodgkin, T. (1983). A medium for germinating brassica pollen *in vitro. Cruciferae Newsl.*, **8**, 63

Hodgkin, T. (1984). The potential for pollen selection in Brassicas. In *Proceedings of Better Brassicas '84 Conference.* Ed. W. Macfarlane Smith & T. Hodgkin, SCRI, Dundee. pp. 51–6

Hodgkin, T. & Lyon, G. D. (1979). Inhibition of *Solanum* pollen germination *in vitro* by the phytoalexin rishitin. *Ann. Bot.*, **52**, 253–5

Hodgkin, T. & Lyon, G. D. (1984). Pollen germination inhibitors in extracts of *Brassica oleracea* L. stigmas. *New Phytol.*, **96**, 293–8

Hodgkin, T. & Lyon, G. D. (1985). Pollen germination inhibitors associated with the self-incompatible response of *Brassica oleracea* L. In *Proceedings of the 8th symposium on sexual reproduction in seed plants, ferns and mosses.* Comp. M. T. M. Willemse & J. L. van Went, pp. 108–10, Pudoc, Wageningen

Hodgkin, T. & Lyon, G. D. (1986). The effect of *Brassica oleracea* stigma extracts on the germination of *B. oleracea* pollen in a thin layer chromatography bioassay. *J. Exp. Bot.*, **37**, 406–11

Hogenboom, N. G. (1983). Bridging a gap between related fields of research: pistil–pollen relationships and the distinction between incompatibility and incongruity in nonfunctioning host–parasite relationships. *Phytopathology*, **73**, 381–3

Horsefall, J. G. & Cowling, E. B. (eds.), (1980). *Plant Disease. An Advanced Treatise. Vol V. How Plants defend Themselves.* Acad. Press, New York, London, Toronto, Sydney, San Francisco, pp. 534

Jones, D. R., Graham, W. G. & Ward, E. W. B. (1975). Ultrastructural changes in pepper cells in an incompatible interaction with *Phytophthora infestans*. *Phytopathology*, **65**, 1274–85

Kanno, T. & Hinata, K. (1969). An electron microscopic study of the barrier against pollen-tube growth in self-incompatible *Cruciferae*. *Plant & Cell Physiol.*, **10**, 213–16

Keen, N. T. & Legrand, M. (1980). Surface glycoproteins; evidence that they may function as the race specific phytoalexin elicitors of *Phytophthora megasperma* F. sp. *glycinea*. *Physiol. Plant Path.*, **17**, 175–92

Keen, N. T. & Yoshikawa, M. (1983). B-1, 3-Endoglucanase from soybean releases elicitor-active carbohydrates from fungus cell walls. *Plant Physiol.*, **71**, 460–5

Kerhoas, C., Knox, R. B. & Dumas, C. (1983). Specificity of the callose response in stigmas of *Brassica*. *Ann. Bot.*, **52**, 597–602

Lee, S. C. & West, C. A. (1981*a*). Polygalacturonase from *Rhizopus stolonifer*, an elicitor of casbene synthetase activity in castor bean (*Ricinus communis* L.) seedlings. *Plant Physiol.*, **67**, 633–9

Lee, S. C. & West, C. A. (1981*b*). Properties of *Rhizopus stolonifer* polygalacturonase, an elicitor of casbene synthetase activity in castor bean (*Ricinus communis*) seedlings. *Plant Physiol.*, **67**, 640–5

Lewis, D. H. (1980). Are there inter-relations between the metabolic role of boron, synthesis of phenolic phytoalexins and the germination of pollen? *New Phytol.*, **84**, 261–70

Linskens, H. F. & Heinen, W. (1962). Cutinase-Nachweis in Pollen. *Zeit. Bot.*, **50**, 338–47

Lyon, G. D. & Albersheim, P. (1982). Host–pathogen interactions XXI. Extraction of a heat-labile elicitor of phytoalexin accumulation from frozen soybean stems. *Plant Physiol.*, **70**, 406–9

Mackay, G. R. (1977). A diallel cross method for the recognition of S-allele homozygotes in turnip, *Brassica campestris spp. rapifera*. *Heredity*, **38**, 201–8

McMahon, D. (1975). Cycloheximide is not a specific inhibitor of protein synthesis *in vitro*. *Plant Physiol.*, **55**, 815–21

McNeil, M., Darvill, A. G., Fry, S. C. & Albersheim, P. (1984). Structure and function of the primary cell walls of higher plants. *Ann. Rev. Biochem.*, **53**, 625–63

Marrewijk, N. P. A. van, & Visser, D. L. (1975). The influences of alternating temperatures and relative humidity on the activity of S-alleles in *Brassica oleracea* L. var *gemmifera* (DC). In A. B. Wills & C. North (eds.) *Proc. Eucarpia Meeting – Cruciferae, 1974*. S.C.R.I. pp. 20–6

Mattsson, O., Knox, R. B., Heslop-Harrison, J. & Heslop-Harrison, Y. (1974). Protein pellicle of stigmatic papillae as a probable recognition site in incompatibility reactions. *Nature (Lond.)*, **247**, 298–300

Nakanishii, T., Esashi, Y. & Hinata, K. (1969). Control of self-incompatibility by CO_2 gas in *Brassica*. *Plant & Cell Physiol.*, **10**, 925–7

Nasrallah, J. B., Kao, T.H., Goldberg, M. L. & Nasrallah, M. E. (1985). A cDNA clone encoding an S-locus-specific glycoprotein from *Brassica oleracea*. *Nature (Lond.)*, **318**, 263–67

Nasrallah, M. E. & Wallace, D. (1967). Immunogenetics of self-incompatibility in *Brassica oleracea* L. *Heredity*, **22**, 519–27

Nettancourt, D. de (1977). *Incompatibility in Angiosperms*. Springer Verlag, Berlin, Heidelberg, New York. 230 pp.
Nettancourt, D. de, Ecochard, R., Perquin, M. D. G., van der Drift, T. & Westerhof, M. (1971). The generation of new S-alleles at the incompatibility locus of *Lycopersicon peruvianum* Mill. *Theor. & App. Gen.*, **41**, 120–9
Nishio, T. & Hinata, K. (1977). Analysis of S-specific proteins in pollen and stigmas of *Brassica oleracea* L. by isoelectric focussing. *Heredity*, **38**, 391–6
Nishio, T. & Hinata, K. (1982). Comparative studies on S-glycoproteins purified from different S-genotypes of self-incompatible *Brassica* species. I. Purification and chemical properties. *Genetics*, **100**, 641–7
Ockendon, D. J. (1972). Pollen tube growth and the site of the incompatibility reaction in *Brassica oleracea. New Phytol.*, **71**, 519–22
Ockendon, D. J. (1974). Distribution of self-incompatibility alleles and breeding structure of open-pollinated cultivars of Brussels sprouts. *Heredity*, **33**, 159–71
Ockendon, D. J. (1975). Dominance relationships between S-alleles in the stigmas of Brussels sprouts (*Brassica oleracea* var *gemmifera*). *Euphytica* **24**, 165–72
Ockendon, D. J. & Currah, L. (1977). Self-pollen reduces the number of cross-pollen tubes in the styles of *Brassica oleracea L. New Phytol.*, **78**, 675–80
Ouchi, S., Oku, H. & Hibino, C. (1976). Localization of induced resistance and susceptibility in barley leaves inoculated with the powdery mildew fungus. *Phytopathology*, **66**, 901–5
Roberts, I. N., Stead, A. D., Ockendon, D. J. & Dickinson, H. G. (1979). A glycoprotein associated with the acquisition of the self-incompatibility system by maturing stigmas of *Brassica oleracea. Planta*, **146**, 179–83
Roberts, I. N., Gaude, T. C., Harrod, G. & Dickinson, H. G. (1983). Pollen–stigma interactions in *Brassica oleracea*; a new pollen germination medium and its use in elucidating the mechanism of self-incompatibility. *Theor. & App. Gen.*, **65**, 231–8
Roberts, I. N., Harrod, G. & Dickinson, H. G. (1984). Pollen–stigma interaction in *Brassica oleracea*. II. The fate of stigma surface proteins following pollination and their role in the self-incompatibility response. *J. Cell Sci.*, **66**, 255–64
Sakai, S., Tomiyama, K. & Doke, N. (1979). Synthesis of a sesquiterpenoid phytoalexin rishitin in non-infected tissue from various parts of potato plants immediately after slicing. *Ann. Phytopath. Soc. Japan*, **45**, 705–11
Sato, N. Kitazawa, K. & Tomiyama, K. (1971). The role of rishitin in localizing the invading hyphae of *Phytophthora infestans* in infection sites at the cut surfaces of potato tubers. *Physiol. Plant Path.*, **1**, 289–95
Sedgely, M. (1974). Assessment of serological techniques for S-allele identification in *Brassica oleracea. Euphytica*, **23**, 543–51
Schmidt, E. L. & Bohlool, B. B. (1981). The role of lectins in symbiotic plant-microbe interactions. In *Encyclopaedia of Plant Physiology, NS. Vol. 13B, Plant Carbohydrates II*. Springer Verlag, Berlin. pp. 658–77
Somssich, I. E., Schmelzer, E., Bollmann, J. & Hahlbrock, K. (1986). Rapid activation by fungal elicitor of genes encoding 'pathogenesis-related' proteins in cultured parsley cells. *Proc. Natl. Acad. Sci. USA*, **83**, 2427–30

Stead, A. D., Roberts, I. N. & Dickinson, H. G. (1980). Pollen–stigma interaction in *Brassica oleracea*: the role of stigmatic proteins in pollen grain adhesion. *J. Cell Sci.*, **42**, 417–23

Taylor, J. P. (1982). Carbon dioxide treatment as an effective aid to the production of selfed seed in kale and brussels sprouts. *Euphytica*, **31**, 957–64

Teasdale, J., Daniels, D., Davis, W. C., Eddy, R. & Hadwiger, L. A. (1974). Physiological and cytological similarities between disease resistance and cellular incompatibility responses. *Plant Physiol.*, **54**, 690–5

Thompson, K. F. (1957). Self-incompatibility in marrow-stem kale, *Brassica oleracea var. acephala*. I. Demonstration of a sporophytic system. *J. Gen.*, **55**, 45–60

Thompson, K. F. & Taylor, J. P. (1966). Non-linear dominance relationships between S-alleles. *Heredity*, **21**, 345–62

Thompson, K. F. & Taylor, J. P. (1971). Self-compatibility in kale. *Heredity*, **27**, 459–71

Vithanage, H. I. M. V. & Knox, R. B. (1976). Pollen-wall proteins: Quantitative cytochemistry of the origins of intine and exine enzymes in *Brassica oleracea*. *J. Cell Sci.*, **21**, 423–35

Verma, S. C., Malik, R. & Dhir, I. (1977). Genetics of the incompatibility system in the crucifer *Eruca sativa L. Proc. Roy. Soc. Lond., B.*, **96**, 131–59

Wade, M. & Albersheim, P. (1979). Host-pathogen interactions XVI. Race-specific molecules that protect soybeans from *Phytophthora megasperma* var. *sojae*. *Proc. Nat. Acad. Sci., USA*, **76**, 4433–7

Williams, E. G., Ramm-Anderson, S., Dumas, C., Mau, S. L. & Clarke, A. E. (1982). The effect of isolated components of *Prunus avium* L. styles on *in vitro* growth of pollen tubes. *Planta*, **156**, 517–19

Wills, A. B. & North, C. (1978). Problems of hybrid seed production. *Acta Hort.*, **83**, 31–6

Wood, R. K. S. & Graniti, A. (1976). *Specificity in Plant Diseases*. NATO Advanced Study Institute Series, Vol. 10. Plenum Press, New York and London. 354 pp.

Yoshikawa, M. (1983). Macromolecules, recognition and the triggering of resistance. In J. A. Callow (ed.) *Biochem. Plant Path.* J. Wiley & Sons Ltd, London, pp. 267–98

Yoshikawa, M., Matama, M. & Masago, H. (1981). Release of a soluble phytoalexin elicitor from mycelial walls of *Phytophthora megasperma* var. *sojae* by soybean tissues. *Plant Physiol.*, **67**, 1032–5

Ziegler, E. & Pontzen, R. (1982). Specific inhibition of glucan elicited glyceollin accumulation in soybeans by an extracellular mannan-glycoprotein of *Phytophthora megasperma* f.sp. *glycinea*. *Physiol. Plant Path.*, **20**, 321–31

Zuberi, M. I. & Dickinson, H. G. (1985). Pollen-stigma interaction in *Brassica*. III. Hydration of the pollen grains. *J. Cell Sci.*, **76**, 321–36

18 Host–parasite recognition in *Ustilago violacea–Silene alba*

Manfred Ruddat and John M. Kokontis[1]

Department of Molecular Genetics and Cell Biology,
Barnes Laboratory,
[1]Department of Biochemistry and Molecular Biology,
The University of Chicago, Chicago, IL, USA

Abstract

The disease-producing dikaryotic hyphae of the anther smut, *Ustilago violacea*, an obligate parasite of the *Caryophyllaceae* (carnation family), can be grown *in vitro* in the presence of extracts from susceptible host plants but revert to avirulent sporidia on laboratory medium. We hypothesise that host-extract substances promoting hyphal growth *in vitro*, termed hyphal growth factors (HGF), regulate fungal development and form the biochemical basis for host recognition. The direct correlation between the hyphal growth response of fungal strains and their host species as the HGF source indicates the function of HGFs as recognition substances. Plants not recognised by the fungus as a host, i.e. resistant plants are either deficient in HGFs or HGFs are sequestered or masked by hyphal growth inhibitors.

Introduction

Host specificity or a limited host range of microbial parasites indicates that a recognition process for successful host–parasite or host–symbiont interactions occurs. Fungal parasites discriminate between host and non-host through mechanisms that have not as yet been fully elucidated but appear to function similarly to those in other recognition systems (Heslop-Harrison, 1978). This is amplified in the brief but encyclopaedic review by Larsen (1986).

Positive selection pressure for the evolution of recognition processes in parasitic or symbiotic interactions may have come from the advantage accruing either to the invader or to the host-species to maximise the reproductive success of either one or both of the interacting organisms (Lippincott & Lippincott, 1984).

The response by a parasite to a host cell surface, its components or a wound product(s) in the host and also the response of the host to the parasite signal-molecules lead to a sequence of biochemical and physiological responses. In certain interacting systems, the responses can be quantitatively determined and serve as a useful measure for the recognition process. Gene expression was proven to occur in the parasite-, elicitor- or wound-induced synthesis of the enzymes involved in phytoalexin synthesis (Bell *et al.*, 1986; Cramer *et al.*, 1985; Ebel *et al.*, 1984; Schmelzer *et al.*, 1984; Callow, 1983; Yoshikawa *et al.*, 1981). Likewise, elicitors from host plants have been shown to induce the expression of genes critical for the host–parasite or host–symbiont interactions (Bolton, Nester & Gordon, 1986; Peters, Frost & Long, 1986; Callow, 1983). Recognition between host and parasite is then the culmination of a complex series of physiological reactions and events. It is unclear, however, whether recognition consists of a single process or involves the entire series of processes. The greater the complexity of the interaction, the larger is the number of events that may be useful to assess the recognition process in a host–microbe interaction (Dazzo & Truchet, 1983).

Discrimination between host and non-host appears to occur only after the parasite has developed structures to penetrate the host surface. In obligate parasites, such as the rust fungi, teliospores will germinate and develop elaborate infection structures on inert replicas of the host epidermis (Staples & Macko, 1980). While the infection structures in rust and smut fungi are intricate and capable of discriminating characteristic epidermal features of potential hosts, they are not host-specific and therefore can not participate in the final host evaluation. It is only after violating the host surface that the fungus encounters the final recognition signal for a compatible/incompatible organism (Yoder & Turgeon, 1985). Chemical substances rather than morphological features serve as signals to identify the fungal host.

For economic reasons and technical advantages, more studies have been conducted on host–pathogen interactions characterising the defence mechanisms of the host that identify the parasite, and relatively few on substances that identify a potential host. It is therefore not surprising that current models emphasise the pathogen eliciting and suppressing host defence mechanisms in the establishment of a compatible or incompatible reaction (Bushnell & Rowell, 1981; Heath, 1981; Keen, 1982). Rapid progress and the many breakthroughs in the investigations on defence substances, such as on the elicitors of phytoalexin synthesis, proteinase inhibitor synthesis, accumulation of hydroxyproline-rich glycoproteins as well as other defence substances have led to view susceptibility as the lack of defence (Dixon, 1986; Ebel, 1986; Ryan *et al.*, 1985; Darvill & Albersheim, 1984; Ouchi, 1983; Sequeira 1983; Bell, 1981; Cruickshank, 1980; Esquerre-Tugaye *et al.*, 1979). Obviously this is an important and successful concept, but it does not, however, account for the frequently observed specificity of pathogens that are restricted to specific species or cultivars.

Lectins, which provide a high degree of specificity in plant–microbe interaction, have been implicated in disease resistance. There is, however, no conclusive evidence that lectins protect plants from pathogens (Bell, 1981; Barondes, 1981; Etzler, 1981, 1985; Pistole, 1981; Sequeira, 1978). Lectins, however, do appear to be specific in establishing symbiontic interactions, such as in clover-*Rhizobium* (Abe *et al.*, 1984) and soybean-*Bradyrhizobium* (Halverson & Stacey, 1986*a*) and thus function as specific recognition molecules.

Flor's (1942) recognition of a 'gene-for gene' system of the transmission of resistance, susceptibility, virulence and avirulence traits has become a favourite model for the host–pathogen relationship; and in flax rust *Melamspora lini* and its host, 29 gene pairs for resistance–avirulence have been identified (Lawrence *et al.*, 1981). The transfer of an avirulence gene from *Pseudomonas syringae* pv *glycinae* to a virulent race of the same pathovar changed it into an avirulent race, whereas, transferred to another bacterium species, it did not transfer the hypersentitivity eliciting reaction (Staskawicz *et al.*, 1984). This is convincing evidence for Flor's gene-for-gene concept and proves the specificity of the incompatibility reaction, even though elicitor molecules or the elicited gene products involved in the defence reaction, e.g. the accumulation of phytoalexins probably causing the hypersensitive reaction, are often commonly occurring molecules that do not seem to be able to convey species or pathovar specificity (Cooksey *et al.*, 1983). The genetic evidence, however, is strong, and it is generally accepted that incompatibility is an active and specific interaction because each member of a resistance–avirulence gene pair, which are normally dominant to susceptibility and virulence, only functions in the presence of its allele (Ellingboe, 1981, 1982).

Susceptibility of plants to pathogens, however, may not result simply from the absence of a defence but may also be related to, or depend on, the presence of signal molecules in the host tissue. Such molecules may have functioned, at one stage, in the co-evolution of host–parasite relationship as defence substances. In a mutant parasite that developed resistance to the defence substance, the defence substance elicited a growth response, which came to control the parasite development. At least in certain plant–fungus interactions, susceptibility may therefore require the presence of unique host substances that are essential to the regulation of the fungal parasite development; and resistance may entail a deficiency in, a sequestration or a masking of these compounds and/or the presence of a specific growth inhibitor (Strange *et al.*, 1974; Staples & Macko, 1980; Lippincott & Lippincott, 1984).

Stimulation of growth and pathogenicity by plant products has been thoroughly documented for auxotrophic mutants (Garber, 1960). The compounds serve as essential nutritional function for the microorganisms, permitting their growth and development only in the presence of the required compounds. They are commonly encountered metabolites and do not determine host specificity.

There are, however, host-identifying substances that have been characterised (Table 18.1). While these compounds elicit highly specific physiological reactions in the parasite, presumably by binding to highly specific receptor molecules, the compounds involved are widely occurring substances that do not seem likely to convey host-specificity.

It is intuitively acceptable that recognition of a host by a pathogen is through a process controlling a critical stage in the development of the parasite, e.g. the development from the saprophytic growth form to the parasitic

Table 18.1. *Signal molecules involved in host recognition*

Elicitor	Source	Response	Reference
luteolin	alfalfa seed exudate	*Rhizobium nod ABC* expression	Peters *et al.*, 1986
acetosyringone	culture medium	*Agrobacterium tumefaciens vir* locus	Stachel *et al.*, 1986
7 phenolics[1]	tobacco	*A. tumefaciens vir* locus	Bolton *et al.*, 1986
galacturonic acid	root	*Phytophthora* spec.	Grant *et al.*, 1985

[1] Catechol, gallic acid, pyrogallic acid, p-hydroxybenzoic acid, protocatechuic acid, β-resorcylic acid, vanillin.

growth form. In most fungi, saprophytic and parasitic stages are not distinguished by different morphologies as they are in the smut fungus, *Ustilago violacea*. We propose that in *U. violacea* host-recognition occurs through chemical compounds obtained from the host that specifically stabilise and promote the development of the parasitic form of the fungus. The instability of the parasitic growth form of this obligate parasite provides a simple experimental process for bioassaying susceptibility compounds. Host-recognition presupposes a susceptible host. In *U. violacea* the interaction with its host plants, e.g. *Silene alba*, is less complex than in other host–parasite systems, e.g. the rusts, and appears to be amenable to analysis.

We will first describe the characteristics of the host–parasite interaction of *U. violacea* and *S. alba*, and then present evidence that the host substance-induced stabilisation and promotion of the parasitic growth-form constitutes the recognition process that identifies the compatible host to the fungus.

Host recognition signal

Life history of Ustilago violacea

The anther smut, *Ustilago violacea*, is a heterobasiodiomycete that systemically parasitises compatible members of the *Carophyllaceae*. In diseased plants, the anthers are filled with purple teliospores instead of pollen and internode growth is stunted. Teliospores, disseminated by wind or insect vectors (Jennersten, 1983), germinate and undergo meiosis on plant surfaces, including floral parts (Fischer & Holton, 1957; Baker, 1947). The resulting meiotic products, four haploid basidiospores, segregate into a_1 and a_2 mating types and bud sporidia. A conjugation tube forms only between sporidia of opposite mating types. The nucleus of one of the sporidia migrates into the other cell forming a dikaryon typical for basidiomycetes. Dikaryotic hyphae developing from conjugated sporidia grow systemically in the host plant.

The interesting question of how sporidia of opposite mating type recognise each other will not be discussed here. Fimbriae on the cell surface are implicated in the process because the morphology of fimbriae, as observed in the electron microscope, is altered in mixtures of a_1/a_2 sporidia (Day & Poon, 1975).

While conjugation is achieved on water agar, the resulting dikaryotic hyphae are characteristically unstable on common laboratory media (Day *et al.*, 1981; Kokontis & Ruddat, 1986). Dikaryotic cells, either obtained by conjugation or isolated from immature anthers of host plants, plated on laboratory medium revert rapidly to haploid and aneuploid sporidia (Day *et al.*, 1981; Garber, Ruddat & Nieb, 1984).

The saprophytic growth form of *U. violacea*, haploid sporidia, grows stably and luxuriantly on a variety of common laboratory media. The parasitic growth form, dikaryotic hyphae of *U. violacea*, does not develop on any of the laboratory media rich in organic material but can be grown *in vitro* in the presence of host plant extracts. (Day *et al.*, 1981; Day & Castle, 1982; Kokontis & Ruddat, 1986). Day *et al.*, 1981 called the active fraction silenin, and Castle & Day (1984) identified the organic solvent soluble compound as alpha-tocopherol. We isolated water-soluble compounds from the leaves of the host plant *Silene alba* that stabilised the dikaryotic cells and induced and promoted hyphal growth in *U. violacea* (Fig. 18.1). We termed these compounds hyphal growth factors (HGF) (Kokontis & Ruddat, 1986).

Do hyphal growth factors function as host recognition substances?

It is our hypothesis that HGFs promoting hyphal growth *in vitro* regulate fungal growth *in vivo* and form the biochemical basis for host-recognition. The availability of HGFs is essential for the parasitic growth form of *U. violacea* to develop and to remain stable. For HGFs to function as host recognition substances, they must meet at least three criteria: (i) HGFs must be restricted in their availability for *U. violacea* only in host species. It follows that (ii) resistant plants are either deficient in HGFs or HGFs are not available to the invading fungus because they are sequestered or their effect is masked by the hyphal growth inhibitor (HGI). (iii) *Ustilago violacea* isolates responding *in vitro* to the HGF of a specific host need not produce the disease in that species but may in another carophyllaceous species.

Our main evidence for the function of the HGFs as host recognition substances consists of a direct correlation between the fungal growth response to the HGFs originating from a susceptible host species and the disease production in such species.

Hyphal growth factors (HGF)

Isolation of HGF and HGI

We obtain HGF from *S. alba*, *S. otites*, *Dianthus deltoides* and *Lychnis flos-cuculi*. The rosette leaves of vegetative plants were homogenised in chloroform/methanol (1 : 2) and the HGF and hyphal growth inhibitor (HGI) partitioned into the aqueous phase; alpha-tocopherol remained in the organic phase.

The aqueous phase was chromatographed on a Bio-Gel P-2 filtration column with an exclusion limit of 1,800–2,500 D. The column, eluted with 20 mM ammonium acetate, yielded four active fractions (Fig. 18.2), HGF-A

Fig. 18.1. *Ustilago violacea* hyphae grown *in vitro*. Conjugated sporidia 1.C449 × 2.C449, grown for 48 h in the aqueous fraction of extracts from *Silene alba*, develop into hyphae (H). A few haploid sporidia (S) can be seen. ×1000.

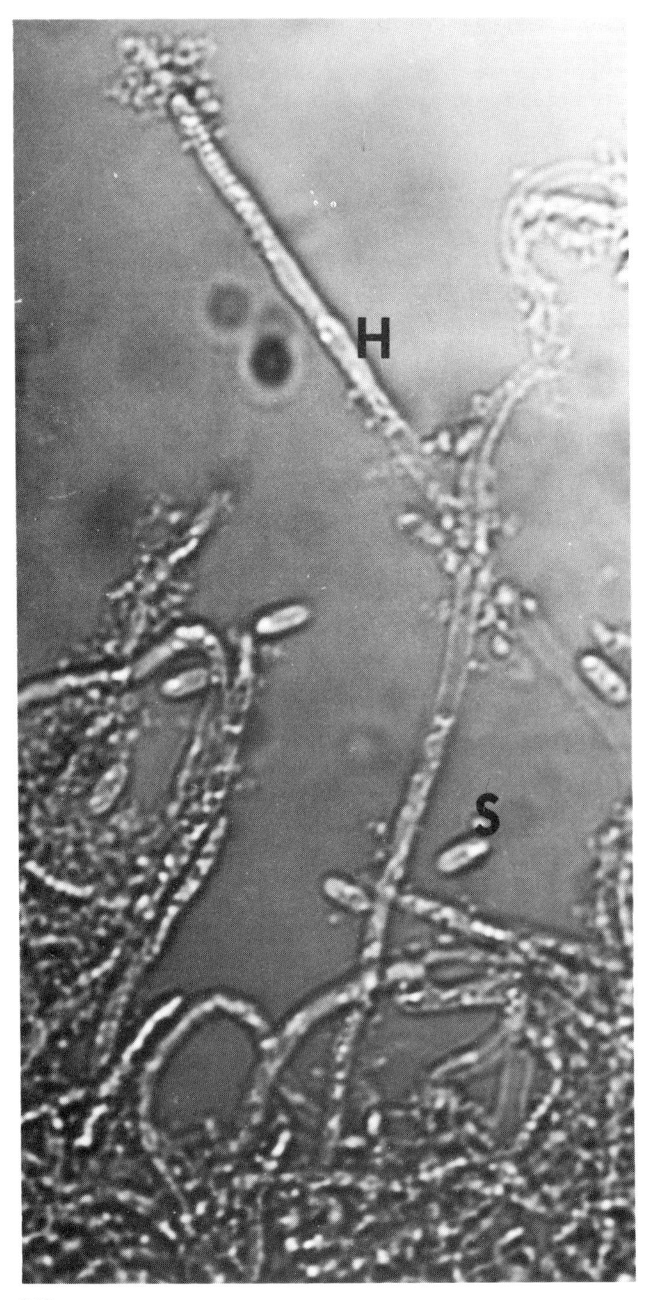

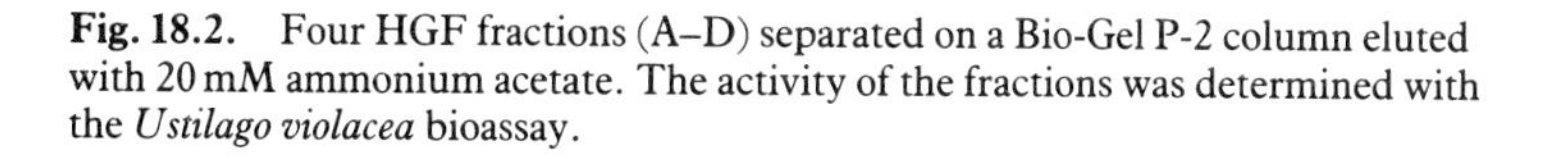

Fig. 18.2. Four HGF fractions (A–D) separated on a Bio-Gel P-2 column eluted with 20 mM ammonium acetate. The activity of the fractions was determined with the *Ustilago violacea* bioassay.

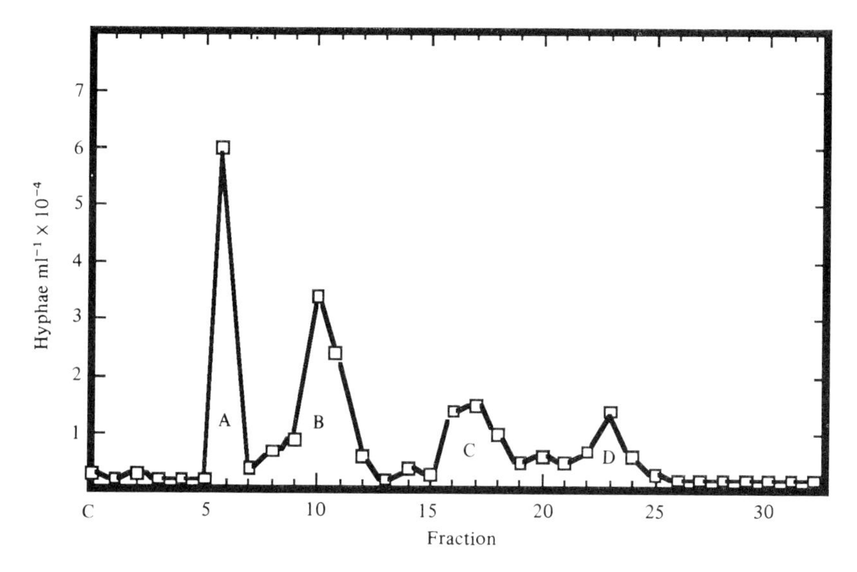

Fig. 18.3. Inhibition of hyphal growth by fraction 2 from a Bio-Gel P-2 column. Filter-sterilised P-2 fractions 1–5 were added to the bioassay medium, which contained unfractionated HGF. The hyphal growth inhibition of fraction 2 was alleviated by alpha-chymotrypsin treatment. Chymotrypsin did not, however, remove the inhibitory activity of fractions 4 and 5.

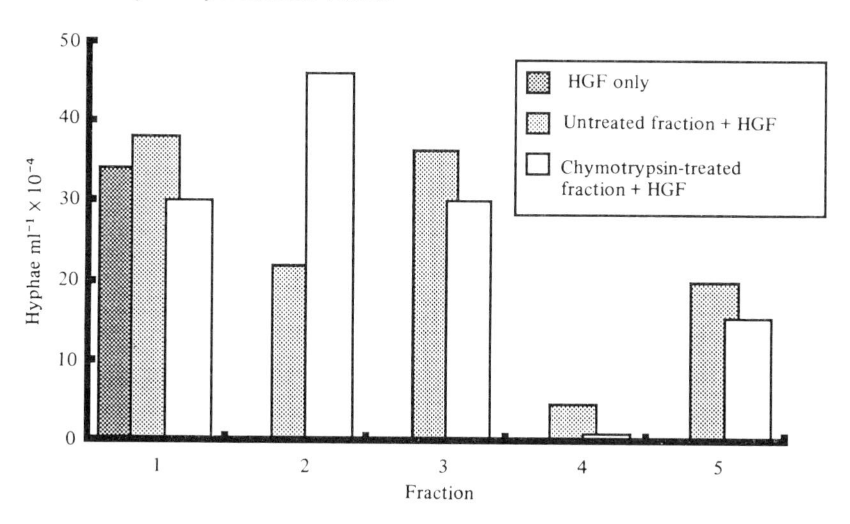

to HGF-D (Kokontis & Ruddat, 1986). The HGI eluted close to the void volume. Further purification was achieved on a C-18 flash column, thin layer chromatography and high performance liquid chromatography.

Preliminary chemical characterisation of the HGFs indicate that HGF-A is a small peptide linked to an aromate glycoside. HGF-B is sensitive to pronase and β-glucoronidase but has not been further characterised. HGF-C and HGF-D are proteolysis products of HGF-A. HGF-D consists of the highly substituted aromate linked to a sugar moiety, like HGF-C, and is sensitive to β-glucoronidase (unpublished results). The HGI which decreases HGF activity in the bioassay is sensitive to proteolysis by alpha-chymotrypsin (Fig. 18.3). The hyphal growth inhibitory activity in fractions 4 and 5, however, was not alleviated by alpha-chymotrypsin.

Bioassay for HGF with *U. violacea*

Ustilago violacea dikaryons grown in liquid culture indicate the activity of the HGFs and also assay the host recognition substances. The number of hyphae provides a sensitive measure of the biological activity of the plant extract added to the fungal culture medium.

Haploid sporidia of opposite mating type, from strain C449, were conjugated overnight on water agar at 14 °C to form dikaryons (Day & Jones, 1968). The sporidia were added to the culture medium at a final density of 10^6 cells/ml and incubated for 42–48 h on a rotatory shaker at 22–24 °C. The culture medium was a supplemented minimal medium adapted from Day & Jones (1968) and Linsmaier & Skoog (1965) and contained minerals, 1% glucose and 2 mg/l thiamine. The number of hyphae was determined with a hemocytometer.

In experiments assessing host recognition substances, HGFs were added as the aqueous partitioning phase. Individual HGF-fractions have also been tested.

Occurrence of HGF in species of Caryophyllaceae

Hyphal growth response of *U. violacea* isolates from different *caryophyllaceous* species

Fungal taxonomists have assigned the anther smut fungi that parasitise the entire family of the *Caryophyllaceae* to a single species, *U. violacea*. While the morphology of smut fungi isolated from different *caryophyllaceous* species is indistinguishable, we find that these isolates can be differentiated on the basis of their hyphal growth response to HGF extracted from various potential host species. *Ustilago violacea* discriminates among the species of the *Caryophyllaceae* as indicated by the higher hyphal growth response to extracts from the plant species from which the fungal strain had been isolated (Table 18.2). Two strains of *U. violacea* isolated from *Lychnis flos-cuculi* showed significantly higher hyphal growth on *L. flos-cuculi* extract than on extracts from any of the other three *Caryophyllaceae*. Similarly, strains C417 isolated from *S. otites* and C413 from *D. deltoides* showed by far the highest hyphal growth on extracts from their own host species. The growth response of C449 from *D. deltoides* and 1.C2 from *S. alba* to all extracts except to that from *L. flos-cuculi* may indicate the presence of alpha-tocopherol in these

Table 18.2. *Hyphal growth factor activity from* Caryophyllaceae *hosts*

Fungal strain	Origin	Relative hyphal growth on extracts from:			
		L. flos-cuculi	*S. otites*	*D. deltoides*	*S. alba*
C431	*Lychnis flos-cuculi*	1.8	0.32	0.79	1
C434	*Lychnis flos-cuculi*	2.2	0.26	0.26	1
C417	*Silene otites*	0.14	2.3	0.17	1
C413	*Dianthus deltoides*	1.0	1.5	37.0	1
C449	*Dianthus deltoides*	0.02	0.58	0.54	1
1.C2	*Silene alba*	0.6	1.1	1.2	1

Number of hyphae per ml is compared with the hyphal growth on extract from *S. alba*, which is normalised to 1.

relatively crude aqueous preparations. The lack of hyphal growth of strain C449 on extracts from *L. flos-cuculi* indicates the presence of an inhibitor to which C449 is uniquely sensitive.

In general, strains of *U. violacea* show the highest response to HGFs extracts from the host species from which they were isolated. One exception is strain C449, isolated from *Dianthus deltoides*, which gave a greater hyphal growth response to HGF obtained from *S. alba*. It is significant that strain C449 also was pathogenic on *S. alba*. Possibly, strain C449, isolated from *D.deltoides*, may have evolved on *S. alba*.

Disease production of *U. violacea* isolates in only those host species that yielded active HGF

It is consistent with our hypothesis that the HGFs function as a host-recognition substance. Fungal strains that respond to the HGF of a given *Caryophyllacean* species *in vitro* also produce disease in this species *in vivo*. We find that this correlation holds for the six *U. violacea* strains so far tested (Table 18.3).

While the development of the infection structures of *U. violacea* is promoted by host extract (Day *et al.*, 1981), it is apparently not host-specific and therefore not involved in the recognition process. It is after breaching the outer protective plant surface that *U. violacea* must determine whether the invaded plant species will support its growth and reproduction (Yoder & Turgeon, 1985). We, therefore, inoculate plants by overlayering the petioles of the rosette leaves with a thick suspension of conjugated cells and puncturing the petiole with a sterile needle (Day & Jones, 1968). In turgid plants the inoculate is soon entirely taken up.

Since the cuticle is not involved in host recognition for *U. violacea*, the use of host extracts of the entire leaf tissue for the identification of HGF

Table 18.3. *Pathogenicity of* Ustilago violacea *isolated from* Caryophyllaceae *species tested on* Silene alba

Strains[1] isolated from	% Diseased *S. alba*
C417 *S. otites*	0
C431 *L. flos-cuculi*	0
C434 *L. flos-cuculi*	0
C413 *D. deltoides*	0
C449 *D. deltoides*	40
1C.2 *S. alba*	60

[1] Conjugated strains.

Table 18.4. *The aqueous partitioning phase of organic extracts from* Silene alba, Phaseolus vulgaris, Raphanus sativus *and* Zea mays *assayed for hyphal growth activity with* Ustilago violacea, *strains 1.C449 × 2.C449.*

	Hyphae/ml
S. alba	$5.52 \times 10^5 \pm 5.0 \times 10^4$
P. vulgaris	$4.72 \times 10^5 \pm 5.0 \times 10^3$
R. sativus	$1.3 \times 10^4 \pm 5.0 \times 10^3$
control	$1.0 \times 10^4 \pm 4.0 \times 10^3$

Ten ml of assay medium contained 50 mg lyophilised extract.

and recognition factors is justified. The use of tissue extracts, in which the integrity of the tissue is lost, does, however, raise the question of whether the extracted substances are available to the invading fungus *in vivo*. (See p. 287.)

Non-host species do not yield HGF

It is equally important for our hypothesis that extracts from non-host species do not promote hyphal growth in *U. violacea*. We find that aqueous extracts from beans and radishes do not support hyphal growth, while corn, *Zea mays*, does (Table 18.4). *Zea mays*, however, is host to *Ustilago maydis*. The host recognition process in *U. violacea–S. alba* may share a hyphal growth factor common to several *Ustilago* species as amplified by the data of Day and his colleagues.

Day & Castle (1982) Day, Castle & Cummins (1981) reported that aqueous extracts from *c.* 100 non-host species for *U. violacea* did not induce hyphal growth in *U. violacea*, except those plant species that also were host to a species of the genus *Ustilago*. Alpha-tocopherol, present in all extracts

Table 18.5. *Exoprotease activity and pathogenicity of strains of* Ustilago violacea.

Strains isolated from	Relative exoprotease activity		% diseased *S. alba*
	Per cell	Per mg protein culture filtrate	
C413 *D. deltoides*	1.0	1.0	0
C431 *L. flos-cuculi*	1.6	3.4	0
C434 *L. flos-cuculi*	1.4	2.9	0
C449 *D. deltoides*	11.4	24.5	40
1.C2 *S. alba*	10.6	26.2	60

Exoprotease determined by azocasein hydrolysis.
Values normalised to those of 413.

obtained with organic solvents from host or non-host species, promoted hyphal growth (Castle & Day 1984; Kokontis & Ruddat, 1986). It is, however, unlikely that alpha-tocopherol controls hyphal growth *in planta* because it is unlikely to be available to the invading hyphae. (See p. 287.)

Fungal exoenzymes

Exoenzymes form an important interface between fungi and their environment. *Ustilago violacea* hyphae probe their environment with exoenzymes and in the process act upon the HGFs. Host recognition, therefore, may involve the action of fungal exoenzymes (Donly & Day, 1984).

We determined the exoprotease levels in the culture filtrates of five strains of *U. violacea* grown under conditions promoting hyphal growth. Only strains C449 and 1.C2 yielded high protease activity as determined by azocasein hydrolysis (Table 18.5). It is significant that the same two strains also were pathogenic in *S. alba*, and this may indicate a relationship between pathogenicity and exoenzyme production.

Exoprotease from strain 1.C449 increased the hyphal growth promoting activity of unfractionated HGF and HGF-A as tested with conjugated C413 sporidia (Table 18.6). This may indicate that HGF is also a substrate for *U. violacea* exoprotease *in vivo*. We have shown that HGF-C and D result from *in vitro* pronase treatment of HGF-A (unpublished).

The proteolysis of HGF-A may be an important step in the recognition process of a suitable host plant. While the four HGFs are independently active, *in planta* they may form a complex that is activated or its uptake facilitated through enzyme action. The presence of HGF-C and HGF-D in untreated *S. alba* extract could have resulted from proteolysis during the isolation procedures.

Since the hyphal growth inhibitor is sensitive to proteolysis, increase in HGF activity following exposure to exoprotease or pronase may have removed or diminished its inhibitory action on hyphal growth. The protease sensitivity of HGI may be involved in the relationship between exoenzyme production and pathogenecity and hence in the recognition process.

Table 18.6. *The effect of* Ustilago violacea *exoprotease and of pronase on the hyphal growth promoting activity of unfractionated HGF and Bio-Gel fraction 6 (HGF-A)*

		Hyphae/ml × 10^{-3} ± SE[1]	
Substrate	Control	Exoprotease	Pronase
Unfractionated HGF	1.1 ± 1.1	11.0 ± 5.4	39.0 ± 7.3[a]
HGF-A	4.4 ± 2.4	10.0 ± 3.7	12.2 ± 2.8[b]

[1] *Ustilago violacea* strains 1.C413 × 2.C413 were used for the bioassay.
[a] $p = 0.01$; [b] $p = 0.1$.

Alpha-tocopherol

Root extracts of *S. alba* and *Pastinaca sativa* (parsnip) yielded a substance that, like the material from *S. alba* leaves, termed silenin (Day *et al.*, 1981; Day & Castle, 1982), promoted hyphal growth of *U. violacea*. Castle & Day (1984) identified silenin as alpha-tocopherol (vitamin E). We found that the hyphal-growth promotion with the chloroform/methanol partioning phase of *S. alba* leaf extracts resulted from the presence of alpha-tocopherol (Kokontis & Ruddat, 1986).

Alpha-tocopherol occurs as a membrane component of chloroplasts and mitochondria and as such is universally distributed (Threlfall, 1980; Wallwork & Crane, 1970). In fact, Castle & Day (1984) found that methanol extracts of host and non-host species promoted hyphal growth of *U. violacea*. Such a compound is not likely to function as a host recognition substance.

Lack of tissue necrosis or of cell damage in *U. violacea*-infected plants indicates that the fungal cells generally do not enter the host cells. The hyphae apparently grow intercellularly (Batcho, Audran & Zambettakis, 1979; Audran & Batcho, 1980). Since alpha-tocopherol is localised in the membranes of cell organelles, *U. violacea* hyphae may not encounter it in the host plant.

Recently, Day *et al.* (1986) reported that the presence of *Ustilago heufleri* fimbriae in host cells may enable the hyphae to reach intracellular substances. It is intriguing that the immunological label is of approximate equal density over the chloroplast as over the cell wall. This interesting observation awaits confirmation of label specificity and a demonstration that the fimbriae function in transporting nutrients or hyphal growth promoting substances. It remains, therefore, open whether the effect of alpha-tocopherol on hyphal growth in *U. violacea* is hormonal or pharmacological in nature.

Reducing compounds appear to promote hyphal growth in *U. violacea*. Ascorbic acid and cysteine promoted, albeit weakly, hyphal growth *in vitro* (Castle & Day, 1984). Based on TLC separation, we excluded ascorbic acid as the active material in the aqueous fraction from *S. alba* leaf extracts eluted from the Bio-Gel P-2 column (Kokontis & Ruddat, 1986). Reducing compounds or a low oxidation-reduction potential induce parasitic growth in dimorphic fungi pathogenic in mammals. Interestingly, the parasitic growth

Table 18.7. *Synergistic effect of HGF and tocopherol on hyphal growth of* Ustilago violacea

Strain			Hyphal growth[1] Tocopherol –	+
1.C431 × 2.C431	HGF	–	1.0	2.5
		+	11.8	45.5
1.C434 × 2.C434	HGF	–	1.0	1.1
		+	3.9	50.0

[1] Relative growth: divided by non-treatment value.
HGF-extract had been treated with chymotrypsin.

form is sporidial in mammalian tissue (Rippon & Scherr, 1959; Rippon, Conway & Domes, 1965; Rippon, 1968; Rippon & Anderson, 1970). As an anti-oxidant, alpha-tocopherol may promote hyphal growth by affecting a mechanism sensitive to oxidation-reduction potential.

Since alpha-tocopherol promotes hyphal growth, one might expect differences in alpha-tocopherol content between the hyphae and sporidia. Extraction of *U. violacea* sporidia and hyphae, however, showed no difference in alpha-tocopherol content. Possibly the test, thin-layer chromatography followed by a colorimetric assay, may not have been sensitive enough to ascertain small, hormonal-level differences.

If alpha-tocopherol were to function as a host recognition substance, then one would expect it to promote hyphal growth in all strains of *U. violacea*. We found that alpha-tocopherol did not induce hyphal growth in two strains: C431 and C434. Simultaneous application of HGF and alpha-tocopherol, however, yielded a synergistic promotion of hyphal growth (Table 18.7).

Synergistic action of the elicitors arachidonic acid and β-glucan was observed in potato phytoalexin synthesis and of oligogalacturonide cell wall fragments and β-glucan soybean phytoalexin synthesis.

Subcellular localisation of HGF

For the HGF and HGI to be active *in vivo*, they must be available to the invading smut fungus. Since homogenisation and solvent extraction disrupt the structural integrity of the cells, biological activity of the fractions may be artefactual. It is therefore essential to identify the active substances at the sites of *in situ* hyphal growth. Chemical identification is prerequisite for their subcellular localisation.

Although alpha-tocopherol is presumably not involved in the host recognition process, it does, however, promote hyphal growth. A comparison of its subcellular distribution with that of HGF might indicate possible availability of HGF. The hyphal growth promoting activity of fractions obtained by differential centrifugation was analysed (Kokontis & Ruddat, 1986). The

Table 18.8. *Relative distribution of hyphal growth promoting activity in differentially centrifuged fractions of aqueous extracts from* Silene alba *leaves*

Fraction	Partition	Hyphae/ml	Relative value[1]
12,000 × g pellet	Aqueous	5.6×10^3	0
	Organic	2.3×10^5	31
100,000 × g pellet	Aqueous	3.3×10^3	0
	Organic	1.2×10^5	17
100,000 × g sup.	Aqueous	3.0×10^5	41
	Organic	8.0×10^4	11

[1] Relative value = $[(\text{hyphae/ml} \times 10\,\text{ml} \times 20)/10^6]/147 \times 100$.
147 = sum of $[(\text{hyphae/ml} \times 10\,\text{ml} \times 20)/10^6]$.
Hyphae/ml $< 10^4$ (control level), Relative value = 0.

entire biological activity of the water soluble fraction, i.e. the HGF, was localised in the high-speed supernatant, while approximately 80% of the biological activity in the fraction soluble in the organic phase was localised in the pellet (Table 18.8). While we cannot exclude the possibility that the HGFs are in the cytoplasm; they could be associated with the cell wall, where they would be available to the invading fungus.

Origin of HGI

Silene alba grown in a greenhouse from seeds collected in the field and inoculated with *U. violacea* yields about 60–70% diseased plants. The remainder are not diseased, apparently resistant against the fungus. Since we must extract batches of plants, rather than individual plants, to obtain sufficient biological active material, it is possible that HGI in our extracts originates from the resistant plants. It is equally plausible that the ratio of HGF/HGI or differential sequestration determines susceptibility and resistance and thus involves HGI along with HGF in the recognition process.

Concluding remarks

Stabilisation of the dikaryon and promotion of hyphal growth, i.e. the parasitic growth form with HGFs *in vitro*, provide an assay that also determines host recognition substances. In the absence of HGF, parasitic development could not proceed; the presence of HGF is a necessary condition for the host–parasite interaction. HGF may therefore carry the double function as a host specific signal. While four HGFs have been isolated from *S. alba*, these may all be active or they could have a precursor-and-product relationship localised at different sites. It is perhaps likely that more than one identifying substance is required. HGFs, as specific, positive signals indicating a potential host may interact with negative signals, such as chemical or mechanical barriers leading to discrimination between host and non-host.

References

Abe, M., Sherwood, J. E., Hollingsworth, R. I. & Dazzo, F. B. (1984). Stimulation of clover root hair infection by lectin-binding oligosaccharides from the capsular and extracellular polysaccharides of *Rhizobium trifolii*. *J. Bact.*, **160**, 517–20

Audran, J. C. & Batcho, M. (1980). Aspects infrastructuraux des alterations des anthères de *Silene dioica* parasitées par *Ustilago violacea*. *Can. J. Bot.*, **58**, 405–15

Baker, H. G. (1947). Infection of species of *Melandrium* by *Ustilago violacea* (Pers.) Fuckel and the transmission of the resultant disease. *Ann. Bot. N.S.*, **11**, 333–46

Barondes, S. M. (1981). Lectin: their multiple endogenous cellular functions. *Am. Rev. Biochem.*, **50**, 207–31

Batcho, M., Audran, J. C. & Zambettakis, Ch. (1979). Sur quelques données de l' évolution de l'*Ustilago violacea* (Pers.) Rouss. dans les ovules du *Silene dioica* (L.) Cl. *Rev. Cytol. Biol. Végét. Bot.*, **2**, 329–46

Bell, A. A., & Mace, M. E. (1981). Biochemistry and physiology of resistance. In: *Fungal Wilt Disease in Plants*. eds M. E. Mace, A. A. Bell & C. H. Beckman, pp 431–86. Academic Press, London

Bell, A. A. (1981). Biochemical mechanisms of disease resistance. *Ann. Rev. Plant Physiol.*, **32**, 21–81

Bell, J. N., Ryder, T. B., Wingate, V. P. M., Bailey, J. A. & Lamb, C. J. (1986). Differential accumulation of plant defence transcripts in a compatible and an incompatible plant–pathogen interaction. *Mol. & Cell. Biol.*, **6**, 1615–23

Bolton, G. W., Nester, E. W. & Gordon, M. P. (1986). Plant phenolic compounds induce expression of the *Agrobacterium tumefaciens* loci needed for virulence. *Science*, **232**, 983–5

Bushnell, W. R. & Rowell, J. B. (1981). Suppressor of defence reactions: a model for roles in specificity. *Phytopathology*, **71**, 1012–14

Callow, J. A. (1983). Recognition in higher plant–fungal pathogen interactions. In: Cellular Interactions in Plants. *Encycloped. Plant Physiol. New Ser.* Vol. 17, eds H. Linskens and J. H. Heslop-Harrison, pp. 212–37, Springer Verlag, Berlin

Castle, A. J. & Day, A. W. (1984). Isolation and identification of α-tocopherol as an inducer of the parasitic phase of *Ustilago violacea*. *Phytopathology*, **74**, 1194–200

Cooksey, C. J., Garratt, P. J., Dahiya, J. S. & Strange, R. N. (1983). Sucrose: a constitutive elicitor of phytoalexin synthesis. *Science*, **220**, 1398–1400

Cramer, C. L., Ryder, T. B., Bell, J. N. & C. J. Lamb. (1985). Rapid switching of plant gene expression induced by fungal elicitors. *Science*, **27**, 1240–3

Cruickshank, I. M. (1980). Defences triggered by the invader: chemical defences in plant disease. In: *Plant Disease: Vol. V. An Advanced Treatise*. eds J. A. Daly & I. Uritani, pp. 115–31

Darvill, A. G. & Albersheim, P. (1984). Phytoalexin and their elicitors-defence against microbial infection in plants. *Ann. Rev. Plant Physiol.*, **35**, 243–75

Day, A. W. & Castle, A. J. (1982). The effect of host extracts on differentiation of the genus *Ustilago*. *Bot. Gaz*, **143**, 188–94
Day, A. W., Castle, A. J. & Cummings, J. E. (1981). Regulation of the parasitic development of the smut fungus *Ustilago violacea* by extracts from host plants. *Bot. Gaz.*, **142**, 135–46
Day, A. W., Castle, A. J. & Cummings, J. E. (1981). Regulation of the parasitic development of the smut fungus *Ustilago violacea* by extracts from host plants. *Bot. Gaz.*, **142**, 135–46
Day, A. W., Gardiner, R. B., Smith, R., Svircev, A. M. & McKeen, W. E. (1986). Detection of fungal fimbriae by protein A-gold immunocytochemical labelling in host plants infected wtih *Ustilago heufleri* or *Peronospora hyoscyami* f. sp. *tabacina*. *Can. J. Microbiol.*, **32**, 577–84
Day, A. W. & Jones, J. K. (1968). The production and characteristics of diploids in *Ustilago violacea*. *Genet. Res.*, **11**, 63–81
Day, A. W. & Poon, N. H. (1975). Fungal fimbriae II. Their role in conjugation in *Ustilago violacea*. *Can. J. Microbiol.*, **21**, 547–57
Dazzo, F. B. & Truchet, G. L. (1983). Interactions of lectins and their saccharide receptors in the Rhizobium–legume symbiosis *J. Membrane Biol.*, **73**, 1–16
Dixon, R. A. (1986). The phytoalexin response: elicitation, signalling and control of host gene expression. *Biol. Rev.* (Cambridge Philos. Soc.), **61**, 239–91
Donly, B. C & Day, A. W. (1984). A survey of extracellular enzymes in the smut fungi. *Bot. Gaz.*, **145**, 135–46
Ebel, J. (1986). Phytoalexin synthesis: biochemical analysis of the induction process. *Ann. Rev. Phytopath.*, **24**, 235–64
Ebel, J., Schmidt, W. E. & Loyal, R. (1984). Phytoalexin synthesis in soybean cells: elicitor induction of phenylalanine ammonia-lyase and chalcone synthase mRNAs and correlation with phytoalexin accumulation. *Arch. Biochem. Biophys.*, **232**, 240–8
Ellingboe, A. H. (1981). Changing concepts in host–pathogen genetics. *Ann. Rev. Phytopath.*, **19**, 125–43
Ellingboe, A. H. (1982). Host resistance in host–parasite interactions: A perspective. In: *Phytopathogenic Prokaryotes*. eds M. S. Mount & G. H. Lacy, pp. 103–17. Academic Press, New York
Esquerre-Tugaye, M. T., Lafitle, C., Mazau, D., Toppan, A. & Touze, A. (1979). Cell surfaces in plant-microorganism interactions. II. Evidence for the accumulation of hydroxyproline-rich glycoproteins in the cell wall of diseased plants as a defence mechanism. *Plant Physiol.*, **69**, 314–19
Etzler, M. E. (1981). Are lectins involved in plant–fungus interactions? *Phytopathology*, **71**, 744–6
Etzler, M. E. (1985). Plant lectins: molecular and biological aspects. *Ann. Rev. Plant Physiol.*, **36**, 209–34
Fischer, G. W. & Holton, C. S. (1957). *Biology Control of the Smut Fungi*. The Ronald Press Co., New York
Flor, H. H. (1942). Inheritance of pathogenicity in *Melamspora lini*. *Phytopathology*, **32**, 653–69
Garber, E. D. (1960). The host as a growth medium. *Ann NY Acad. Sci.*, **88**, 1187–94

Garber, E. D., Ruddat, M. & Nieb, B. (1984). Genetics of *Ustilago violacea*. XVII. Chromosome transfer as a source of recombinant sporidia recovered from immature infected anthers. *Bot. Gaz.*, **145**, 145–9
Grant, B. R., Irving, H. R. & Radda, M. (1985). The effect of pectin and related compounds on encystment and germination of *Phytophthora palmivora zoospores*. *J. Gen. Microbiol.*, **131**, 669–76
Halverson, L. J. & Stacey, G. (1986). Signal exchange in plant-microbe interactions. *Microbiol. Rev.*, **50**, 193–225
Heath, M. C. (1981). A generalized concept of host–parasite specificity. *Phytopathology*, **71**, 1121–23
Heslop-Harrison, J. (1978). *Cellular Recognition Systems in Plants*. Edward Arnold, London.
Jennersten, O. (1983). Butterfly visitors as vectors of *Ustilago violacea* spores between caryophyllaceous plants. *Oikos*, **40**, 125–30
Keen, N. T. (1982). Specific recognition in gene-for-gene host–parasite systems. *Adv. Plant Path.*, **1**, 35–82
Kokontis, J. & Ruddat, M. (1986). Promotion of hyphal growth in *Ustilago violacea* by host factors from *Silene alba*. *Arch. Microbiol.*, **144**, 302–6
Larsen, K. (1986). Cell–cell recognition and compatibility between heterogenic and homogenic incompatibility. *Hereditas*, **105**, 115–33
Lawrence, G. J., Shepherd, K. W. & Mayo, G. M. E. (1981). Fine structure of genes controlling pathogenicity in flax rust, *Melamspora lini*. *Heredity*, **46**, 397–13
Linsmaier, E. M. & Skoog, F. (1965). Organic growth factor requirements of tobacco tissue cultures. *Physiol. Plant*, **18**, 100–7
Lippincott, J. A. & Lippincott, B. B. (1984). Concepts and experimental approaches in host-microbe recognition. In: *Plant–Microbe Interactions Molecular and Genetic Perspectives*. Vol. 1, ed. T. Kosuge & E. W. Nester, pp. 195–214. Macmillan Publ. Co., New York.
Long, S. R. (1984). Genetics of *Rhizobium* nodulation. In: *Plant–Microbe Interactions*, ed. T. Kosuge & E. Nester, pp. 265–306. New York: Macmillan
Ouchi, S. (1983). Induction of resistance or susceptibility. *Ann. Rev. Phytopathology*, **35**, 82–112
Peters, N. K., Frost, J. W. & Long, S. R. (1986). A plant flavone, luteolin, induces expression of *Rhizobium meliloti* nodulation genes. *Science*, **233**, 977–80
Pistole, T. G. (1981). Interaction of bacteria and fungi with lectins and lectin-like substances. *Ann. Rev. Microbiol.*, **35**, 85–122
Rippon, J. W. (1968). Monitored environmental system to control cell growth, morphology, and metabolic rate in fungi by oxidation–reduction potentials. *Appl. Microbiol.*, **16**, 114–21
Rippon, J. W. & Anderson, D. N. (1970). Metabolic rate of fungi as a function of temperature and oxidation–reduction potential (eh). *Mycopathol. Mycol. Appl.*, **40**, 349–52
Rippon, J. W., Conway, T. P. & Domes, A. L. (1965). Pathogenic potential of *Aspergillus* and *Penicillium* species. *J. Infec. Dis.*, **115**, 27–32
Rippon, J. W. & Scherr, G. (1959). Induced dimorphism in dermatophytes. *Mycologia*, **51**, 902–14

Ryan, C. A., Bishop, P. D., Walker-Simmon, M., Brown, W. E. & Grahams, J. (1985). Pectic fragments regulate the expression of proteinase inhibitor genes in plants. In: *Cellular and Molecular Biology of Plant Stress*, pp. 319–34. Alan R. Liss, New York

Schmelzer, E., Borner, H., Grisebach, H., Ebel, J. & Hahlbrock, K. (1984). Phytoalexin synthesis in soybean (*Glycine max*): similar time course of mRNA induction in hypocotyls infected with a fungal pathogen and in cell cultures treated with fungal elicitor. *FEBS Lett.*, **172**, 59–63

Sequeira, L. (1978). Lectins and their role in host–pathogen specificity. *Ann. Rev. Phytopath.*, **16**, 453–81

Sequeira, L. (1983). Mechanisms of induced resistance in plants. *Ann. Rev. Microbiol.*, **51–79**

Stachel, E., Nester, E. W., Zambryski, P. (1986). A plant cell factor induces *Agrobacterium tumefaciens vir* gene expression. *Proc. Nat. Acad. Sci. USA*, **83**, 379–83

Staples, R. C. & Macko, V. (1980). Formation of infection structures as a recognition response to fungi. *Exp. Mycol.*, **4**, 22–6

Staskawicz, B. J., Dahlbeck, D. & Keen, N. T. (1984). Cloned avirulence gene of *Pseudomonas syringae* pv *glycinea* determined race-specific incompatibility on *Glycine max* (L) *Merr., Proc. Natl Acad. Sci. USA*, **81**, 6024–8

Strange, J., Majer, J. R. & Smith, S. (1974). The isolation and identification of choline and betaine as the two major components in anthers and wheat germ that stimulate *Fusarium graminearum in vitro*. *Physiol. Plant Path.*, **4**, 277–90

Threlfall, D. R. (1980). Polyprenols and terpenoid quinones and cjromanols. In: *Secondary Plant Products*. Encyclopaedia of Plant Physiology, eds E. A. Bell & B. V. Charlwoold, pp. 288–308. Springer-Verlag, Berlin

Wallwork, J. C. & Crane, F. L. (1970). The nature, distribution and biosynthesis of prenyl phytoquinones and related compounds. *Prog. Phytochem.*, **2**, 267–341

Yoder, O. C. & Turgeon, B. G. (1985). Molecular analysis of the Plant-fungus interaction. In: *Molecular Genetics of Filamentous Fungi. Proc. UCLA Symposium*, ed. W. E. Timberlake, pp. 383–403. Alan R. Liss, New York

Yoshikawa, M., Matama, M., Masago, H. (1981). Release of a soluble phytoalexin elicitor from mycelial walls of *Phytophthora megasperma* var. *sojae* by soybean tissues. *Plant Physiol.*, **67**, 1032–5

INDEX

Author index

Organism index

Subject index